COMPUTER VISION, MODELS and INSPECTION

WORLD SCIENTIFIC SERIES IN ROBOTICS AND AUTOMATED SYSTEMS

Editor-in-Charge: Prof T M Husband
(*Vice Chancellor, University of Salford*)

Vol. 1: Genetic Algorithms and Robotics — A Heuristic Strategy for Optimization (*Y Davidor*)

Vol. 2: Parallel Computation Systems for Robotics: Algorithms and Architectures (*Eds. A Bejczy and A Fijany*)

Vol. 3: Intelligent Robotic Planning Systems (*P C-Y Sheu*)

Vol. 4: Computer Vision, Models and Inspection (*A D Marshall and R R Martin*)

Vol. 5: Advanced Tactile Sensing for Robotics (*Ed. H R Nicholls*)

Vol. 6: Intelligent Control: Aspects of Fuzzy Logic and Neural Nets (*C J Harris, C G Moore and M Brown*)

Vol. 7: Visual Servoing (*Ed. K Hashimoto*)

Vol. 8: Modelling and Simulation of Robot Manipulators: A Parallel Processing Approach (*A Y Zomaya*)

Forthcoming volumes:

Vol. 9: Advanced Guided Vehicles – Aspects of the Oxford AGV Project (*S Cameron and P Probert*)

Vol. 10: Recent Trends in Mobile Robots (*Ed. Y F Zheng*)

Vol. 11: Cellular Robotics and Micro Robotic Systems (*T Fukuda and T Ueyama*)

World Scientific Series in Robotics and Automated Systems – Vol. 4

COMPUTER VISION, MODELS and INSPECTION

A. D. Marshall and R. R. Martin
Department of Computing Mathematics
University of Wales
Cardiff CF2 4YN, UK

World Scientific
Singapore • New Jersey • London • Hong Kong

Published by

World Scientific Publishing Co. Pte. Ltd.
P O Box 128, Farrer Road, Singapore 9128
USA office: Suite 1B, 1060 Main Street, River Edge, NJ 07661
UK office: 73 Lynton Mead, Totteridge, London N20 8DH

Library of Congress Cataloging-in-Publication Data
Marshall, A. D. (A. Dave)
   Computer vision, models, and inspection / by A.D. Marshall and R.R. Martin.
      p.   cm. -- (Series in robotics and automated systems; vol. 4)
   ISBN 9810207727
   1. Quality control--Optical methods--Automation. 2. Computer vision-- Industrial applications. 3. Engineering inspection--Automation. 4. Three-dimensional display systems. I. Martin, R. R. (Ralph R.) II. Title. III. Series.
TS156.2.M37   1992
670.42'5--dc20   91-47664
                                                                      CIP

Copyright © 1992 by World Scientific Publishing Co. Pte. Ltd.

Reprinted 1993.

All rights reserved. This book, or parts thereof, may not be reproduced in any form or by any means, electronic or mechanical, including photocopying, recording or any information storage and retrieval system now known or to be invented, without written permission from the Publisher.

For photocopying of material in this volume, please pay a copying fee through the Copyright Clearance Center, Inc., 27 Congress Street, Salem, MA 01970, USA.

Printed in Singapore by Utopia Press.

Preface

Three-dimensional computer vision has advanced a great deal in the last decade. Although industrial inspection is one of the areas where computer vision has been most successfully applied, mainly two-dimensional inspection problems have been solved to date. Three-dimensional geometric inspection to typical engineering tolerances is just becoming possible with computer vision techniques, and a main objective of this book is to show how this can be achieved. An important sub-task in inspection is object recognition, and a substantial part of this book is devoted to a consideration of various current object recognition strategies. The book is rounded out by showing what other types of (two-dimensional) inspection problems can also be solved currently by computer vision methods. Furthermore, an introduction to a selection of basic concepts and ideas in computer vision is provided.

Thus, we hope that this book will be of use to a wide range of readers. Firstly, the introductory material should provide a suitable text for taught (or self-study) courses in basic computer vision, at the final year undergraduate or postgraduate level. As an aid, exercises are provided at the ends of the relevant Chapters. Throughout the book we have tried to keep the level of mathematics assumed to a minimum. Certain details are explained as we proceed, others are provided in Appendices. However, we do assume a basic knowledge of matrix algebra, geometry, trigonometry and calculus. Secondly, the book as a whole should be of interest to researchers in both industry and academia who are working in computer vision based inspection, object recognition, and related areas. This book both surveys existing methods and presents new results in these fields.

As well as being an application of considerable importance in its own right, three-dimensional geometric inspection in many ways provides a good means of assessing the current capabilities of computer vision technology. Such inspection processes imposes high accuracy requirements both on the data acquisition hardware and subsequent processing methods used by a vision system. When designing algorithms for our inspection system, these accuracy requirements will be of considerable importance.

In writing this book we have tried to keep the material as self-contained as possible. In order to achieve this we have divided the book into three Parts.

The subject matter of the first Part of the book and the early Chapters of the second and third Parts of the book is based on lecture courses which we have given to final year undergraduate and postgraduate students. Elsewhere in the book we detail research which we have conducted over the last few years in collaboration with British Aerospace's Sowerby Research Centre at Bristol.

The first Part is a basic introduction to the fundamentals of computer vision. Here, we provide an overview of many important aspects of computer vision, ranging from forming various kinds of two-dimensional and three-dimensional images, to low-level processing, to methods of higher level reasoning aimed at extracting information from images. This Part provides a foundation for many of the topics discussed in the second and third Parts of the book. Although in this Part we discuss both two-dimensional and three-dimensional computer vision techniques, we shall sometimes shift the balance of the discussion towards the three-dimensional aspects, as they will be required later in the book.

In more detail, the Chapters in the first Part are as follows. After a general introduction, Chapter 2 deals with the acquisition of images. It details both two- and three-dimensional imaging hardware and associated processing techniques. One important step is camera calibration. We provide details of one popular calibration method and assess its performance in our vision system. An introduction to our research vision system for three-dimensional inspection is also provided in this Chapter. Chapter 3 then deals with image processing techniques which are designed to either enhance features in the image or reduce errors arising from noise and camera distortion. It also deals with methods of storing the acquired images efficiently. Chapter 4 considers what features can be extracted from two- and three-dimensional images and gives details of how they can be extracted. The important features dealt with are edges, planes and other higher order surfaces. Finally Chapter 5 deals with higher order reasoning, where the basic features can be grouped together into clusters of features which form objects or parts of an object. This Chapter also deals with methods of determining other object properties, such as shape and texture of the surface of an object. A preliminary discussion of object recognition strategies is provided in preparation for Part 2.

Part 2 of the book addresses the problems of object recognition from three-dimensional data. In order to recognise an object we need a stored model of the object to which the data extracted from a scene can be compared or matched. Such model based matching is far from being easy to perform even for relatively simple images containing only one or just a few objects. Chapter 6 provides a basic introduction to the whole area of model based matching. Chapter 7 discusses how to represent models of three-dimensional objects in order to achieve efficient model based matching. Chapter 8 provides a survey of many currently popular three-dimensional model based matching strategies. Chapter 9 goes on to consider algorithmic details of the various stages involved in these matching strategies. Finally, a matching strategy is developed for the purpose of visual three-dimensional geometric inspection of objects, the main topic of Part 3.

Part 3 of the book discusses the application of computer vision to inspection. Primarily we deal with the task of inspecting the geometric properties of a three-dimensional object although other important areas such as surface finish inspection are also considered. Chapter 10 defines what we mean by inspection in general, and in particular, three-dimensional geometric inspection. It also addresses the application of vision systems to this area, outlining the requirements of any automated inspection system. In Chapter 11 we discuss how geometric tolerances are defined and represented using the object models developed in Chapter 7. Chapters 12 and 13 go on to develop strategies for performing the complete inspection of an object using the specified vision system. In Chapter 12 the emphasis is placed on what can be inspected in a single scene. In Chapter 13 we develop methods for overall object inspection by rotating the object using single axis turntable and by using a robot arm to reposition the object, to obtain a set of views sufficient for inspection of all object features. To provide a balanced treatment, various other computer vision inspection methods and applications are considered in Chapter 14. Such topics as surface finish inspection and crack detection are considered, which can be inspected using two-dimensional image data. The application of such methods to two particular areas, printed circuit board inspection and the food industry, is discussed. Finally, Chapter 15 provides a critical assessment of the research described in this book and also points to future developments.

Finally, we would like to thank the many people who have contributed to the creation of this book. Firstly, we would like to thank the staff at the Advanced Image Processing Department of the Sowerby Research Centre, British Aerospace, Bristol who were responsible for assembling the vision and inspection system, and for providing the data used throughout the book. Dave Marshall would like to thank all of them for making him most welcome at all times and especially during periods of his secondment there. Particularly, he would like to thank Dave Hutber for providing him with much practical advice, assistance and guidance over our many years of association. We would also like to thank Andy Page, Michael Gay, Simon Tomlinson and Chris Hopkins who have contributed to the development of the system and provided camera calibration and segmentation routines in particular.

Throughout the book for simplicity we shall refer to this particular vision system as "our" system. We wish to make it clear that in some cases work being described may have been largely, or even entirely, performed by these people at the Sowerby Research Centre and not by ourselves. Such references to "our" are purely for conciseness of explanation, and we wish to make it quite clear that we are not trying to steal the credit for the work done in such cases. References provided at the appropriate points in the text should make it clear who has done the work on various components of the system.

Secondly, we would like to thank Bruce Batchelor and John Chan of the Electrical, Electronic and Systems Engineering Department of the University of Wales College of Cardiff, for much useful information about current industrial inspection strategies. They also very kindly provided the photographs for Chapter 14.

Lastly, we would like to thank our Department for the facilities to enable us to complete the book.

Dave Marshall
Ralph Martin

Department of Computing Mathematics
University of Wales College of Cardiff
September 1991

Contents

I Fundamentals of Computer Vision — 1

1 Introduction — 3
1.1 Introduction 3
1.2 Computer Vision 5
1.3 Related Topics 7
1.4 Industrial Inspection 9
1.5 Layout of this Book 9
1.6 Scope of this Book 10
1.7 Our Visual Inspection System 11
    1.7.1 Inspection Aims and Limitations 11
    1.7.2 System Components 12

2 Image Acquisition — 15
2.1 Introduction 15
2.2 Image Input Devices 16
2.3 Camera Geometry 18
    2.3.1 A Camera Model 19
    2.3.2 Camera Calibration 22
    2.3.3 Analysis of the Calibration 25
2.4 Three-Dimensional Imaging 26
    2.4.1 Depth Maps 26
    2.4.2 Introduction to Stereo Imaging 27
    2.4.3 Methods of Acquisition 32
2.5 Our Vision Acquisition System 36
    2.5.1 Description 36
    2.5.2 Errors in Our Vision Acquisition System 40
    2.5.3 Performance of Our Vision System 43

2.6	Simulating Depth Data	45
2.7	Exercises	45

3 Image Preprocessing   49
  3.1 Introduction 49
  3.2 Fourier Methods 50
      3.2.1 Introduction 50
      3.2.2 Theory 51
      3.2.3 Convolutions 55
      3.2.4 The Dirac Delta Function 58
      3.2.5 Other Properties of Fourier Transforms 59
      3.2.6 The Fast Fourier Transform Algorithm 60
  3.3 Smoothing Noise 63
      3.3.1 Introduction 63
      3.3.2 Real Space Smoothing Methods 63
      3.3.3 Fourier Space Smoothing Methods 65
  3.4 Contrast Enhancement 68
      3.4.1 Introduction 68
      3.4.2 Histogram Equalisation 68
  3.5 Correction of Camera Errors 72
      3.5.1 Introduction 72
      3.5.2 Geometric Correction 73
      3.5.3 Correcting Blurred Images 74
  3.6 Image Compression 77
      3.6.1 Introduction 77
      3.6.2 Real Space Compression 78
      3.6.3 Transform Encoding 79
      3.6.4 Colour Space Compression 81
  3.7 Exercises 82

4 Segmentation and Feature Representation   85
  4.1 Introduction 85
  4.2 Segmentation Goals 87
  4.3 Extracting and Describing Features in an Image 88
  4.4 Point Extraction 89
      4.4.1 Oriented Surface Points 89
      4.4.2 Problems with Points as Primitives 90

4.5 Edge Extraction ... 91
- 4.5.1 Representing Lines ... 91
- 4.5.2 Extracting Edges from Images ... 92
- 4.5.3 Detecting Edge Points ... 93
- 4.5.4 Edge Linking ... 102
- 4.5.5 Problems with Edges as Primitives ... 108

4.6 Surface Extraction ... 110
- 4.6.1 Representing Surfaces ... 110
- 4.6.2 Extracting Surfaces from Images ... 120
- 4.6.3 Region Splitting ... 121
- 4.6.4 Region Growing ... 123
- 4.6.5 Similarity Constraints ... 126
- 4.6.6 A Method for Segmenting Planes ... 128
- 4.6.7 Segmenting Higher Order Surfaces ... 129
- 4.6.8 Problems with Surface Segmentation ... 135

4.7 Statistical Region Description ... 137
- 4.7.1 Statistical Descriptions ... 137
- 4.7.2 Invariant Measures ... 140

4.8 Exercises ... 141

5 Reasoning With Images 143
5.1 Introduction ... 143
5.2 Line Labelling ... 143
- 5.2.1 Introduction ... 143
- 5.2.2 Assumptions ... 145
- 5.2.3 Junction Types ... 146
- 5.2.4 Labelling an Image ... 148

5.3 Relaxation Labelling ... 149
- 5.3.1 Introduction ... 149
- 5.3.2 Statistical Relaxation Techniques ... 151

5.4 Shape from Shading ... 157
- 5.4.1 Introduction ... 157
- 5.4.2 Radiance and Irradiance ... 158
- 5.4.3 Bidirectional Reflectance Distribution Function . 158
- 5.4.4 Radiance of a Surface ... 160
- 5.4.5 Surface Orientation ... 162
- 5.4.6 Reflectance Map ... 164

       5.4.7   Photometric Stereo 165
  5.5 Object Recognition 167
       5.5.1   Introduction 167
       5.5.2   Extended Gaussian Images 167
  5.6 Optical Flow 169
       5.6.1   Introduction 169
       5.6.2   Optical Flow Constraint Equation 170
       5.6.3   Further Constraints 171
       5.6.4   Finding the Optical Flow 172
  5.7 Texture 173
       5.7.1   Introduction 173
       5.7.2   Texture Methods 174
       5.7.3   Shape Grammars 175
       5.7.4   Texture Recognition 179
  5.8 Exercises 180

II Model-Based Matching Techniques   183

6 Introduction to Model Based Matching   185
  6.1 Introduction 185
  6.2 Outline of Part 2 187

7 Geometric Modelling for Computer Vision   189
  7.1 Introduction 189
  7.2 Solid Model Representations 191
       7.2.1   Introduction 191
       7.2.2   Spatial Enumeration 192
       7.2.3   Set-Theoretic Modelling 195
       7.2.4   Boundary Representation 198
       7.2.5   Model Validity 201
  7.3 Solid Models for Computer Vision 203
       7.3.1   Desirable Model Properties for Vision 204
       7.3.2   Which Representation is Best? 205

8 Model Based Matching   207
  8.1 Introduction 207
  8.2 The Nevatia and Binford Method 209

	8.3	The Oshima and Shirai Method 212
		8.3.1 The Learning Phase 212
		8.3.2 The Matching Phase 213
	8.4	The *3DPO* Method 216
	8.5	The *ACRONYM* Method 217
	8.6	The Grimson and Lozano-Perez Method 219
		8.6.1 Generating Feasible Interpretations 220
		8.6.2 Model Testing 223
	8.7	The *TINA* Method 224
	8.8	The Faugeras and Hebert Method 226
	8.9	Relaxation Labelling Methods 228
	8.10	The *SCERPO* Method 229
	8.11	The *IMAGINE* Method 231
	8.12	Fan's Method 233
	8.13	Hough Transform Methods 235
	8.14	Extended Gaussian Image Methods 235
	8.15	Summary 235

9 Practical Model Based Matching   239

	9.1	Introduction 239
	9.2	Estimation of the Transformation 242
		9.2.1 Estimation of the Rotation 243
		9.2.2 Estimation of the Translation 246
	9.3	Improving the Method's Accuracy 247
		9.3.1 Replacing the Pseudo-Inverse 247
		9.3.2 Using the Centres of Gravity of Regions 249
	9.4	Improving the Search Efficiency 251
		9.4.1 Order of Matching Pairs 251
		9.4.2 An Additional Local Constraint 254
		9.4.3 Extension to Primitives Other Than Planes ... 255
		9.4.4 Further Search Control 256
		9.4.5 Rogue Faces 258
		9.4.6 Position Recovery From Insufficient Data 259
	9.5	A Practical Assessment of our Vision System 262
		9.5.1 Our Vision Acquisition System Reviewed 262
	9.6	Segmentation Algorithms in Practice 263
		9.6.1 Our Segmentation Approach 263

             9.6.2 Results From Artificial Data 264
             9.6.3 Results from Real Data 268
             9.6.4 Conclusions 272
       9.7   Matching Algorithm Assessment 273
             9.7.1 Matching Segmented Artificial Data 273
             9.7.2 Matching Segmented Real Data 276
       9.8   Conclusions 278

III Inspection 281

10 Introduction to Inspection 283
   10.1 Introduction 283
   10.2 The Geometric Inspection Problem 284
         10.2.1 Introduction 284
         10.2.2 Current Industrial Inspection 285
   10.3 Current Industrial Vision Systems 286
         10.3.1 Introduction 286
         10.3.2 Three-Dimensional Automated Visual Inspection 286
   10.4 Three-Dimensional Data 288
         10.4.1 Introduction 288
         10.4.2 Requirements of a Vision Acquisition System . . . 289
         10.4.3 Possible Data Acquisition Methods 289
         10.4.4 Our Vision Acquisition System 291
   10.5 Matching for Inspection 292
         10.5.1 Introduction 292
         10.5.2 Suitability of Matching Methods 292
   10.6 Outline of Part 3 295

11 Geometric Tolerances 297
   11.1 Introduction 297
   11.2 Tolerance Representation 298
   11.3 Implementing Datum Systems 304
   11.4 Measuring Conformance to Desired Tolerances 306
   11.5 Points Lying Outside Tolerance 307

12 Geometric Inspection and Single Scenes — 311
- 12.1 Introduction 311
- 12.2 Geometric Inspection Strategies 312
- 12.3 Single Scene Inspection 314
- 12.4 Errors in our Inspection System 315
- 12.5 Single Scene Inspection Results 316
  - 12.5.1 Results on Artificial Data 316
  - 12.5.2 Results on Real Data 319
- 12.6 Conclusions 322

13 Complete Object Geometric Inspection — 325
- 13.1 Introduction 325
- 13.2 An Ideal Visual Inspection System 326
  - 13.2.1 Choice of Vision System for Inspection 328
- 13.3 Single Axis Inspection 328
  - 13.3.1 Choosing Sets of Features for Each View 329
  - 13.3.2 Obtaining the Best View of a Set of Features . . . 333
- 13.4 The Single Axis Inspection Scheme 338
  - 13.4.1 The Registration Phase 338
  - 13.4.2 The Inspection Phase 340
  - 13.4.3 Diagnostics 342
- 13.5 Complete Inspection 343
- 13.6 Results 344
- 13.7 Conclusions 344

14 Other Types of Inspection — 349
- 14.1 Introduction 349
- 14.2 Inspection in Two Dimensions 350
- 14.3 Surface Finish and Crack Detection 354
- 14.4 Applications of Visual Inspection 360
  - 14.4.1 Printed Circuit Board Inspection 360
  - 14.4.2 Inspection in the Food and Agriculture Industries 364
  - 14.4.3 Future Applications of Visual Inspection Systems 368

15 Conclusions and Future Work — 371
- 15.1 Conclusions 371
- 15.2 Future Work 375

Appendices 381

A Matrix Definitions 383
A.1 Matrix Signature and Rank 383
A.2 Eigenvectors and Eigenvalues 384
A.3 Singular Value Decomposition 384

B Least Squares Approximation 387
B.1 Approximation of a Planar Surface 387
B.2 Approximation of a Quadric Surface 389

C Quaternions 393
C.1 Rotation by Quaternions 394

D Outliers 395
D.1 Significance Test For a Single Outlier 395

Bibliography 399

Index 423

Part I

Fundamentals of Computer Vision

Chapter 1

Introduction

1.1 Introduction

The human sense of vision can be employed to perform a very wide range of tasks which would otherwise be very difficult or impossible. Although we seem to take our visual processing abilities for granted, they are probably amongst the most complex operations we humans have to perform. If you find this hard to believe, then after reading this book, or any other book on computer vision for that matter, reflect on the current visual capabilities of today's computer controlled machines and robots. Even though the field of computer vision has been studied seriously for about thirty/years, with quite significant progress made in the last ten, we are still a very long way off competing with human visual capabilities.

Let us reflect on the everyday tasks that humans perform with the aid of vision. You are reading this book right now, recognising characters, and probably whole words or even sentence fragments. You can do this irrespective of the typeface or size of the type. We can find our way around in many environments even if we have not been there before. We can pick up and put down objects using visual (as well as tactile) feedback. One particular application that humans frequently employ vision for is *inspection* — testing the suitability of an object for a purpose. We can quickly tell which shoe fits on which foot and if it is roughly the correct size. We can often tell if food has gone bad just by looking at it. We can also see, at least in some ways, if a car body panel has been acceptably repaired after an accident, just by looking at the reflections from its surface.

Man has always sought to automate human processes and in particular, "making machines that can see" has been a subject of widespread interest in recent years. We make the point here that, realistically, work has not sought to impart the total capability of the human visual system to a single machine, but rather to construct machines that perform a specific visual task.

Clearly, a machine with intelligent vision capabilities — being able to reason with data without the aid of human assistance — could be applied in many areas. There are various industrial and military tasks, especially of a repetitive or a hazardous nature, that could benefit. Indeed, machines with limited intelligent vision capabilities have been employed in industry for some time as we shall see in Chapter 14. However, recent advances in both the relevant hardware and software have made computer vision potentially applicable and affordable for a much wider range of industrial problems.

One such industrial task is the geometric inspection of manufactured mechanical parts for quality assurance. The main thrust of this book examines the feasibility of applying computer vision to perform three-dimensional geometric inspection of manufactured parts, and a practical working system has been built as a testbed for evaluating the ideas proposed. The various issues concerning the choices of hardware, and the algorithms and ideas used in the software, will be explained in this book. The book also provides a general tutorial introduction to the field of computer vision, and briefly considers other types of inspection problem, such as two-dimensional inspection and surface finish inspection.

The other major topic addressed in this book, which in fact has to be solved as a sub-task of the inspection process, is *object recognition*. One particularly important by-product of this step is information about the pose (position and orientation) of the object. The manufactured parts we are interested in will often be represented by some form of geometric model , typically generated by a computer aided design (CAD) system. Because we expect to be looking at known objects in inspection tasks, such models can be used by the vision system to help recognise what is seen in the observed scene. Thus, by using the model, the system can

1.2. COMPUTER VISION

decide what parts of the object are visible in the scene and where they are, and hence the position and orientation of the whole object. Indeed model based approaches, using a stored description in some form of what is expected to be present in the scene have been applied to many vision problems as will be seen throughout the book.

Qualitative information provided by a model based vision system can be used to verify the presence or absence in the observed object of features from the model, and hence to perform some basic types of inspection. The major aim of this book, however, is to examine the use of model based vision techniques to perform quantitative three-dimensional geometric inspection of manufactured mechanical parts.

1.2 Computer Vision

For many years the field of computer vision has been developing as a research theme in artificial intelligence. The main purpose of computer vision is to allow a computer to understand aspects of its environment using information provided by "visual" sensors. The term visual is applied loosely, since although the majority of techniques have used visible light images of some sort (including two-dimensional monochrome or colour data and three-dimensional data), a variety of other techniques have been developed using X-rays, infrared and ultraviolet light, thermal images, sonar and radar, for example.

Most successful vision techniques have been constrained to operate in a particular type of environment where certain assumptions can be made about the scene. Examples include both outdoor scenes such as those provided by aerial photography and many indoor scenes provided by industrial manufacturing processes.

Most early visual techniques [203] involved reasoning with one or more two-dimensional images and attempted to infer three-dimensional information from them. Indeed, most commercially available practical vision systems built to date deal only with two-dimensional data [216]. Computer vision using two-dimensional data and models is in some ways fairly straightforward, as there exists a direct one-to-one correspondence between features in the model and the scene. However, in three-dimensional vision only certain features in the model can be visible at any one time, and correspondence problems become important.

When we think of a two-dimensional image we usually think of two-dimensional array of grey scale intensity values (see Section 2.2). We can think of a three-dimensional image as a similar array containing depth values (*i.e.* distances from points on the object to a reference plane) instead, as discussed in Section 2.4. Usually, however, valid depth values will only be contained in a certain subset of array elements due to the nature of the sensing methods used. Throughout the last decade three-dimensional information has become increasingly available with many techniques being developed to produce such data. Some of these techniques are described in Chapter 2. This increase in availability is mainly due to advances in reliability of the hardware. In particular, advances in camera technology (especially charge coupled device — CCD — cameras) have facilitated the production of three-dimensional data. In many cases, two cameras are used to provide three-dimensional data from a scene by combining, using software, the two different two-dimensional views from each camera. This is not a straightforward task, as difficulties arise in deciding which features visible in each of the two views correspond to the same real three-dimensional entity. Furthermore, because of camera inaccuracies, errors in the calculation of three dimensional coordinates on combining the two views can be large enough to prevent reliable three-dimensional measurements from being possible unless care is taken.

If three-dimensional data can be reliably provided, three-dimensional vision techniques have a significant advantage over two-dimensional methods in many applications. This is because three-dimensional geometry is explicitly represented in the data in terms of three-dimensional coordinates, rather than only being implicitly available from projections in two-dimensional images. For example, finding the length of an edge of a solid object becomes a straightforward measurement between two three-dimensional points rather than a quantity which must be inferred from two-dimensional data. As well as relying on possibly questionable assumptions, such inferences can also be very prone to numerical error. For example, consider the case where the z-axis of the world coordinate system of the vision system lies nearly along the line of sight of a single camera. Even large changes in the z coordinate of a point will lead to little or perhaps no observable change in the image.

Such issues may obviously be important in detailed three-dimensional assembly or inspection tasks. Similar considerations also affect navigation tasks, for example, where the distance from an obstacle may be the most important single piece of information required.

As we have already noted, in order to recognise an occurrence of an object in a scene it is necessary to have some sort of predefined description or model of the object to refer to. Many different model representations have been developed, both in two and three dimensions. The simplest may record relatively little qualitative geometric information, perhaps just a few statistics about an object, while more sophisticated approaches, such as those originally developed for computer aided design (CAD) applications, can record both complete low-level geometric details for the object and higher-level relationships. Many model-based recognition methods have been developed, and a range of such techniques and suitable model representation methods are discussed later in the book.

1.3 Related Topics

Computer vision originally started out as a branch of artificial intelligence. However, it has grown into such a large field in recent times that it may now be considered to be a separate field of study in its own right. Probably one of the main attractions of computer vision is that it involves ideas from such a broad range of disciplines. For example, computer vision uses concepts and results taken from optics, signal processing, psychology and psychophysics, many branches of mathematics and statistics, computer graphics and computer aided design, and many other branches of computer science.

Some people regard *image processing* as a closely related but separate field to computer vision whilst others consider that the former topic has been subsumed by the latter. Both points of view have their relative merits. If a distinction is to be made, we may regard image processing as mainly consisting of the manipulation of data contained within an image to improve or enhance certain properties of the image. For example, this may involve reducing the noise in an image or changing the contrast of the image. The aim of image processing is often to improve the appearance of an image for subsequent human viewing. Nevertheless, as we will see later in the book, some of the same techniques can also be used, often as preprocessing steps, to aid subsequent computer vision tasks. Finally, one other important aspect of image processing is image compression.

An important distinction which can be used to separate image processing from computer vision is that the former generally is not concerned with what the image represents. On the other hand, *computer vision* is concerned with the extraction and use of information contained within, or represented by, the image.

Another area closely related to computer vision is that of *pattern recognition*. The main aim here is to classify data considered as forming some sort of pattern into one of several predetermined categories. The classification is usually based on a selection of statistics measured from the pattern, such as height, width, and so on. One example where pattern recognition is widely used is in optical character recognition (OCR). Many early two-dimensional vision systems employed pattern recognition techniques to identify objects in pictures. However, in general pattern recognition is not flexible enough for many computer vision tasks. The numeric classifiers are too restrictive and prone to errors; having to put an object into one of several predetermined classes is clearly very restrictive.

Many branches of artificial intelligence, for example natural language understanding, have used *symbolic reasoning* strategies to recognise and understand occurrences of certain groups of symbols. Indeed many computer languages, such as *LISP* and *PROLOG*, are particularly suited to these strategies. Such techniques make it possible to reason with incomplete, inconsistent and changing models of the world. Clearly, certain computer vision tasks may well have such needs. Some of the matching methods discussed in Chapter 8 rely on a form of symbolic reasoning.

One further area related to computer vision which is worth mentioning is *computer graphics*. This area is concerned with producing artificial images by computer, which can be regarded in some sense as the opposite of computer vision. However, since both subjects deal with images of some kind, there is obviously some common ground. For example, input to a graphics systems is, or is used to create, a model of some sort, from which the image is generated. Models used in computer graphics and CAD have already been mentioned above.

1.4 Industrial Inspection

Industry often requires that a sample or perhaps all of the products output by a given manufacturing process should be inspected in order to provide measures of quality and reliability. For a complex manufactured object the inspection process may have many stages, involving the testing of each component through to the testing of performance and acceptability of the finished product.

There is an obvious need for a totally automated inspection system which can perform all tasks without any human aid during the process, as production costs both in time and money can be reduced [9]. However, to date, totally automated inspection systems are rare and their scope limited. Current industrial inspection techniques have their drawbacks and computer vision can help to overcome many of them. We shall discuss these points in more detail throughout the book.

1.5 Layout of this Book

In order to aid the reader we have laid this book out in three Parts. Each of these Parts is more-or-less self-contained and a reader with sufficient knowledge could skip one Part and read the others without too much difficulty. Each Part however, does build substantially on topics covered in previous Parts. We shall now briefly summarise what is to come in the rest of this book.

Part 1 — Fundamentals of Computer Vision Basic concepts of image acquisition, image processing, and the extraction of and reasoning with higher-level information from images are discussed in this Part of the book. Throughout this Part the emphasis is on three-dimensional techniques although many relevant two-dimensional techniques are also given.

Part 2 — Model Based Matching Techniques The geometric modelling of three-dimensional objects is discussed, as are many techniques for recognising objects by matching higher-level image data to stored object models. Practical implementation details of matching algorithms are given together with such an algorithm developed specifically for geometric inspection purposes.

Part 3 — Inspection The ideas from Part 2 are extended to cover the complete three-dimensional geometric inspection of objects and strategies to achieve this task are discussed. Also, to round off the coverage of inspection, other inspection tasks than geometric inspection are discussed together with relevant applications such as inspection of surface finish and surface fractures. The inspection of two-dimensional geometric properties is also addressed.

1.6 Scope of this Book

The first Part of the book (and perhaps some of the second and third Parts) can be used as a basis for final year undergraduate and postgraduate courses. The rest of the book surveys current research as well as describing our own work over the last few years in the field of visual inspection.

The purpose of the research detailed in this book was to investigate the feasibility of a three-dimensional computer vision based system for performing geometric inspection. The inspection system is intended to inspect single manufactured components or small assemblies of about 0.5m in size to typical dimensional tolerance accuracies of about 0.25mm (0.01inch). The tolerances addressed in this book are geometric tolerances [38], which is to say that the inspection strategies are mainly concerned with firstly verifying the presence or absence of features (such as holes), and secondly measuring relationships between features such as distances and angles to specified accuracies. The shape of the features (such as the roundness of holes) also needs inspecting. The inspection of surface defects such as surface finish and fractures is not a goal of the implemented system. However, methods for inspecting such defects are discussed in Chapter 14.

An overview of the adopted visual system is given in the following Section. The relevant techniques used at each stage in the system are subsequently described in detail throughout the book.

It should be noted that although the general problem studied is that of automatic inspection, the methods described may have useful results that contribute to many other related fields of study. These fields include industrial assembly, general computer integrated manufacturing (CIM), object recognition and computer navigation where similar matching and object measurement techniques could be applied, perhaps with less strict requirements.

In summary, then, the major part of this book aims to show that the goal of automatic three-dimensional geometric tolerance inspection can be achieved, by giving details of the hardware and algorithms. This conclusion is justified by results from simulations and from real experimental data.

1.7 Our Visual Inspection System

This Section gives an overview of the architecture of the complete inspection system discussed in this book. This prototype inspection system has been assembled at the British Aerospace Sowerby Research Centre in collaboration with whom the work detailed in this book was performed. The image acquisition, camera calibration and segmentation routines described in this book were developed at the Centre. Detailed descriptions of each of the components of the system will be presented throughout the book. The system has been purposely developed using a black box approach for each of the various stages so that if and when better techniques are developed for any particular stage they can be readily incorporated into the system.

1.7.1 Inspection Aims and Limitations

The inspection task that the proposed system is intended to perform is to check geometric tolerances [38]. Clearly, the inspection of every possible three-dimensional object is impossible, and so we require that the part must lie within a cube of side 0.5m. This limitation is imposed so that the field of view of the available cameras is sufficiently narrow to the achieve the necessary accuracy in the acquired data.

1.7.2 System Components

The major components of our vision-based inspection system and the data passed between them are outlined in Fig. 1.1. In accordance with the black-box approach described above, the interfaces between these components are carefully defined.

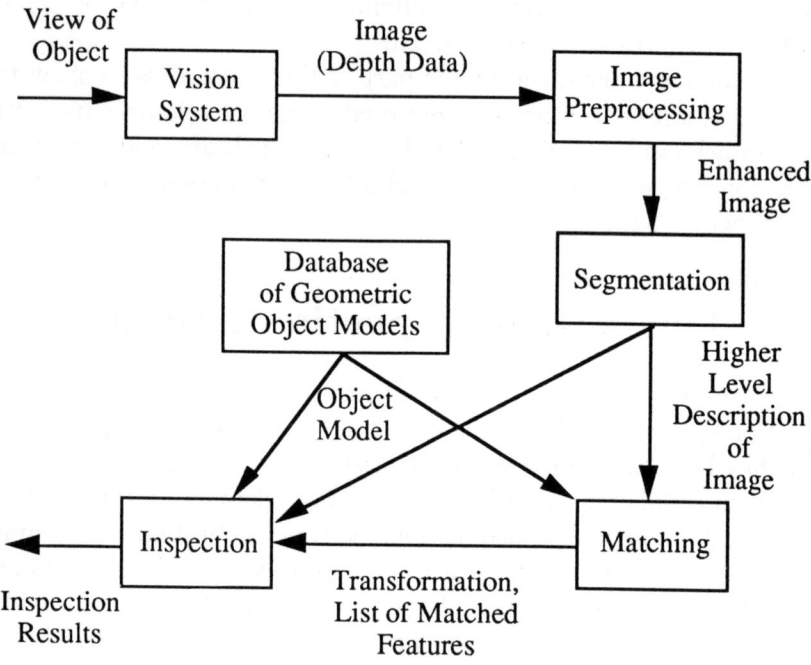

Figure 1.1: Schematic diagram of outlined inspection system

The first stage of the vision system acquires three-dimensional data from a scene. Techniques for acquiring depth data are discussed in Chapter 2. In this system many three-dimensional samples of data are taken over the image and hence the data is said to *dense*. The dense depth data are acquired to a high accuracy using triangulation from a pair of cameras and a laser to illuminate the scene (see Chapter 2).

The next stage in the vision process is the segmentation stage. The raw depth data is processed to yield higher level geometric descriptions which are more efficient to use in the matching and inspection stages.

1.7. OUR VISUAL INSPECTION SYSTEM

Segmentation is discussed in detail in Chapter 4. Typically it yields one or more sets of data describing the object, which may comprise

- oriented surface points (points embedded in a surface with an associated surface normal),
- edges, or
- surface patch descriptions such as those belonging to planes, cylinders, spheres, quadrics and so on.

In our system, segmentation produces object descriptions of the third type.

Thirdly, the geometric description obtained from the segmentation stage is then matched to a stored geometric model of the viewed object. This ensures that the correct object is being viewed, and more importantly, gives information about the position and orientation of the real object. Geometric modelling of objects is discussed in Chapter 7 while the matching of models and scenes is discussed in Chapters 8 and 9.

Fourthly, having obtained a description of the location and orientation of the observed object, the actual positioning of various features relative to those predicted by the model can then be verified to a high precision. This is the *inspection* part of the system.

Clearly, not all features of an object are visible in a single view of an object and so multiple views are required in order to perform complete inspection. As indicated by Fig. 1.2, the object under inspection is initially placed in front of the cameras on a turntable which is capable of being rotated accurately. The inspection of the object about a single axis is possible by choosing a set of angular positions for the turntable, and by taking a series of views. After such a single axis inspection is complete the robot can then pick up and reorient the object on the turntable to provide any further views required to give information which was not visible during the rotation of the turntable about its axis. Complete inspection strategies are outlined in Chapters 12 and 13.

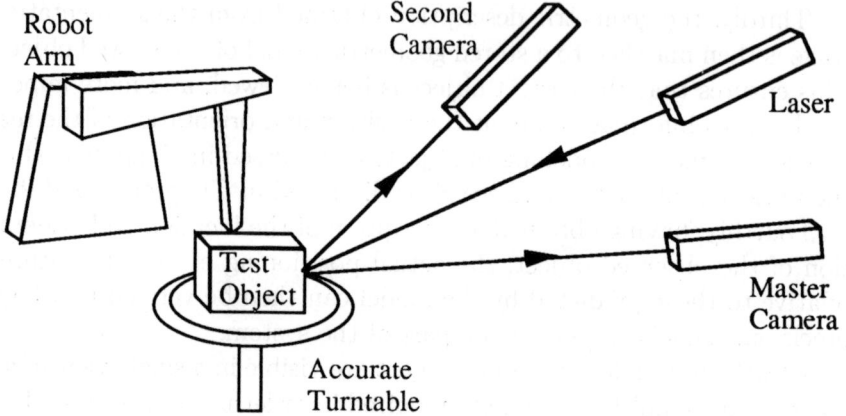

Figure 1.2: Outlined inspection system

Chapter 2

Image Acquisition

2.1 Introduction

The first stage of any vision system is the image acquisition stage. After the image has been obtained, various methods of processing can be applied to the image to perform the many different vision tasks required today. However, if the image has not been acquired satisfactorily then the intended tasks may not be achievable, even with the aid of image enhancement (see Chapter 3).

This Chapter addresses many aspects of capturing an image in a form suitable for analysis by computer. Firstly we define the problem of two-dimensional image acquisition and discuss means of obtaining such images. Secondly, the description of camera geometry is considered together with the problem of its determination and calibration for any particular hardware setup. This is discussed with reference to both two-dimensional and three-dimensional data. Thirdly, the problem of acquiring three-dimensional images is discussed. The techniques used in acquiring such data and suitable data representations are addressed.

It should be noted that the nature of the data captured by a three-dimensional image acquisition system is generally rather different to that obtained in two-dimensional images. Firstly, pixel values in two-dimensional images are often light intensities (or similar values) reflected by points on an object, whilst in three-dimensional images just coordinates of points are stored. Secondly, although in two-dimensional images, every pixel usually contains a valid value, this may well not be the case for three-dimensional images.

Finally in this Chapter, our three-dimensional image acquisition system is described and its performance analysed.

2.2 Image Input Devices

This Section provides a brief overview of two-dimensional image capture devices. The basic two-dimensional image is a monochrome (greyscale) image which has been *digitised*. Thus, we have a two-dimensional light intensity function $f(x,y)$ where x and y are spatial coordinates and the value of f at any point (x, y) is proportional to the brightness or grey value of the image at that point. A digitised image is one in which the spatial and greyscale values have been made discrete. Samples of light intensity are measured across a regularly spaced grid in x and y directions; the sample values themselves are also recorded as one of a set of perhaps 256 different possible intensities. For computational purposes, we may think of a digital image as a two-dimensional array where x and y index an image point. Each element in the array is called a *pixel* (picture element). An example of a digitised image and corresponding pixel values in a highlighted region are shown in Figs. 2.1 and 2.2.

The most obvious choice for a two-dimensional image input device is a television camera, the output of which is a video signal. The image of the scene is focused by a lens onto a photoconductive target. The target is scanned line by line horizontally by an electron beam and produces an electric current as the beam passes over it. The current from the target is proportional to the intensity of light incident at each point. The output video signal is obtained directly from this electric current. A single complete scan of the image is called a *frame*.

This form of device has several disadvantages. It has a finite number of scan lines (about 625) and frame rate (30 or 60 frames per second) which limit its resolution and the number of images that can be obtained in a given time. The camera itself is also subject to a number of distorting factors that affect the image. For example, there can be some unwanted persistence between one frame and the next and the video output may not be linear with respect to light intensity. Also, the target in the tube may not be exactly flat.

2.2. IMAGE INPUT DEVICES

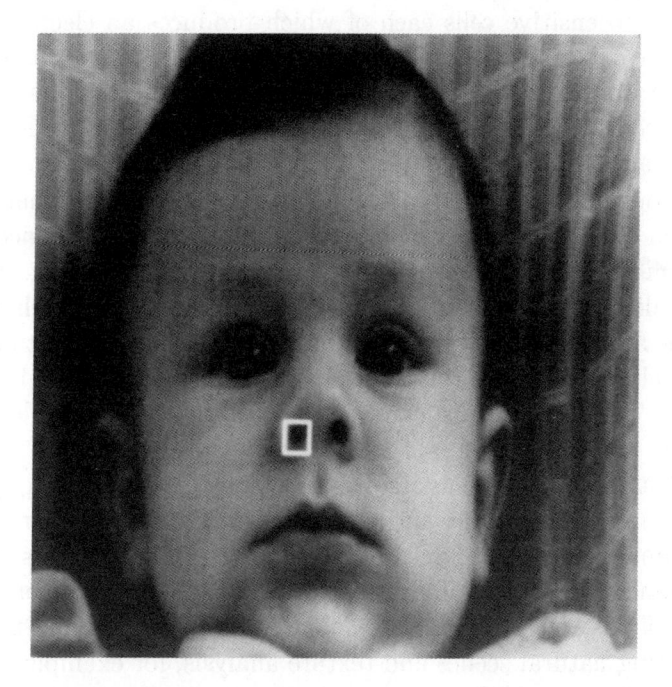

Figure 2.1: Greyscale image and highlighted region

99	71	61	51	49	40	35	53	86	99
93	74	53	56	48	46	48	72	85	102
101	69	57	53	54	52	64	82	88	101
107	82	64	63	59	60	81	90	93	100
114	93	76	69	72	85	94	99	95	99
117	108	94	92	97	101	100	108	105	99
116	114	109	106	105	108	108	102	107	110
115	113	109	114	111	111	113	108	111	115
110	113	111	109	106	108	110	115	120	122
103	107	106	108	109	114	120	124	124	132

Figure 2.2: Pixel values in highlighted region

By far the most popular two-dimensional imaging device nowadays is the charge-coupled device (CCD) camera. This consists of an array of photosensitive cells each of which produces an electric current dependent on the incident light falling on it. All cells exist on a single integrated circuit chip, producing a device which suffers from less geometric distortion than the larger television camera tubes. It also provides a more linear output with respect to light intensity.

The output signal from a CCD camera is similar to that from a television camera and in both cases must be digitised. A device known as a *frame store* or *frame grabber* usually performs this task. A frame grabber digitises the incoming video signal into discrete pixels by sampling the analogue signal at appropriate intervals and converting it to a digital value. As it does this, it stores the values line by line in its own memory. The stored frame can then be easily transferred to computer memory.

Whilst the above discussion has considered greyscale images, other types of image also exist and can be digitised in a similar manner. *Colour* images can be obtained from colour television cameras or colour CCD cameras to obtain images with red, green and blue components instead of just one grey value. Colour information can be useful for interpreting natural scenes and texture analysis, for example.

Three-dimensional imaging systems typically employ a camera, as above, in conjunction with a controlled or structured light source or a second camera to provide three-dimensional information (see Section 2.4).

Other devices can be used to provide X-ray, infrared, ultraviolet or even thermal image information.

2.3 Camera Geometry

In order to deduce anything about an object's position and orientation from an image, we need details of the camera's position and orientation in space relative to some reference coordinate system, called the *world coordinate system*. Furthermore, we must also have some geometric model of the camera arrangement, sometimes called the *imaging geometry* or *camera model*, and then some process of finding the various

2.3. CAMERA GEOMETRY

parameters present in the model. This latter process is known as *camera calibration*. A suitable model and its calibration, due to Tsai [228, 229], are described in the rest of this Section.

The choice of camera model depends on the particular application of the vision system. Since the main topic addressed in this book requires three-dimensional information, we will describe a camera model suitable for providing this type of information. In particular, we will use two cameras together to form a *stereo* system. The geometry of each camera is determined independently and then the knowledge of the position of a particular feature in the image formed by each camera enables the three-dimensional coordinates of the feature to be calculated. This is discussed later in Section 2.5.

The model we describe is equally useful for single camera or stereo use. If only two-dimensional image information is required from the model then the z dimension (depth) information in our discussion can be disregarded.

Usually the choice of world coordinates is arbitrary and if there is only one camera we can choose coordinate systems so that world and camera axes are aligned. This is not appropriate if there is more than one camera, however.

As are most camera models, this model is based on a *pin-hole* camera. Here, we ignore the extent of the aperture of the camera, and consider it to be negligible with respect to the size of the whole system.

We will now consider in turn the camera model and its calibration.

2.3.1 A Camera Model

As well as describing the geometry (position and orientation with respect to some coordinate system) of the camera, a camera model will also describe various internal camera characteristics such as focal length and lens distortion. The internal characteristics are called *intrinsic parameters*, while the parameters describing the geometry are said to be *extrinsic*.

Lens distortion deteriorates the geometric quality of an image and hence the ability to measure positions of objects in it. Lens distortion may be classified as being either *radial* or *tangential* [240]. Both types of distortion arise because light rays do not emerge from the lens in the directions predicted by simple geometric optics. Radial lens distortion causes image points to be shifted along radial lines from the optical axis. Its main cause is faulty grinding of lens elements. Tangential distortion occurs at right angles to radial lines from the optical axis. Its main cause is faulty centring of lens elements that make up a compound lens. Tangential distortion effects are usually much less significant than radial distortion effects, and so tangential distortion is not included in the camera model described below. Any gain in accuracy in modelling it would be negligible compared to sensor quantisation, whilst incorporating it would also cause numerical instability in the calibration methods [228, 229, 240]. We shall discuss methods that attempt to *correct* errors due to lens distortion and other causes in Section 3.5.

We shall firstly outline the camera model and then derive equations describing the model. The relationship between some of the parameters involved is illustrated in Fig. 2.3.

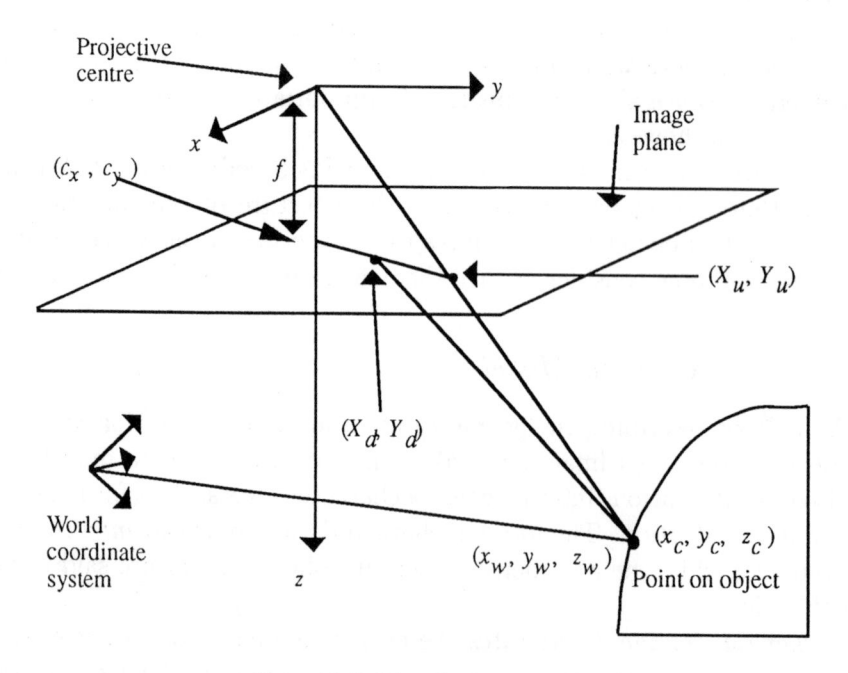

Figure 2.3: The camera model

The **intrinsic parameters** in this model are:

2.3. CAMERA GEOMETRY

- The *effective focal length*, f, which is the distance between the image plane and the projective or optical centre.

- The *lens distortion* parameters, k_1 and k_2, which model the radial lens distortion in the x and y directions.

- The *uncertainty scale factor*, s_x, between computed coordinates for pixels and actual pixel (CCD element) coordinates, arising due to errors in the digitisation of the incoming video signal. These arise because of slight hardware timing mismatches between image acquisition hardware and camera scanning hardware for successive pixels on a scan line [229].

- The *computer image coordinates* of the point at the centre of the image array, in pixel coordinates, (c_x, c_y).

The **extrinsic parameters** are:

- The parameters of the *transformation* between world and camera coordinates, expressed as a 3×3 rotation matrix, **R**, and a translation vector, **t**.

The transformation between three-dimensional world coordinates

$$\mathbf{v}_w = (x_w, y_w, z_w) \qquad (2.1)$$

and three-dimensional camera coordinates

$$\mathbf{v}_c = (x_c, y_c, z_c) \qquad (2.2)$$

is given by:

$$\mathbf{v}_c = \mathbf{R}\mathbf{v}_w + \mathbf{t}. \qquad (2.3)$$

The transformation between three-dimensional camera coordinates and an ideal undistorted image (X_u, Y_u) is given by the perspective projection:

$$\begin{aligned} X_u &= f x_c / z_c, \\ Y_u &= f y_c / z_c. \end{aligned} \qquad (2.4)$$

Allowing for radial distortion of the lenses gives image coordinates in the plane (X_d, Y_d):

$$\begin{aligned} X_d + D_x &= X_u, \\ Y_d + D_y &= Y_u, \end{aligned} \quad (2.5)$$

where

$$\begin{aligned} D_x &= X_d(k_1 r^2 + k_2 r^4), \\ D_y &= Y_d(k_1 r^2 + k_2 r^4), \\ r &= \sqrt{X_d^2 + Y_d^2}. \end{aligned} \quad (2.6)$$

The computer image coordinates (X_f, Y_f) of the actual image formed by the camera, allowing for the uncertainty scale factor, are then given by:

$$\begin{aligned} X_f &= \frac{s_x X_d}{d_x''} + c_x, \\ Y_f &= \frac{Y_d}{d_y'} + c_y, \end{aligned} \quad (2.7)$$

where

(X_f, Y_f) is the row and column number of the pixel image,
$d_x'' = d_x' N_{c_x}/N_{c_y}$,
d_x' is the distance between adjacent sensor elements in the x direction,
d_y' is the distance between adjacent sensor elements in the y direction,
N_{c_x} is the number of sensor elements in the x direction, and
N_{c_y} is the number of sensor elements in the y direction.

2.3.2 Camera Calibration

Camera calibration is the process of determining the internal geometric and optical characteristics of the cameras (intrinsic parameters) and the three-dimensional position and orientation of the cameras relative to the chosen world coordinate system (extrinsic parameters).

Camera calibration is probably the most important phase in the data acquisition stage since without accurate calibration there is no

2.3. CAMERA GEOMETRY

hope of obtaining data to a high degree of accuracy. Unfortunately, camera calibration is a very difficult problem. In stereo vision the problem is compounded since two cameras need to be calibrated. Any errors in the calibration procedures will lead to errors in the sensed position of points in each image and hence to errors in depth measurements, as will be further discussed in Section 2.5.2. The calibration method described here is based on the techniques proposed by Tsai [228, 229]. The method has proved to be popular in recent years and is the calibration method used by our vision system described in Section 2.5 as well as other vision systems [191].

Tsai's technique is designed to be accurate, efficient, versatile and to work with "off the shelf" cameras and lenses. Most other existing calibration techniques [240] do not satisfy all of these criteria and require one or more of the following:

- Special cameras, special processing equipment, or both — This prohibits large scale automation and versatility. It may also be uneconomical.

- Non-linear optimisation — The determination of the internal and external camera parameters using non-linear methods may require inefficient iterative searches, which also normally require good initial guesses to achieve reliable results. It is better to use linear parameters.

- Ignoring or assuming important camera properties — Some linear methods of estimating the parameters require that lens distortions and uncertainty scale factors are ignored. Others assume certain camera information such as focal length. These lead to inaccurate results.

The calibration of each camera is performed as follows. A calibration chart is placed in the view of the cameras. The calibration chart contains a lattice of identical black squares of known dimension $(dx \times dy)$ and known repeat distances $(dx' \times dy')$, as shown in Fig. 2.4. The corners of the squares are taken as calibration points. Since all points on the calibration chart are coplanar, the world coordinate system can be chosen so that the plane of the calibration chart is $z_w = 0$.

In order to improve the efficiency of the method the world coordinate system is also chosen such that (c_x, c_y) is not close to the origin in world coordinates and the y axis of the camera coordinate system does not pass close to the origin [229]. This ensures that not all of the components of the translation vector t are zero simultaneously and thus the computation procedure can be simplified.

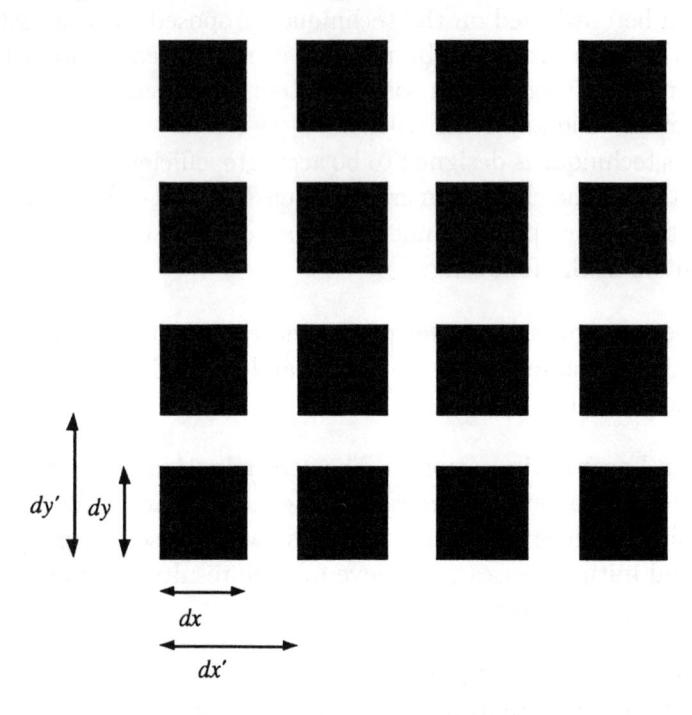

Figure 2.4: A calibration chart

Having obtained an image of the calibration chart, the locations of the calibration points need to be found. The dark areas in the image are located and are iteratively grown in all directions. The resulting dark regions correspond to the black squares on the chart. The calibration points at the corners of the squares can then be obtained by tracing around the boundaries of the dark regions.

Knowing the row and column number of each calibration point in the computer frame store and $N_{c_x}, N_{c_y}, d_x, d_y$ from camera and frame store specifications, values for (X_{d_i}, Y_{d_i}) for each calibration point (where

2.3. CAMERA GEOMETRY

$i = 1\ldots n$, the number of calibration points), can be found. As noted, the point (c_x, c_y) is taken to be at the centre of the pixel array and s_x may be assumed to be constant for similar cameras after an initial calibration (details are given in [133, 229]).

A linear system of equations is set up using Eqns. 2.3 – 2.5 for the components of **R** and **t** and solved (solving for **t** first). Having found solutions for **R** and **t** then estimates for the focal length and lens distortion coefficients are easily derived. It should be noted that there are many more calibration points than parameters in the system of equations. This overdetermines the equations given above and least squares or other approximation methods can be used to produce more accurate results.

2.3.3 Analysis of the Calibration

It is difficult to analytically assess the calibration accuracy since all measurements taken by the system are dependent on the calibration results. Therefore the calibration is usually assessed by how well it can sense the real world.

Three types of measures can be applied for this type of analysis [229]:

Error in locating calibration points — Having calibrated both the cameras, the errors in the sensed location of the calibration points can be measured by projecting the calibration points into the respective camera frame and measuring the distance between the sensed and calibrated position of the points. Thus a radius of ambiguity for the positions of points can be estimated.

Error in locating known points in the real world — Having calibrated the cameras, by using stereo triangulation to measure the three-dimensional coordinates of known points on test objects whose positions and dimensions are accurately known, the goodness of the calibration can be assessed.

General accuracy of 3D measurement — The calibrated cameras are used to perform many three-dimensional measurements whilst gathering data. The goodness of camera calibration can be measured in terms of how well the cameras perform such tasks as part of a complete system. Since this type of assessment is dependent on other processes, such as segmentation and matching, discussion is deferred until such processes have been described.

Our system will described in Section 2.5 and its calibration will be assessed there.

2.4 Three-Dimensional Imaging

An image containing explicit three-dimensional information (*depth data*) has many advantages over its two-dimensional counterpart. From a two-dimensional image only limited information can be deduced about the physical shape and size of an object in a scene (see Chapter 5). However with a three-dimensional image the geometry of the scene is represented in terms of three-dimensional coordinates. Thus, for example, measurements of the size of an object in a scene can be straightforwardly computed from its three-dimensional coordinates.

Recent technological advances (*e.g.* in camera optics, CCD cameras and laser rangefinders) have made the production of reliable and accurate three-dimensional depth data possible, and consequently many three-dimensional data acquisition systems have been developed [28, 65, 76, 130, 132, 191]. This Section assesses various means for producing depth data. An actual three-dimensional system using one of the described methods will then described in the following Section.

2.4.1 Depth Maps

The simplest and most convenient way of representing and storing the depth measurements taken from a scene is a *depth map*. A depth map is a two-dimensional array where the x and y distance information corresponds to the rows and columns of the array as in an ordinary image, and the corresponding depth readings (z values) are stored in the array's elements (pixels). Essentially, the depth map is similar to a grey scale image except that the z information replaces the intensity information.

2.4. THREE-DIMENSIONAL IMAGING

In practice, methods of capturing three dimensional data may result in a partially filled depth map, where only certain pixels contain valid depth values, and other pixels are set to some signal value such as zero to indicate there is not a valid reading for that pixel. Some acquisition methods produce a relatively small number of depth readings over a large region, some only produce values for pixels corresponding to edges present in a scene [191], whilst others produce a complete scan of at least some area of the scene giving a more or less complete array of depth readings. In the latter case, the depth map is said to be dense. A typical dense depth map, produced by the system described in Section 2.5, may have individual depth readings at every millimetre for a $300 \times 300\text{mm}^2$ area. Some examples of the type of depth maps produced by this system are presented in Figs. 2.5 and 2.6. In order to give the depth maps a realistic three-dimensional appearance in these pictures, the depth maps are viewed under three-dimensional projection, and only show a sample of the pixels in each depth map. Fig. 2.6 shows depth maps obtained from the actual vision system described later in this Chapter, while Fig. 2.5 shows artificial depth maps simulated by adding noise which is discussed in Section 2.5.2.

2.4.2 Introduction to Stereo Imaging

Before considering actual methods of three-dimensional imaging let us consider a simplified approach to the mathematics of the problem in order to aid understanding of the tasks involved. The approach given below closely follows that in Horn's book [123], which the reader may wish to consult for further details.

We will consider a set up using two cameras in *stereo*. However a similar approach can be adopted when considering other methods that involve stereo. For example, as we will see shortly, structured lighting and related methods involve a single camera and a fixed light source, the positions, orientations and physical properties of which are known. This information is then combined to obtain depth information.

Fig. 2.7 shows two cameras with their optical axes parallel and separated by a distance d. The line connecting the camera lens centres is called the *baseline*. Let this baseline be perpendicular to the line of sight of the cameras. Let the x axis of the three-dimensional world

(a) A test object "Widget"

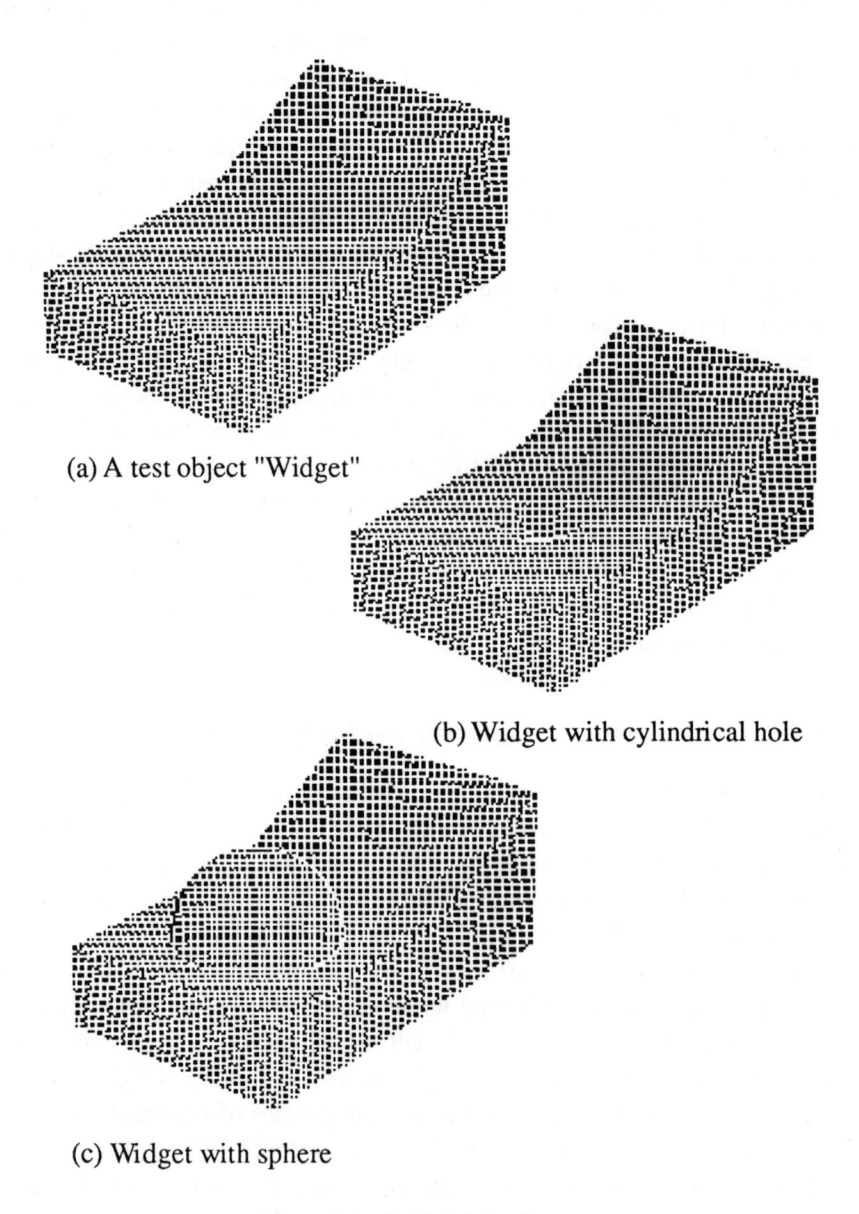

(b) Widget with cylindrical hole

(c) Widget with sphere

Figure 2.5: Artificial depth maps

2.4. THREE-DIMENSIONAL IMAGING

(a) Widget with small coin

(b) Another test object

Figure 2.6: Real depth maps

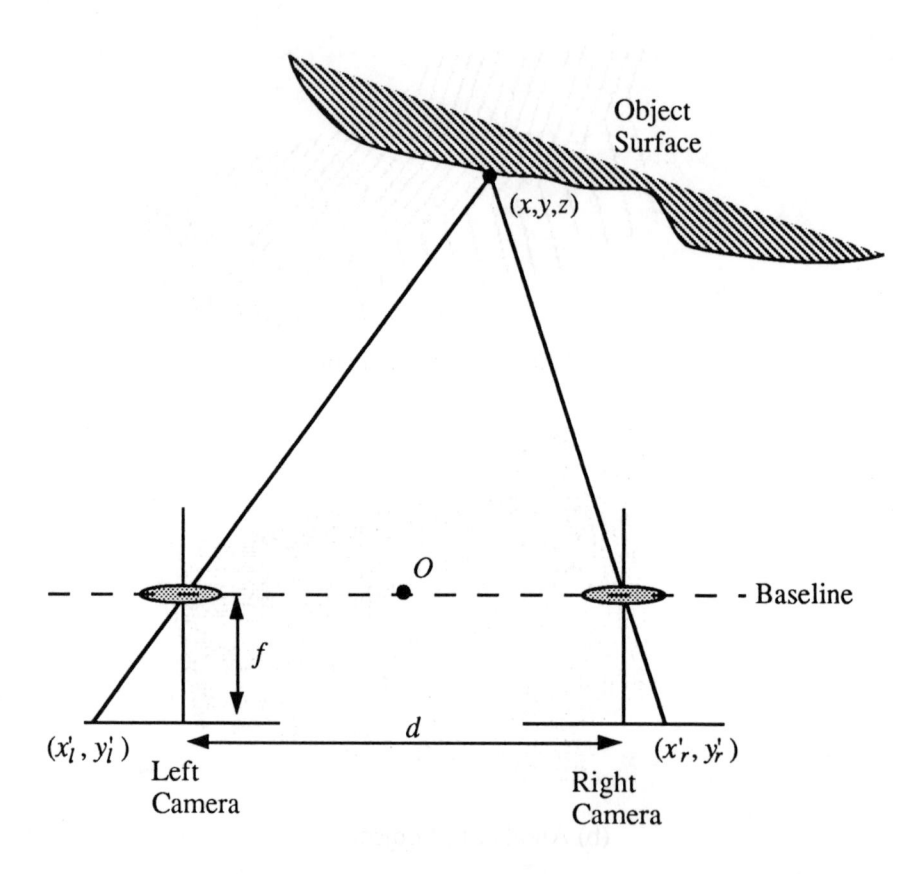

Figure 2.7: A simplified stereo imaging system

2.4. THREE-DIMENSIONAL IMAGING

coordinate system be parallel to the baseline and let the origin O of this system be mid-way between the lens centres.

Consider a point (x, y, z), in three-dimensional world coordinates, on an object. Let this point have image coordinates (x'_l, y'_l) and (x'_r, y'_r) in the left and right image planes of the respective cameras.

Let f be the focal length of both cameras, the perpendicular distance between the lens centre and the image plane. Then by similar triangles:

$$\frac{x'_l}{f} = \frac{x + d/2}{z}, \qquad (2.8)$$

$$\frac{x'_r}{f} = \frac{x - d/2}{z}, \qquad (2.9)$$

$$\frac{y'_l}{f} = \frac{y'_r}{f} = \frac{y}{z}. \qquad (2.10)$$

Solving for (x, y, z) gives:

$$x = \frac{d(x'_l + x'_r)}{2(x'_l - x'_r)}, \qquad (2.11)$$

$$y = \frac{d(y'_l + y'_r)}{2(x'_l - x'_r)}, \qquad (2.12)$$

$$z = \frac{df}{x'_l - x'_r}. \qquad (2.13)$$

The quantity $x'_l - x'_r$ which appears in each of the above equations is called the *disparity*.

Unfortunately as we shall soon see, things are not usually this simple when considering practical stereo imaging systems:

- With the given set up it is easy to measure the position of near objects accurately but impossible for far away objects. Normally, d and f are fixed. However, distance is inversely proportional to disparity. Disparity can only be measured in pixel differences.

- It can also be seen that disparity is proportional to the camera separation d. This implies that if we have a fixed error in determining the disparity then the accuracy of depth determination will increase with d. However as the camera separation becomes large difficulties arise in correlating the two camera images. In order to measure the depth of a point it must be visible to both cameras and we must also be able to identify this point in both images. As the camera separation increases so do the differences in the scene as recorded by each camera. Thus it becomes increasingly difficult to match corresponding points in the images. This problem is known as the *stereo correspondence problem*.

- It is impossible in practice to set up a vision system as described above where the axes and baseline are perfectly perpendicular.

Many of the problems mentioned above will be described in more detail in subsequent Sections. A method for measuring depth using a more general stereo set up will be discussed in Section 2.5.

2.4.3 Methods of Acquisition

The main methods for acquiring depth data fall into one of the following categories.

Laser Ranging Systems

Laser ranging works on the principle that the surface of the object reflects laser light back towards a receiver which then measures the time (or phase difference) between transmission and reception in order to calculate the depth.

Most laser rangefinders are intended for use at long distances (greater than 15m) and consequently their depth resolution is inadequate for detailed vision tasks. Some shorter range systems have been developed [13] but even these have an inadequate depth resolution (1cm at best) for most practical industrial vision purposes.

Structured Light Methods

These methods project patterns of light (grids, stripes, elliptical patterns *etc.*) onto an object. Surface shapes are then deduced from the distortions of the patterns that are produced when the light is reflected

2.4. THREE-DIMENSIONAL IMAGING

from the surfaces of the object. Knowing relevant camera and projector geometry, depth can be inferred by triangulation.

Many methods have been developed using this approach [16, 13]. The major advantage of these methods is that they are simple to use. However since the patterns are optically produced the patterns become sparser with distance resulting in quite low spatial resolutions. Some close range (4cm) sensors have been developed that produce good depth resolution (around 0.05mm) but these are not very practical for general purpose vision use due to their very narrow field of view and close range of operation.

Moire Fringe Methods

The advent of digital image processing equipment has stimulated the development of systems which employ Moire fringe methods [193]. The essence of the method is that a grating is projected onto an object and an image is formed in the plane of some reference grating as shown in Fig. 2.8. The image then interferes with the reference grating to form

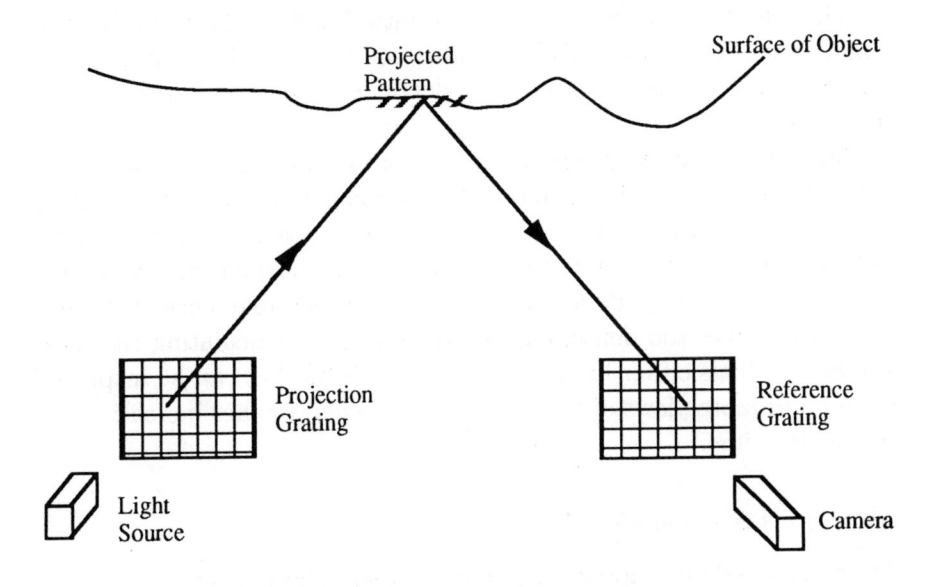

Figure 2.8: A moire projection system

Moire fringe contour patterns which appear as dark and light stripes, as demonstrated by Fig. 2.9. Analysis of the patterns then gives accurate descriptions of changes in depth and hence shape.

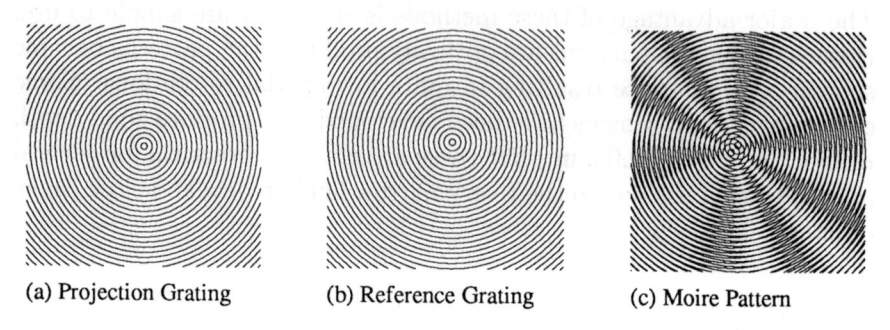

(a) Projection Grating   (b) Reference Grating   (c) Moire Pattern

Figure 2.9: Moire fringe patterns

However, it should be noted that ambiguities arise in interrogating the fringe patterns. It is not possible to determine whether adjacent contours are higher or lower in depth. This can be resolved by moving one of the gratings and taking multiple Moire images, noting the phase shift between patterns in the sequence of images [221].

The reference grating can also be omitted and its effect can be simulated in software. Methods based on this technique [135] can produce accurate depth resolutions but are dependent on the digitised image resolution.

Moire fringe methods are capable of producing very accurate depth data (resolution to within about 10 microns) but the methods have certain drawbacks. Firstly, Moire fringe methods are relatively computationally expensive for finding the depth information. Secondly, surfaces at a large angle are sometimes unmeasurable since the fringe density becomes too dense, causing difficulties in calculating the measurements. This either limits the class of objects that can be inspected or requires more views of the object to be taken, further increasing the processing time.

Shape from Shading Methods

Methods based on shape from shading employ photometric stereo techniques [24, 123, 207] to produce depth measurements, as will be dis-

2.4. THREE-DIMENSIONAL IMAGING

cussed in Section 5.4.7. Using a single camera, two or more images are taken of an object in a fixed position but under different lighting conditions. By studying the changes in brightness over a surface and employing constraints in the orientation of surfaces, certain depth information may be calculated.

However, shape from shading methods are mostly used when it is desired to extract surface shape information. Methods based on these techniques are not suited for general three-dimensional depth data acquisition for the following reasons:

- The methods are sensitively dependent on the illumination and surface reflectance properties of objects present in the scene.

- The methods only work well on objects with uniform surface texture.

- It is difficult to infer absolute depth, and only surface orientation is easily inferred.

Passive Stereoscopic Methods

We have already discussed stereoscopy as a technique for measuring range by triangulation to selected locations in a scene imaged by two cameras; further detail will be given in Section 2.5. The primary computational problem of stereoscopy is to find the correspondence of various points in the two images [188, 189].

This requires reliable extraction of certain features (such as edges or points) from the separate two-dimensional images and matching of corresponding features between images. Both of these tasks are non-trivial and can be computationally complex as will be seen in subsequent Chapters. Passive stereo may not be able to produce depth maps within a reasonable time. Furthermore, the depth data produced by passive stereo systems is typically sparse since the matching is performed between *high level* features, such as edges rather than points.

Problems in finding and accurately locating features in each image can be serious. These then contribute errors to the three-dimensional edge positions calculated from the two images. Such systems produce at best depth measurements accurate to a few millimetres. Thus, depth maps produced by these techniques, although suitable for some purposes, are not accurate enough for inspection, for example. One such passive stereo vision system is *TINA* [191], developed at Sheffield University.

Active Stereoscopic Methods

The problems of passive stereoscopic techniques may be overcome by illuminating the scene with a strong source of light (in the form of a point or line of light) which can be observed by both cameras to provide known corresponding points in each image. Depth maps can then be produced by sweeping the light source across the whole scene. The light source typically employed is a laser.

Clearly active stereoscopy can only be applied in controlled environments where external light sources can be limited or eliminated to achieve the desired results. Thus such techniques are particularly suited to industrial vision applications, such as *automatic inspection*.

2.5 Our Vision Acquisition System

2.5.1 Description

This Section describes the active stereoscopic subsystem which provides the three-dimensional data to our system for automatically inspecting mechanical parts. The system was developed at the British Aerospace Sowerby Research Centre (SRC) [130, 132, 133]. Whilst this Section considers some specific active stereo problems, many of the other issues discussed are not specific to any particular three-dimensional data acquisition technique, and will be of general interest.

A photograph of this vision system is shown in Fig. 2.10 and its main components are illustrated by the schematic diagram in Fig. 2.11.

The vision system consists of a matched pair of cameras and a laser scanner all mounted on an optical bench to reduce vibration. The cameras are a high sensitivity CCD type, each mounted on accurate turntables capable of independent rotation. The cameras are interfaced to a computer via a frame store that is able to grab frames from both cameras.

2.5. OUR VISION ACQUISITION SYSTEM

Figure 2.10: Vision system

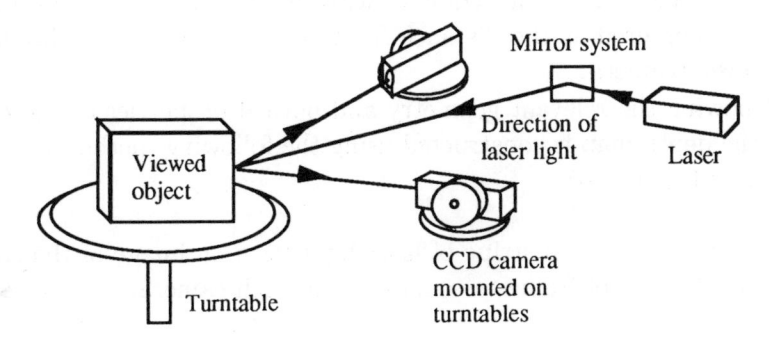

Figure 2.11: Schematic diagram of vision system

The scanner consists of a laser and two mirrors that can be moved independently to steer the beam of laser light. The mirrors are manipulated in such a way as to project a vertical stripe of light onto the object and to move the stripe across it.

The whole system is controlled by the computer which also has access to a database of solid geometric models.

Initially the cameras of the system must be calibrated as described in Section 2.3.2. However, the basic method for the determination of camera parameters has been extended to use multiple views of the calibration chart taken by rotating the cameras on their turntables (see Fig. 2.11). The angular positions of these turntables can be accurately determined and hence described described by suitable rotation matrices. All of these orientations will be relative to the initial angular position of each camera. This allows the camera positions and orientations in world coordinates to be estimated, as well as the other camera parameters, using least-squares or similar methods from the whole set of views. Using multiple views can give more accurate estimates of the intrinsic camera parameters in particular, since the calibration chart is more likely appear across the whole field of view of the cameras. This is especially relevant for the lens distortion parameters. More generally, using several views for calibration gives more data for least-squares fitting than using a single view, and hence better results.

Once the system has been calibrated, depth maps can then be extracted from the scene. The system operates by moving the laser stripe across the scene to obtain a series of vertical columns of pixels that are processed to give the required dense depth map. The depth of a point is measured as the distance from one of the cameras, chosen as the *master* camera.

Knowing the relevant geometry and optical properties of the cameras the depth map is constructed using the following method (refer to Fig. 2.12 for details):

1. For each vertical stripe of laser light form an image of the stripe in the pair of frames associated with each camera.

2. For each row in the master camera image, search until the stripe is found at point $P(i,j)$, say.

2.5. OUR VISION ACQUISITION SYSTEM

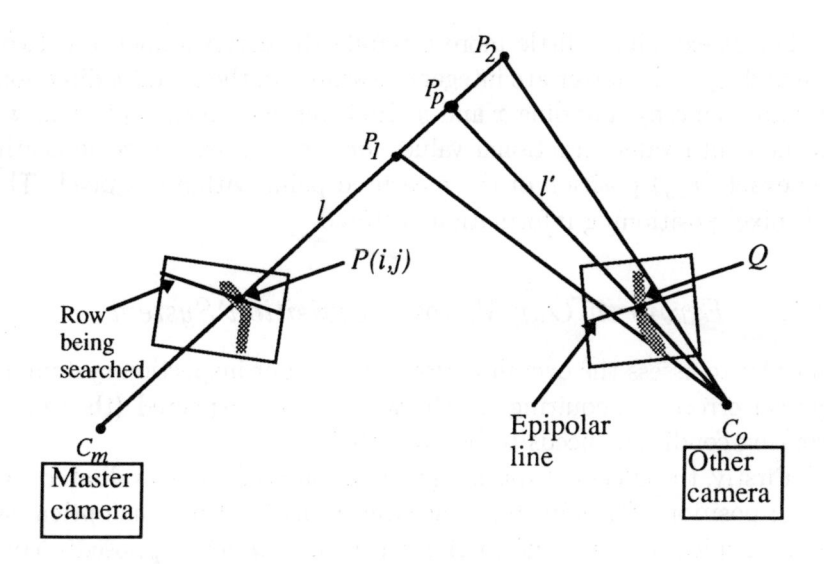

Figure 2.12: Measuring a depth value

3. Form a three-dimensional line l passing through the centre C_m of the master camera and $P(i,j)$.

4. Construct the epipolar line which is the projection of the line l into the image formed by the other camera. This is achieved by projecting two arbitrary points P_1 and P_2 into the image and constructing a line between the two projected points.

5. Search along the epipolar line for the laser stripe. If it is not found, no value is written. If it is found at Q, proceed to Step 6.

6. Find the point P_p on line l which corresponds to Q. Calculate the (x, y, z) coordinates of P_p, and store the z value at position (i, j) corresponding to x and y in the depth map.

The position of the point P_p is easily found by projecting a line l' from the centre C_o of the secondary camera passing through Q. The intersection of the lines l and l' gives the coordinates of P_p.

The depth map is formed by using a world coordinate system fixed on the master camera with its origin at C_m.

Let us examine a little more carefully the correspondence between x, y and i, j. The latter are integers measured in the x and y directions, and are found by rounding x and y. However, within each pixel, as well as the depth value, fractional values for x and y are stored indicating the exact (x, y) position of the measured point within the pixel. Thus sub-pixel positioning information is stored.

2.5.2 Errors in Our Vision Acquisition System

In order to assess the overall performance of our inspection system, the typical errors in acquiring depth values under expected (theoretical) working conditions needs to be evaluated.

Firstly, the effects of spatial quantisation need to be evaluated. The (x, y) position of a point has inaccuracies arising from the optical system. Typically, a z value in the depth map which supposedly corresponds to a position (x, y) may actually be a depth reading for a point whose (x, y) coordinates are up to 0.2 pixels away from the stored position. This result is based on observed differences in depth maps obtained from identical setups. It depends on many things, including surface reflectance properties of objects, the optical properties of the lenses and camera system, and the resolution of the framestore. It also depends on the distance from the vision system to the object as will be explained below. It should also be noted the laser spot size is finite and this implies that the depth reading is spatially averaged over the spot.

Thus an *area of uncertainty* in the precise location of the sensed data point arises, as illustrated in Fig. 2.13. Both the angular and depth resolution can be readily evaluated.

In this schematic view of the system let

γ = field of view of cameras in radians,
n = number of resolvable points across field of view,
d = separation of cameras in metres,
β = angle between optical axis and image point in radians,
z = depth in metres,
δ = angular resolution of the camera,
δ_z = depth resolution.

2.5. OUR VISION ACQUISITION SYSTEM

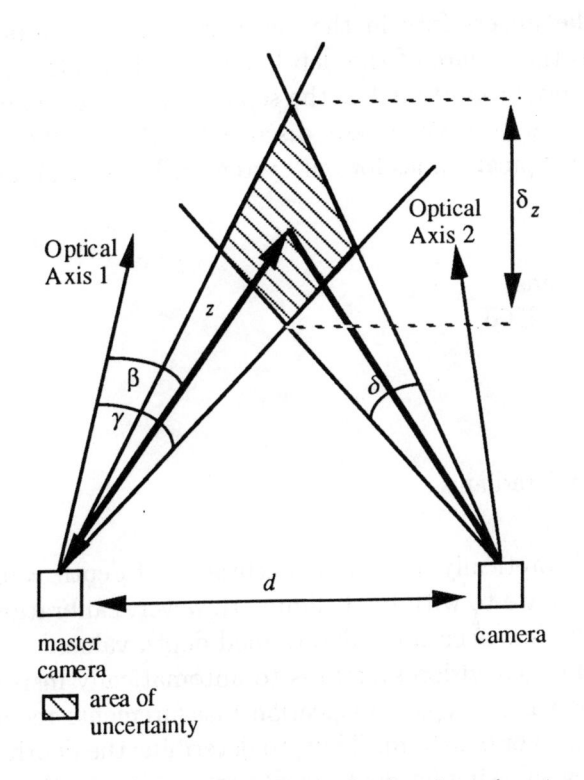

Figure 2.13: Error in measuring a depth point using stereo

Then the angular resolution may be expressed as

$$\delta = 0.5 \tan^{-1}(\frac{\tan \gamma}{n}), \tag{2.14}$$

while the depth resolution can be shown to be [130]

$$\delta_z = \frac{2z^2 \tan \delta}{d}(\tan^2 \beta + \frac{d}{z}\tan \beta + 1) \tag{2.15}$$

Thus, the uncertainty in the depth reading is to first order proportional to the square of the depth value z. It is also, again to first order, inversely proportional to the separation of the cameras, d, causing problems as we have already discussed in Section 2.4.2.

Inserting typical values for our system in Eqns. 2.14 and 2.15:

$\beta = 0.0$ radians,
$d = 0.3$m,
$\gamma = 0.08$ radians,
$n = 512 \times 5 = 2560$,
$z = 0.7$m

gives

$\delta = 4.9 \times 10^{-5}$ radians,
$\delta_z = 0.052$mm.

Thus, the theoretically achievable accuracies of depth values acquired by the system are to within 0.052mm. However, calibration errors will lead to rather less accurately determined depth values.

The aim of our vision system is to automatically inspect manufactured components. Typical inspection measurements require an accuracy of 0.01inch or 0.254mm. Thus, to determine the depth of a point to this accuracy, allowing for depth resolution errors, depth readings must be determined by the system to within $0.254 - 2 \times 0.052 = 0.15$mm under the above assumptions. This is the maximum error due to incorrect calibration which can be tolerated.

A different problem which also needs to be considered is that greater spatial quantisation errors arise near sharp edges of the object under observation, due to laser light being scattered there. Consequently, the intensity of the images of edges produced in the CCD cameras is less, which leads to errors in accurately locating edges in images.

2.5.3 Performance of Our Vision System

Our vision system was calibrated using the techniques described in Section 2.3.2. However, it is difficult to assess the accuracy of the calibration of the system except for its performance in sensing the real world, as discussed in that Section. An analysis can be made from the following point of view:

- Error in locating calibration points in the camera frame — After calibration using least-squares fitting, the locations of the calibration points in camera coordinates are calculated. The resulting root-mean-square error in position is called the *radius of ambiguity*. The current camera setup gives an average error of the order of 0.5 of a pixel. With a typical field of view of 200mm and taking 256×256 pixel depth maps gives an error of about 0.4mm in position.

- Errors in locating sensed points in the real world — Here points lying on the ground plane ($z = 0$) are measured and their respective errors in z coordinate noted. Typical root-mean-square errors of this type using the current camera setup are of the order of 0.5mm. Since the line of sight of the cameras typically makes an angle of about 30°–40°, under projection this error is in agreement with the previous paragraph. Errors of this type are illustrated in Fig. 2.14 where an acquired depth map has been superimposed onto the image of the test object (*the widget*) obtained from the master camera. Clear misplacements around the edges are visible. These are particularly noticeable on the faces furthest from the camera.

- General performance of three-dimensional measurement — The overall performance of our complete vision system is described in Part 3 of this book where the use of the system use to perform three-dimensional inspection tasks is considered. Other performance aspects are also considered in Chapter 9 of Part 2.

Considering the above results and the theoretical system errors as derived in Section 2.5.2, it can be concluded that the current camera calibration is not sufficiently accurate for inspection to tolerances of

Figure 2.14: Error in placement of depth map on image

2.6. SIMULATING DEPTH DATA

0.01inch. Errors in the calibration procedure need to be investigated and overcome before a fully operational vision acquisition system suitable for inspection is available. However, the calibration errors *are* small enough to allow inspection to tolerances not much greater than those required in practice. It is realistic to expect future advances both in the camera calibration method and the hardware to achieve the necessary accuracy.

2.6 Simulating Depth Data

Throughout the book, and in particular in Parts 2 and 3 of the book, reference is made to experiments conducted with artificial or simulated depth data of the type shown in Fig. 2.5. In order to assess the respective performance of the various algorithms used in our vision and inspection processes as described later, we need to simulate artificial depth data. Using such data also lets us investigate the suitability of the various algorithms for other tasks and hardware. Thus we may simulate input hardware of differing accuracies to find the resulting effect on the algorithms and systems under consideration. Depth data of the sort supplied by our vision system is simulated.

Depth maps are simulated with errors in depth reading by adding Gaussian noise [55, 112]. The standard deviation of the Gaussian distribution is taken as the size of the error. Thus, to simulate depth maps which will test the system working at the desired accuracy a standard deviation of 0.15mm is used — the largest allowable working depth resolution error for our vision system. Differing distances between camera and object can also be modelled by noting that the depth error is approximately proportional to the square of the distance so increasing the distance by a factor of $\sqrt{2}$ requires doubling the standard deviation of the added noise. For a definition of Gaussian distribution see Section 4.5.3.

2.7 Exercises

Exercise 2.1 *Explain which means of acquiring an image you would choose to use to perform the following tasks:*

(a) Inspecting printed circuit boards. (b) Traffic surveillance. (c) Production of three-dimensional aerial images. (d) General purpose two-dimensional medical imaging. (e) Obtaining three-dimensional profiles of internal human body organs. (f) Obtaining three-dimensional profiles of external body features such as face, back, and chest. (g) Three-dimensional imaging of natural scenes. (h) Guiding a robot manipulator arm in an industrial application.

Exercise 2.2 *What means of image acquisition are possible other than those discussed so far?*

A manufacturer wishes to build an autonomous guided vehicle that can be guided solely by visual means. The manufacturer recognises that adopting a single visual acquisition method might not be sufficient to enable his vehicle to be adapted to many different environments. He therefore wants to incorporate several different visual sensing methods.

Discuss the methods that you would suggest for use in the vehicle, stating clearly why each would be of use. (Your answer should not necessarily limit itself to the methods that have been described in this book). Briefly discuss how all the information present could be interpreted efficiently.

Exercise 2.3 *Suggest various applications where the use of* **colour** *imaging could be useful or essential.*

Exercise 2.4 *A Moire projection system projects a grating consisting of 1mm vertical lines 1mm apart onto an object and uses a reference grating consisting of similar stripes angled at 10° to the vertical. The projector and camera are mounted side by side so that the lines of sight to both are almost parallel.*

Predict the spacing of the fringes in the interference pattern obtained from the system when the grating is projected onto a flat surface perpendicular to the line of sight of the camera. At what angle to the line of sight does the fringe pattern projected onto the flat surface become too dense to be measured? Assume that the camera has a resolution of 256×256 pixels, a focal length of 8mm and a field of view of 30°, and that the whole system is placed 1m away from the object.

2.7. EXERCISES

Exercise 2.5 *A stereo camera system uses lenses with a focal length of 55mm. Assuming that the cameras are placed 0.5m apart, what is the greatest spacing of the pixels in the sensors which will allow distances of 5mm to be measured to within ±10% ?*

Exercise 2.6 *It is proposed to make a night-time automatic landing system for commercial aircraft, with a camera placed at each wingtip using the geometry shown in Fig. 2.15. The cameras are rotated to point directly at a distant landing beacon. The distance between the wingtips*

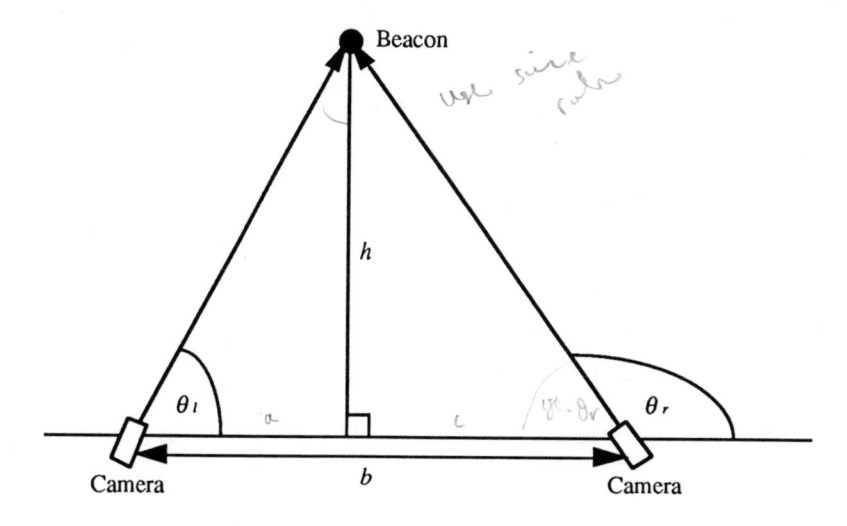

Figure 2.15: Stereo for automatic landing

is b, while h is the distance to the landing beacon (greatly shortened in the diagram).

1. Show that the distance h to the landing beacon is given by

$$h = \frac{b \sin \theta_r \sin \theta_l}{\sin(\theta_r - \theta_l)};\qquad(2.16)$$

2. By making suitable approximations, show that if the beacon is far away, and the aircraft is pointing almost towards it, a small percentage error in $\theta_r - \theta_l$ leads to a similar percentage error in h, but of the opposite sign;

3. By making suitable assumptions, estimate how close the aircraft must be to the beacon before its distance can be found to within 5%, given that θ_r and θ_l can each be measured to within $1/100°$.

Exercise 2.7 *By referring to Fig. 2.13, prove Eqn. 2.15.*

Chapter 3

Image Preprocessing

3.1 Introduction

Image processing is in many cases concerned with taking one array of pixels as input and producing another array of pixels as output which in some way represents an improvement to the original array. In many cases, the aim of this processing is to make the image more intelligible and useful to the human eye and brain. For example, this processing may remove noise, it may improve the contrast between light and dark features in the image, it may remove blurring in an image caused by movement of the camera during image acquisition, or it may correct for geometrical distortions caused by the lens. Nevertheless, such enhancements of the image may also be useful as a preprocessing step for a computer vision system. An image that is more intelligible to a human being will also often be a better starting point for computer-based extraction of information. When used specifically for this purpose, image processing is often referred to as *image preprocessing*.

Usually, the array of pixel values is rectangular, and each pixel represents the brightness level of a given part of the image. However, the pixels may contain other data, for example temperatures if the image has been captured using a heat-sensitive camera. One particular type of non-intensity image we have already met is the *depth map*, where the pixel values are distances from each pixel to the corresponding point on the surface of an object under inspection.

Image processing methods may be broadly divided into *real space* methods and *Fourier space* methods. The former work by directly processing the input pixel array to produce the output array. The latter work by firstly deriving a new representation of the input data by performing a *Fourier transform*, which is then processed, and finally, an *inverse Fourier transform* is performed on the resulting data to give the final output image. Although extra work is involved in computing the Fourier transform and its inverse, this is compensated for by the fact that certain processes applied to the image can be described in more straightforward and simple terms when the image is represented in its Fourier form. Thus this Chapter commences with an introduction to Fourier methods before proceeding to describe various image preprocessing methods.

3.2 Fourier Methods

3.2.1 Introduction

Consider a complicated sound such as the noise of a car horn. It is possible to describe this sound in two related ways. The first of these samples the amplitude of the sound many times a second, which gives an approximation to the sound as a function of time. Alternatively, it is possible to analyse the sound in terms of the pitches of the notes, or frequencies, which make the sound up, recording the amplitude of each frequency. In practice, there will be a continuous range of frequencies, rather than a discrete set of them, but again, the sound would be sampled for a set of equally spaced frequencies. Both of these descriptions contain equivalent information, and as will be seen later, either of these descriptions can be converted into the other.

In just the same way brightness along a line, information in space rather than time, can be recorded as a set of values measured at equally spaced distances apart, or equivalently, at a set of spatial frequency values. Each of these frequency values is referred to as a *frequency component*.

In image processing, we are interested in a two-dimensional array of pixel measurements on a uniform grid, but the principle is the same, and this information can also be described in terms of a two-dimensional grid of spatial frequencies. A given frequency component now speci-

3.2. FOURIER METHODS

fies what contribution to the overall image is made by data which is changing with specified x and y direction spatial frequencies.

Let us try to obtain an intuitive feeling for how the frequency data is related to the original pixel array. If an image has large values at *high* frequency components, it means that the data is changing rapidly on a short distance scale. A typical example might be a page of text, where there are many changes from black to white and back again over short distances. On the other hand, if the image has large *low* frequency components, it means that the large scale features of the picture are more important. For example, we might have a single fairly simple object which occupies most of the image, with relatively little internal detail, such as a picture of the planet Mars.

The tool which converts a spatial (real space) description of an image into one in terms of its frequency components is called the *Fourier transform*, and the new version is usually referred to as the Fourier space description of the image. The corresponding inverse transformation which turns a Fourier space description back into a real space one is called the *inverse Fourier transform*.

3.2.2 Theory

We give here only a brief description of the most important properties of the Fourier transform which are useful in image processing. Further details and proofs can be found in the book by Gonzalez and Wintz [105], for example.

Let us commence by considering a continuous function $f(x)$ of a single variable x representing distance, such as the brightness along a line. Then the Fourier transform of that function is denoted $F(u)$, where u represents spatial frequency. $F(u)$ can be computed from $f(x)$ using the formula

$$F(u) = \int_{-\infty}^{\infty} f(x) e^{-2\pi i x u} \, dx. \tag{3.1}$$

Note that in general $F(u)$ will be a complex quantity, because of the presence of i in the exponential, even though the original data is purely real. The meaning of this is that not only is the magnitude of each frequency present important, but that its phase relationship is too, *i.e.*

the point in its cycle at which each frequency component is considered to commence.

The inverse Fourier transform for regenerating $f(x)$ from $F(u)$ is given by

$$f(x) = \int_{-\infty}^{\infty} F(u)e^{2\pi i x u} \, du, \qquad (3.2)$$

which is rather similar, except that the exponential term has the opposite sign. One benefit of this is that with slight modifications the same algorithm can be used for computing the Fourier transform and its inverse.

As an example, consider a particular function $f(x)$ defined as

$$f(x) = \begin{cases} 1 & \text{if } |x| \leq 1 \\ 0 & \text{otherwise,} \end{cases} \qquad (3.3)$$

shown in Fig. 3.1, which is sometimes called a *top hat* function on

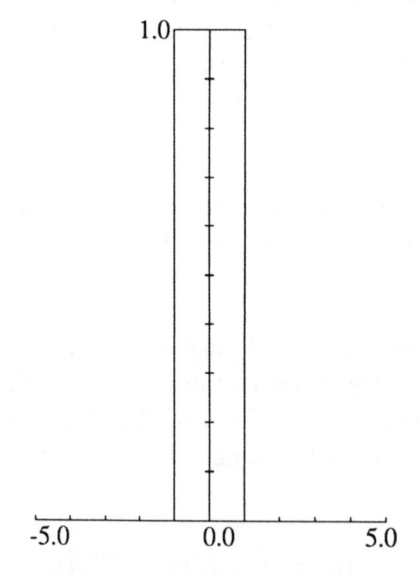

Figure 3.1: A top hat function

account of its shape. We can compute its Fourier transform as follows.

$$F(u) = \int_{-\infty}^{\infty} f(x)e^{-2\pi i x u} \, dx$$

3.2. FOURIER METHODS

$$= \int_{-1}^{1} 1 \times e^{-2\pi i x u} \, dx$$

$$= \frac{-1}{2\pi i u}(e^{2\pi i u} - e^{-2\pi i u})$$

$$= \frac{\sin 2\pi u}{\pi u}. \qquad (3.4)$$

In this case $F(u)$ is purely real, which is a consequence of the original data being symmetric in x and $-x$. A graph of $F(u)$ is shown in Fig. 3.2. This function is often referred to as the Sinc function.

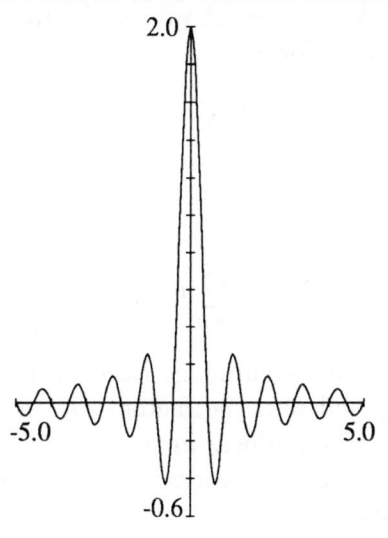

Figure 3.2: Fourier transform of a top hat function

The Fourier transform of a function of two variables is really no more complicated as each variable is treated separately. Thus, if $f(x,y)$ is such a function, for example the brightness in an image, its Fourier transform is given by

$$F(u,v) = \int_{-\infty}^{\infty}\int_{-\infty}^{\infty} f(x,y) e^{-2\pi i(xu+yv)} \, dx \, dy, \qquad (3.5)$$

and the inverse transform, as might be expected, is

$$f(x,y) = \int_{-\infty}^{\infty}\int_{-\infty}^{\infty} F(u,v) e^{2\pi i(xu+yv)} \, du \, dv. \qquad (3.6)$$

In practice when we are carrying out image processing, we do not have a continuous function to work with, but rather a set of regularly spaced data values, the pixel values. Thus, we need a discrete formulation of the Fourier transform, which takes such regularly spaced data values, and returns the value of the Fourier transform for a set of values in frequency space which are equally spaced. This is done quite naturally by replacing the integral by a summation, to give the *discrete Fourier transform* or DFT for short. Taking the one-dimensional case first, it is most convenient now to assume that x goes up in steps of 1, and that there are N samples, at values of x from 0 to $N-1$. In this case the DFT takes the form

$$F(u) = \frac{1}{N} \sum_{x=0}^{N-1} f(x) e^{-2\pi i x u/N}, \tag{3.7}$$

while the inverse DFT is

$$f(x) = \sum_{x=0}^{N-1} F(u) e^{2\pi i x u/N}. \tag{3.8}$$

Minor changes from the continuous case are a factor of $1/N$ in the exponential terms, and also the factor $1/N$ in front of the forward transform which does not appear in the inverse transform.

The two dimensional DFT works in exactly the same way, so for an $N \times M$ grid in x and y we have

$$F(u,v) = \frac{1}{NM} \sum_{x=0}^{N-1} \sum_{y=0}^{M-1} f(x,y) e^{-2\pi i(xu/N + yv/M)}, \tag{3.9}$$

and

$$f(x,y) = \sum_{u=0}^{N-1} \sum_{v=0}^{M-1} F(u,v) e^{2\pi i(xu/N + yv/M)}. \tag{3.10}$$

Often $N = M$, and it is then it is more convenient to redefine $F(u,v)$ by multiplying it by a factor of N, so that the forward and inverse transforms are more symmetrical:

$$F(u,v) = \frac{1}{N} \sum_{x=0}^{N-1} \sum_{y=0}^{N-1} f(x,y) e^{-2\pi i(xu+yv)/N}, \tag{3.11}$$

3.2. FOURIER METHODS

and
$$f(x,y) = \frac{1}{N} \sum_{u=0}^{N-1} \sum_{v=0}^{N-1} F(u,v) e^{2\pi i(xu+yv)/N}. \quad (3.12)$$

3.2.3 *Convolutions*

As will be described later, several important optical effects can be described in terms of convolutions. Indeed, the effect that an imperfect observing instrument has on an ideal image can often be modelled as a convolution with some function which describes the instrument's characteristics. The relationship which exists between convolutions and Fourier transforms gives us a method by which we can try to compensate for the effect of the instrument, and reconstruct the ideal image.

Let us examine the concepts using one-dimensional continuous functions; the ideas extend directly to the discrete two-dimensional case.

The convolution of two functions $f(x)$ and $g(x)$, written $f(x) * g(x)$, is defined by the integral

$$f(x) * g(x) = \int_{-\infty}^{\infty} f(\alpha) g(x-\alpha) \, d\alpha. \quad (3.13)$$

For example, let us take two top hat functions of the type described earlier. Let $f(\alpha)$ be the top hat function shown in Fig. 3.1,

$$f(\alpha) = \begin{cases} 1 & \text{if } |\alpha| \leq 1 \\ 0 & \text{otherwise,} \end{cases} \quad (3.14)$$

and let $g(\alpha)$ be as shown in Fig. 3.3, defined by

$$g(\alpha) = \begin{cases} 1/2 & \text{if } 0 \leq \alpha \leq 1 \\ 0 & \text{otherwise.} \end{cases} \quad (3.15)$$

Then $g(-\alpha)$ is the reflection of this function in the vertical axis, while $g(x-\alpha)$ is the latter shifted to the right by a distance x. Thus for a given value of x, $f(\alpha)g(x-\alpha)$ integrated over all α is the area of overlap of these two top hats, as $f(\alpha)$ has unit height. An example is shown for x in the range $-1 \leq x \leq 0$ in Fig. 3.4. If we now consider x moving from $-\infty$ to $+\infty$, we can see that for $x \leq -1$ or $x \geq 2$, there is no overlap; as x goes from -1 to 0 the area of overlap steadily increases

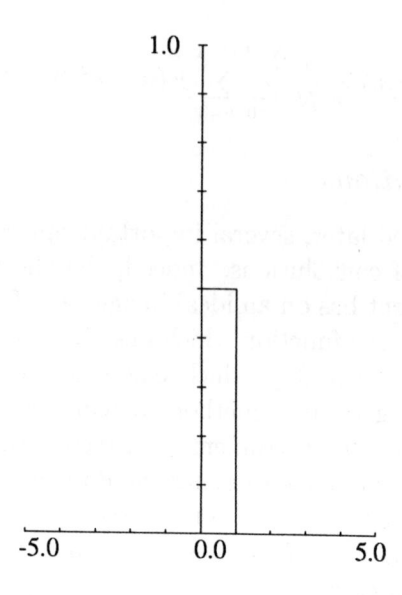

Figure 3.3: Another top hat: $g(\alpha)$

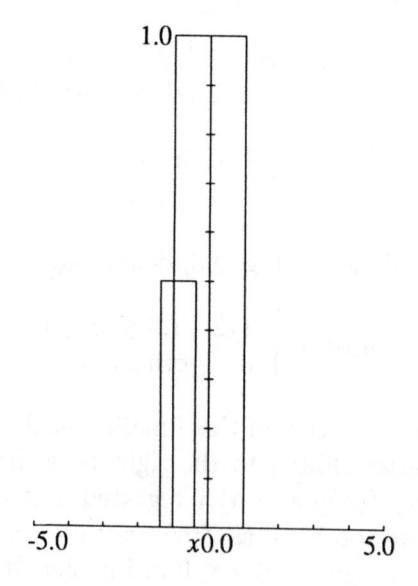

Figure 3.4: Convolving two top hats

3.2. FOURIER METHODS

from 0 to 1/2; as x increases from 0 to 1, the overlap area remains at 1/2; and finally as x increases from 1 to 2, the overlap area steadily decreases again from 1/2 to 0. Thus the convolution of $f(x)$ and $g(x)$, $f(x) * g(x)$, in this case has the form shown in Fig. 3.5, or expressed

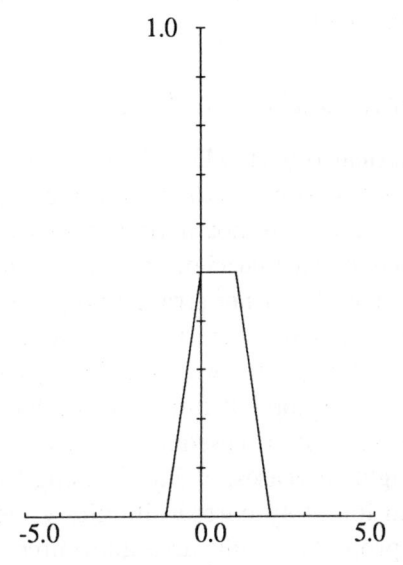

Figure 3.5: Convolution of two top hats

symbolically,

$$f(x) * g(x) = \begin{cases} (x+1)/2 & \text{if } -1 \leq x \leq 0 \\ 1/2 & \text{if } 0 \leq x \leq 1 \\ 1 - x/2 & \text{if } 1 \leq x \leq 2 \\ 0 & \text{otherwise.} \end{cases} \quad (3.16)$$

One major reason that Fourier transforms are so important in image processing is the *convolution theorem* which states that

Theorem 3.1 *If $f(x)$ and $g(x)$ are two functions with Fourier transforms $F(u)$ and $G(u)$, then the Fourier transform of the convolution $f(x) * g(x)$ is simply the product of the Fourier transforms of the two functions, $F(u)G(u)$.*

Thus in principle we can undo a convolution, for example to compensate for a less than ideal image capture system, by taking the Fourier transforms of the imperfect image and of the function describing the effect of the system, and by dividing the former by the latter to obtain the Fourier transform of the ideal image. We can then take the inverse Fourier transform to recover the ideal image. This process is sometimes referred to as *deconvolution*.

3.2.4 The Dirac Delta Function

A useful mathematical object when dealing with Fourier transforms and convolutions is the *Dirac delta function* $\delta(x)$, also called the *impulse function*. In fact, it is not a proper mathematical function at all, but rather what mathematicians call a *generalised function*. The delta function $\delta(x)$ has the rather strange property that it is zero for all values of x except at $x = 0$, where it becomes infinite. Nevertheless it does so in a controlled way. One way of defining the delta function is as the limit of a sequence of top hat functions, each of which is centred on $x = 0$ and has unit area. As this sequence of top hat functions becomes narrower, their height increases, always keeping their unit area, in the limit passing to an infinitely narrow, infinitely high top hat function. Nevertheless, the property of unit area guarantees that

$$\int_{-\infty}^{\infty} \delta(x)\,dx = 1. \tag{3.17}$$

More usefully, it can be seen that

$$\int_{-\infty}^{\infty} \delta(x_0 - x) f(x)\,dx = f(x_0), \tag{3.18}$$

as the delta function is non-zero only when $x = x_0$, so only the value of f at $x = x_0$ matters, while the total area under the delta function is 1.

More generally, if we consider a delta function located not at the origin, but at $x = t$, in other words $\delta(x - t)$, and convolve it with some other function $f(x)$, using the above result, we have

$$\begin{aligned} f(x) * \delta(x - t) &= \int_{-\infty}^{\infty} f(\alpha) \delta(\alpha - x + t)\,d\alpha \\ &= f(x - t). \end{aligned} \tag{3.19}$$

Thus, the result is a copy of the original function shifted from the origin to $x = t$.

3.2. FOURIER METHODS

One example of the usefulness of delta functions will be seen in Section 3.5.3, which considers correcting blurring in images.

3.2.5 Other Properties of Fourier Transforms

We give here just a few of the other properties of Fourier transforms which are useful when reasoning or computing with them.

The Fourier transform is a *linear operator*. This means that

Theorem 3.2 *If $f(x)$ and $g(x)$ are two functions with Fourier transforms $F(u)$ and $G(u)$, then the Fourier transform of $af(x) + bg(x)$ where a and b are constants is simply $aF(u) + bG(u)$.*

Shifting the real space data through a fixed distance x_0 has the effect that

Theorem 3.3 *If $f(x)$ is a function with Fourier transform $F(u)$, then the Fourier transform of $f(x - x_0)$ is given by $e^{-2\pi i x_0 u} F(u)$.*

Note that the exponential term here has unit magnitude for all values of u. Thus, the magnitude of the resulting Fourier transform is left unchanged for all values of u, although the relative sizes of the real and imaginary components are altered.

If the spacing of the real space data in distance is scaled by a factor a, we have that

Theorem 3.4 *If $f(x)$ is a function with Fourier transform $F(u)$, then the Fourier transform of $f(ax)$ where a is a real constant is given by $\frac{1}{|a|} F(\frac{u}{a})$.*

In other words, if we spread out the data in real space, the data is compressed into a more compact region of Fourier space. This is what we expect — if we double the spacing of a grid pattern, its spatial frequency is halved. Note however that the magnitude of the Fourier representation is also affected.

One useful property in two dimensions is that if we rotate the real space data, its Fourier transform is rotated by the same angle, or more exactly

Theorem 3.5 *If $f(x,y)$ is a function with Fourier transform $F(u,v)$, on expressing these functions in terms of polar coordinates r, θ, ρ, ϕ where $x = r\cos\theta, y = r\sin\theta, u = \rho\cos\phi, v = \rho\sin\phi$ so that $f(x,y)$ and $F(u,v)$ become $f(r,\theta)$ and $F(\rho,\phi)$ respectively, the Fourier transform of $f(r, \theta + \omega)$ where ω is a constant is given by $F(\rho, \phi + \omega)$.*

Basically this means that the two-dimensional Fourier transform is an intrinsic property of the data which is independent of our choice of axis directions.

A final property is that the zeroth component of the Fourier space representation is just the average data value (apart from a factor of $1/N$, in two dimensions). This is shown below for the two-dimensional DFT case:

$$F(u,v) = \frac{1}{N} \sum_{x=0}^{N-1} \sum_{y=0}^{N-1} f(x,y) e^{-2\pi i(xu+yv)/N}, \qquad (3.20)$$

so

$$F(0,0) = \frac{1}{N} \sum_{x=0}^{N-1} \sum_{y=0}^{N-1} f(x,y) = N\overline{f}(x,y). \qquad (3.21)$$

3.2.6 The Fast Fourier Transform Algorithm

We shall now consider how the discrete Fourier transform (DFT) may be computed efficiently. To compute the DFT in one dimension,

$$F(u) = \frac{1}{N} \sum_{x=0}^{N-1} f(x) e^{-2\pi i xu/N} \qquad (3.22)$$

has to be evaluated for N values of u, which if done in the obvious way clearly takes N^2 multiplications. (Note that the evaluation of terms of the form $e^{-2\pi i xu/N}$ is relatively unimportant. These are N^{th} roots of unity and so there are only N different such values, which can be computed once at the beginning of the calculation and stored in an array.)

It is possible to calculate the DFT more efficiently than this, using the *fast Fourier transform* or FFT algorithm, which reduces the number of operations to $O(N \log_2 N)$. In what follows, we shall assume for simplicity that N is a power of 2, $N = 2^n$, although other versions

3.2. FOURIER METHODS

of fast transform algorithms exist for other input data sizes. A useful reference for these, which also gives a little of the history of the FFT algorithm, is Blahut [23].

If we define ω_N to be the Nth root of unity given by $\omega_N = e^{-2\pi i/N}$, and set $M = N/2$, we have

$$F(u) = \frac{1}{2M} \sum_{x=0}^{2M-1} f(x) \omega_{2M}^{xu}. \qquad (3.23)$$

This can be split apart into two separate sums of alternate terms from the original sum,

$$F(u) = \frac{1}{2}\left(\frac{1}{M}\sum_{x=0}^{M-1} f(2x)\omega_{2M}^{(2x)u} + \frac{1}{M}\sum_{x=0}^{M-1} f(2x+1)\omega_{2M}^{(2x+1)u}\right). \qquad (3.24)$$

Now, since the square of a $2M^{th}$ root of unity is an Mth root of unity, we have that

$$\omega_{2M}^{(2x)u} = \omega_M^{xu} \qquad (3.25)$$

and hence

$$F(u) = \frac{1}{2}\bigg(\underbrace{\frac{1}{M}\sum_{x=0}^{M-1} f(2x)\omega_M^{xu}}_{F_{even}(u)} + \underbrace{\frac{1}{M}\sum_{x=0}^{M-1} f(2x+1)\omega_M^{xu}\omega_{2M}^{u}}_{F_{odd}(u)}\bigg). \qquad (3.26)$$

If we call the two sums demarcated above $F_{even}(u)$ and $F_{odd}(u)$ respectively, then we have

$$F(u) = \frac{1}{2}\left(F_{even}(u) + F_{odd}(u)\omega_{2M}^u\right). \qquad (3.27)$$

Note that each of $F_{even}(u)$ and $F_{odd}(u)$ for $u = 0, \ldots, M-1$ is in itself a discrete Fourier transform over $N/2 = M$ points. How does this help us? Well, if we also observe, using the properties of roots of unity, that

$$\omega_M^{M+u} = \omega_M^u \quad \text{and} \quad \omega_{2M}^{M+u} = -\omega_{2M}^u, \qquad (3.28)$$

we can also write

$$F(u+M) = \frac{1}{2}\left(F_{even}(u) - F_{odd}(u)\omega_{2M}^u\right). \qquad (3.29)$$

Thus, we can compute an N-point DFT by dividing it into two parts. The first half of $F(u)$ for $u = 0, \ldots, M-1$ can be found from Eqn. 3.27, whilst the second half for $u = M, \ldots, 2M - 1$ can be found simply be reusing the same terms differently, without further evaluation, as shown by Eqn. 3.29. This is obviously a divide and conquer method.

To show how many operations this requires, let $T(n)$ be the time taken to perform a transform of size $N = 2^n$, measured by the number of multiplications performed. The above analysis shows that

$$T(n) = 2T(n-1) + 2^{(n-1)}, \quad (3.30)$$

the first term on the right hand side coming from the two transforms of half the original size, and the second term coming from the multiplications of F_{odd} by ω_{2M}^u. Induction can be used to prove that

$$T(n) = 2^{(n-1)} \log_2 2^n = \frac{1}{2} N \log_2 N, \quad (3.31)$$

as inserting this result into both sides of Eqn. 3.30 gives us a tautology, whilst for $n = 0$, the DFT of a single point is just the same value, no multiplications being required, which also agrees with Eqn. 3.31.

A similar argument can also be applied to the number of additions required, to show that the algorithm as a whole takes time $O(N \log_2 N)$.

It should be pointed out that the same algorithm can be used with a little modification to perform the inverse DFT too. Going back to the definitions of the DFT and its inverse,

$$F(u) = \frac{1}{N} \sum_{x=0}^{N-1} f(x) e^{-2\pi i x u / N} \quad (3.32)$$

and

$$f(x) = \sum_{x=0}^{N-1} F(u) e^{2\pi i x u / N}, \quad (3.33)$$

if we take the complex conjugate of the second equation, we have that

$$f^*(x) = \sum_{x=0}^{N-1} F^*(u) e^{-2\pi i x u / N}. \quad (3.34)$$

This now looks (apart from a factor of $1/N$) like a forward DFT, rather than an inverse DFT. Thus to compute an inverse DFT, we take the

3.3. SMOOTHING NOISE

conjugate of the Fourier space data, put it through a forward DFT algorithm, and take the conjugate of the result, at the same time multiplying each value by N.

Exactly the same fast Fourier transform algorithm can also be used directly for computing two-dimensional DFTs, using the separability property of the two-dimensional transform. We can rewrite the definition of the two-dimensional DFT as

$$\begin{aligned} F(u,v) &= \frac{1}{N} \sum_{x=0}^{N-1} \sum_{y=0}^{N-1} f(x,y) e^{-2\pi i(xu+yv)/N} \\ &= \frac{1}{N} \sum_{x=0}^{N-1} e^{-2\pi ixu/N} \sum_{y=0}^{N-1} f(x,y) e^{-2\pi iyv/N}. \end{aligned} \quad (3.35)$$

The right hand sum is basically just a one-dimensional DFT if x is held constant. The left hand sum is then another one-dimensional DFT performed with the numbers that come out of the first set of sums. Putting it more directly, we can compute a two-dimensional DFT by performing a one-dimensional DFT for each value of x, i.e. for each column of $f(x,y)$, followed by performing a one-dimensional DFT in the opposite direction (for each row) on the resulting values.

This requires a total of $2N$ one dimensional transforms, so the overall process takes time $O(N^2 \log_2 N)$.

3.3 Smoothing Noise

3.3.1 Introduction

Here, the idea is to reduce various spurious effects of a local nature in the image, caused perhaps by noise in the image acquisition system, or arising as a result of transmission of the image, for example from a space probe utilising a low-power transmitter. The smoothing can be done either by considering the real space image, or its Fourier transform.

3.3.2 Real Space Smoothing Methods

The simplest approach is *neighbourhood averaging*, where each pixel is replaced by the average value of the pixels contained in some neighbourhood about it. The simplest case is probably to consider the 3 × 3 group of pixels centred on the given pixel, and to replace the central pixel value by the unweighted average of these nine pixels. For example, the central pixel in Fig. 3.6 is replaced by the value 13 (the nearest integer to the average). If any one of the pixels in the neighbourhood

10	12	11
11	23	12
10	14	15

Figure 3.6: Neighbourhood averaging

has a faulty value due to noise, this fault will now be smeared over nine pixels as the image is smoothed, thus making it much less noticeable. Unfortunately, other information in the image is also smoothed out, with the result that the image becomes blurred. As the size of the neighbourhood is increased, so does the blurring.

This approach can be improved by using averaging with *thresholding*. The strategy now is

$$p_{new}(x,y) = \begin{cases} A & \text{if } |p_{old}(x,y) - A| \leq T \\ p_{old}(x,y) & \text{otherwise,} \end{cases} \quad (3.36)$$

where A is the average pixel value for the region, T is some threshold value and $p(x,y)$ is a pixel value. In words, if the value of the pixel is not too far from the neighbourhood average we replace it by the average, otherwise we keep it unchanged. Thus, if we are near to some sort of edge, there will be a large change in pixel values, and pixels on both sides of it will not be close to the average value. Overall, the process still smooths pixel values in regions where there are only small changes, but avoids smoothing edges. Much less blurring occurs, and a larger neighbourhood can now be used.

Even better is to use a *median filter* rather than averaging. Again, a neighbourhood around the pixel under consideration is used, but this time the pixel value is replaced by the median pixel value in the neighbourhood. Thus, if we have a 3 × 3 neighbourhood, we write the 9 pixel values in sorted order, and replace the central pixel by the fifth

3.3. SMOOTHING NOISE

highest value. For example, again taking the data shown in Fig. 3.6, the central pixel is replaced by the value 12.

This approach has two advantages. Firstly, occasional spurious high or low values are not averaged in, but are ignored in the new image as they will be at one end or the other of the sorted values. Secondly, the sharpness of edges is preserved. To see the latter, consider the pixel data shown in Fig. 3.7. When the neighbourhood covers the left hand

10	10	20	20
10	10	20	20
10	10	20	20

Figure 3.7: An edge in pixel data

nine pixels, the median value is 10; when it covers the right hand ones, the median value is 20, and the edge is preserved.

If there are large amounts of noise in an image, more than one pass of median filtering may be useful to further reduce the noise.

A rather different real space technique for smoothing is to average multiple copies of the image. The idea is that over several images, the noise will tend to cancel itself out if it is independent from one image to the next. Statistically, we expect the effects of noise to be reduced by a factor $1/\sqrt{n}$ if we use n images. However, alignment of consecutive images can be a problem in practice. One particular situation where this technique is of use is in low lighting conditions.

3.3.3 Fourier Space Smoothing Methods

If there is a lot of noise in an image, there are many rapid transitions in intensity from high to low and back again or vice versa, as faulty pixels are encountered. As these occur on a short distance scale, such noise will contribute heavily to the high frequency components of the image when it is considered in Fourier space.

Turning this argument around, if we reduce the high frequency components, we should reduce the amount of noise in the image. We thus create a new version of the image in Fourier space by computing

$$G(u,v) = H(u,v)F(u,v) \qquad (3.37)$$

where $F(u,v)$ is the Fourier transform of the original image, $H(u,v)$ is a filter function, designed in this case to reduce high frequencies, and $G(u,v)$ is the Fourier transform of the improved image.

The simplest sort of filter to use is an *ideal lowpass filter*, which in one dimension appears as shown in Fig. 3.8. It is, in fact, a top hat

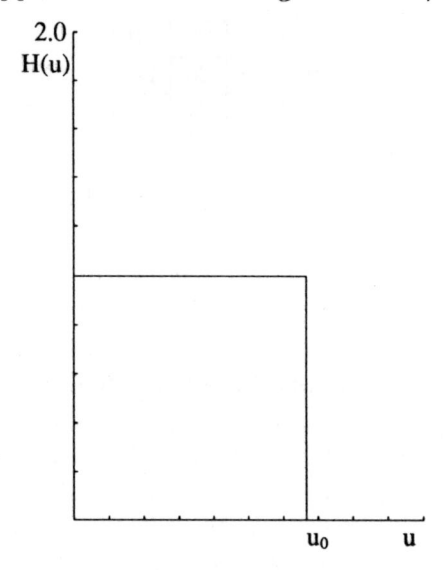

Figure 3.8: Lowpass filter

function which is unity for u between 0 and u_0, the *cut-off frequency*, and zero elsewhere. All frequency space space information above u_0 is thrown away, and all information below u_0 is kept.

The two dimensional analogue of this is the function

$$H(u,v) = \begin{cases} 1 & \text{if } \sqrt{u^2+v^2} \leq w_0 \\ 0 & \text{otherwise,} \end{cases} \qquad (3.38)$$

where w_0 is now the cut-off frequency. Thus, all frequencies inside a radius w_0 are kept, and all others discarded. Note that this filter has radial symmetry, so there will be no preferred directions in the result.

3.3. SMOOTHING NOISE

The problem with this filter is that as well as the noise, edges (places of rapid transition from light to dark) also significantly contribute to the high frequency components. Thus, an unwanted side-effect of applying an ideal lowpass filter is that edges become blurred as high frequency information is thrown away. The lower the cut-off frequency is made, the more pronounced this effect becomes. In respect of this blurring, Fourier smoothing has similar disadvantages to real space smoothing carried out by averaging.

Another filter sometimes used is the *Butterworth lowpass filter*. In this case, $H(u,v)$ takes the form

$$H(u,v) = \frac{1}{1 + [(u^2 + v^2)/w_0^2]^n}, \qquad (3.39)$$

where n is called the order of the filter. This keeps some of the high frequency information, as illustrated by the second order one dimensional Butterworth filter shown in Fig. 3.9, and consequently reduces the blurring.

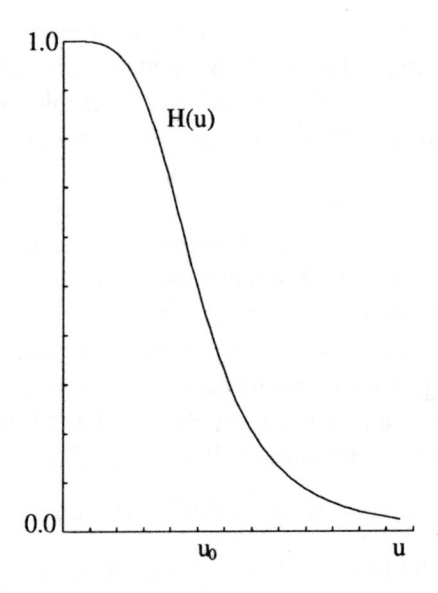

Figure 3.9: A Butterworth filter

3.4 Contrast Enhancement

3.4.1 Introduction

The human eye has a limited sensitivity to differences in intensity, and sometimes images which have been captured display only a limited range of intensities due to poor lighting, incorrect set-up of equipment, or various other causes. One class of image processing methods aims to make images more understandable by enhancing the contrast, *i.e.* increasing the differences in intensity between pixel values, to make these differences more visible to the human eye. In general, however, such uses are not particularly appropriate for computer vision, as the computer can just as well tell a difference in intensity of one unit as of many units! Furthermore, contrast enhancement methods often improve (in intensity terms) some part of the image at the expense of a loss of information in other parts, which is usually not desirable.

3.4.2 Histogram Equalisation

If we count the number of pixels of each intensity (or grey level) in an image, and display the result as a histogram, this gives us some overall information about the image. For example, an image with the histogram shown in Fig. 3.10 has most of its pixels in the range black to medium grey, with rather fewer light grey to white ones. Thus the overall image is quite dark. (For simplicity, in this histogram, and for most of the discussion, we shall assume that the intensities take on a continuous range of values. We shall see at the end how to modify the ideas for a discrete set of intensity levels.)

Let us assume that the original pixel intensities lie in the range $0 \leq r \leq 1$. We are interested in carrying out transformations of the type $s = T(r)$ to give a new intensity s for each old intensity r. For $T(r)$ to be a reasonable transformation, it should satisfy two conditions:

- $T(r)$ should increase monotonically as r goes from 0 to 1.
- $T(r)$ should lie between 0 and 1 for r between 0 and 1.

The first of these two conditions ensures that the order from black to white is preserved in the new image. In other words, any pixel which

3.4. CONTRAST ENHANCEMENT

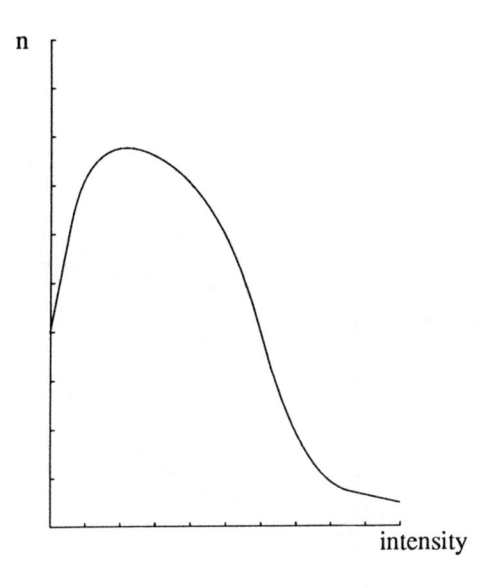

Figure 3.10: A histogram of image intensity

is originally darker than another will remain so. The second condition ensures that the new pixel values also lie in the correct range. A possible function for $T(r)$ is shown in Fig. 3.11.

Now, let $p_r(r)\,dr$ be the probability that an original pixel has intensity in the range r to $r+dr$ (the area of a vertical strip of the histogram between r and $r+dr$, divided by the total area of the histogram), and let the equivalent probability for a new pixel be $p_s(s)\,ds$. As each old pixel of intensity r is turned into a new pixel of intensity s, we have the relationship

$$p_s(s) = p_r(r)\left.\frac{dr}{ds}\right|_{r=T^{-1}(s)}. \qquad (3.40)$$

The derivative dr/ds is to be evaluated at r such that $s = T(r)$.

The aim of the technique known as *histogram equalisation* is to take whatever pixel intensity distribution we have initially, and to produce a new image which has equally many pixels of every shade of grey from black to white. This means that $p_s(s)$, the new pixel intensity distribution, must be a constant. For example, in an image like the one whose histogram is shown in Fig. 3.10, most of the pixels are quite dark, and so it will be difficult for the eye to see details, which may

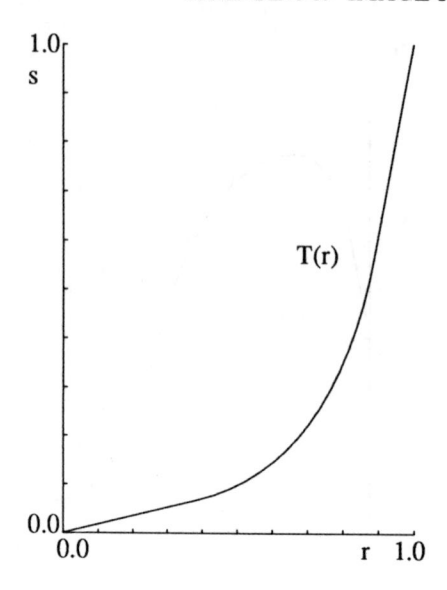

Figure 3.11: A histogram modification function

be hidden by quite small differences in shade. The effect of histogram equalisation will be to make the lighter dark grey pixels even lighter. In general, there will be an increase in the relative differences in brightness of pixel values near intensity values where there are many pixels of those values, although there will be a decrease in brightness differences near intensity values where there are few pixels of those intensities. As there are fewer such pixels, the result is that on the whole, the contrast of the image is improved.

How can we choose $T(r)$ to make $p_s(s)$ constant? Suppose we take

$$s = T(r) = \int_0^r p_r(\alpha) \, d\alpha. \tag{3.41}$$

Thus, $T(r)$ is the area under the graph of $p_r(r)$ from 0 up to r. Because $p_r(r)$ is a probability distribution function,

$$\int_0^1 p_r(\alpha) \, d\alpha = 1 \tag{3.42}$$

(a pixel is certain to have an intensity between 0 and 1), and also, obviously,

$$\int_0^0 p_r(\alpha) \, d\alpha = 0. \tag{3.43}$$

3.4. CONTRAST ENHANCEMENT

Thus $T(r)$ defined in this way satisfies the second of the conditions given earlier. Also, because the area under the graph of $p_r(r)$ increases with r, $T(r)$ also satisfies the first condition.

Now, from Eqn. 3.41 we have that

$$\frac{ds}{dr} = p_r(r), \qquad (3.44)$$

and on substituting this into Eqn. 3.40 we find that

$$p_s(s) = p_r(r) \frac{1}{p_r(r)} \bigg|_{r=T^{-1}(s)} = 1, \qquad (3.45)$$

and hence the form for $T(r)$ given by Eqn. 3.41 makes $p_s(s)$ a constant as required to equalise the histogram.

In practice, we have only a discrete set of pixel intensities, rather than a continuous range, so the method must be modified. In this case, the calculated histogram can not exactly be equalised, but the result is the nearest that can be achieved given that every pixel of each old (discrete) intensity must be mapped to the same discrete new intensity.

The probability $p_r(r)$ that a pixel has a given value r, is now just n_r/n, where n_r is the number of pixels of intensity r, and n is the total number of pixels. The discrete equivalent of Eqn. 3.41 is

$$s = T(r) = \sum_{j=0}^{r} p_r(j) = \sum_{j=0}^{r} \frac{n_j}{n}. \qquad (3.46)$$

Thus, the intensity level r is in principle mapped to the new intensity level s where s is the fraction of pixels in the original image with intensities less than or equal to r. However, in practice s can only take on a discrete set of values, so it must be rounded to the nearest permissible value. A further minor modification of this formula will be required if the old and new pixel intensities are desired to be in a range other than 0 to 1.

Many other similar techniques can be devised for histogram modification for particular purposes, by choosing an appropriate $T(r)$. For example, we may only wish to improve the contrast of pixels in a given range of intensities, and not care about the others — we might have a car registration plate in heavy shadow in an otherwise bright image, and wish to improve the contrast of these few dark pixels for identification purposes. Note that in this case applying histogram equalisation to the *whole* image will have just the opposite of the desired effect. Histogram equalisation is a useful general purpose technique when no prior assumptions are made about the image contents.

3.5 Correction of Camera Errors

3.5.1 Introduction

The image reconstruction or correction techniques discussed so far have been concerned with enhancing images that have been degraded by the working environment. For example, images could have either been acquired under adverse working conditions (for example, with poor lighting) or images could have been affected by noise during image acquisition, transmission or storage (for example, because of electrical interference).

However, some images have errors introduced into them by faults in the very vision system used to acquire them. These faults fall broadly into two categories. Firstly, errors may arise due to physical defects in the image capture apparatus. Such errors give rise to a distorted appearance of the image and clearly any direct measurements of lengths or positions from an image of this type will be in error. For example, distortions of this type arise when images are taken using a television camera (see Section 2.2) because the photosensitive target in the camera tube on which the image is formed is never perfectly flat. Also as we have seen in Section 2.3 errors arise due to lens distortions no matter what type of camera is used.

The second type of errors which may arise occur due to user error. These errors usually give rise to a blurred image and may happen if the lens system is not focussed properly or if the camera is moving as the image is taken.

We shall consider techniques for correcting both types of errors described above. However, errors of the second kind are often not reproducible or difficult to model accurately, and should be avoided if possible. For example, strategies such as the use of an optical bench can avoid vibration problems.

3.5.2 Geometric Correction

Distortions in images are resolved using *geometric correction* techniques. Such distortions in an image generally affect the spatial relationships between image pixels. Thus geometric transformations can be applied to reposition pixels to their 'ideal' places in the image.

We first need a model of the geometric transformation. Let $f(x,y)$ be the ideal, undistorted image, and $g(x,y)$ be the actually observed, distorted image. Let a pixel in f at position (x,y) have been distorted to a position (x',y') in g. We can describe the distortion by

$$\begin{aligned} x' &= d_x(x,y) \\ y' &= d_y(x,y) \end{aligned} \qquad (3.47)$$

where $d_x(x,y)$ and $d_y(x,y)$ are spatial transformations. These transformations can be modelled by linear functions when considering distortion caused by perspective transformation of the scene onto an image, quadratic functions when considering camera tube distortion or even higher order functions if we wish to describe tangential lens distortions.

If these distortion transformations are known then compensating for the distortion is simply a matter of inverting the transformations.

However, generally we do not know the exact parameters of the distortion *a priori*, and so these must be found by calibration. This is done by viewing a set of points whose locations are known precisely and noting their positions in the resultant (distorted) image. If enough test points are used, a system of linear equations can be set up to find the values of the parameters. In practice, more than the minimum number of test points is used, and least-squares methods are then used to give better estimates of the parameters. This was described in Section 2.3.

The whole ideal image can now be obtained from Eqn. 3.47 by using the inverse mapping to compute for all pixels in $f(x,y)$ their positions from the mappings from the distorted image $g(x',y')$. One problem remains however. Since pixel positions in both images are at discrete locations, the distortion of pixels means that the calculated coordinates (x,y) do not in general lie exactly on discrete pixel positions. This means that the value of $f(x,y)$ at each discrete pixel location has to be interpolated from the values of $f(x,y)$ at surrounding points by considering the values of the four (or more) nearest discrete pixel values of $g(x', y')$.

3.5.3 Correcting Blurred Images

The approach we shall take to this problem is to attempt to model the degradation as best we can, and then to try to apply its inverse to the image to give us a restored image.

A fairly general model we can use is that

$$g(x,y) = \Theta f(x,y) + n(x,y) \qquad (3.48)$$

where $g(x,y)$ is the captured, degraded image, $f(x,y)$ is the original image, Θ is an operator describing the degradation, and $n(x,y)$ represents random noise in the image.

Let us ignore n for the present and assume that it is zero; we shall see its importance later.

Normally, to make things simpler, we assume Θ is a *linear* operator, which simply means that

$$\Theta\left[k_1 f_1(x,y) + k_2 f_2(x,y)\right] = k_1 \Theta f_1(x,y) + k_2 \Theta f_2(x,y) \qquad (3.49)$$

where k_1 and k_2 are constants, and $f_1(x,y)$ and $f_2(x,y)$ are any two input functions.

Another property which such an operator may have is position independence, which is to say that the effect of Θ depends *only* on the input at the point in question, *not* on where that point is. This would clearly be the case, for example, if the camera moved sideways during the motion. In such a case

$$g(x - \alpha, y - \beta) = \Theta f(x - \alpha, y - \beta) \qquad (3.50)$$

for any α, β and f.

Whether the assumption that Θ is a position independent, linear operator is a particularly good one or not will depend on what is the exact cause of the blurring in the image.

Assuming it is, we can express $f(x,y)$ as

$$f(x,y) = \int_{-\infty}^{\infty} \int_{-\infty}^{\infty} f(\alpha, \beta) \delta(x - \alpha, y - \beta) \, d\alpha \, d\beta, \qquad (3.51)$$

3.5. CORRECTION OF CAMERA ERRORS

which is just the definition of the two-dimensional delta function (see Section 3.2.4).

Thus,

$$g(x,y) = \Theta f(x,y) = \Theta \int_{-\infty}^{\infty}\int_{-\infty}^{\infty} f(\alpha,\beta)\delta(x-\alpha,y-\beta)\,d\alpha\,d\beta, \quad (3.52)$$

and as Θ is a linear operator,

$$g(x,y) = \int_{-\infty}^{\infty}\int_{-\infty}^{\infty} \Theta\left[f(\alpha,\beta)\delta(x-\alpha,y-\beta)\right] d\alpha\,d\beta. \quad (3.53)$$

Now Θ as an operator operates on a function of x and y, so $f(\alpha,\beta)$ is a constant as far as x and y are concerned and

$$g(x,y) = \int_{-\infty}^{\infty}\int_{-\infty}^{\infty} f(\alpha,\beta)\Theta\left[\delta(x-\alpha,y-\beta)\right] d\alpha\,d\beta. \quad (3.54)$$

Now, if Θ is position independent, we may write

$$\Theta\delta(x-\alpha,y-\beta) = h(x-\alpha,y-\beta) \quad (3.55)$$

where h is called the *impulse response*, or *point spread* function. Let us consider its physical meaning. Remember that a delta function is zero everywhere except at a given point, where its integrated value is one. Optically, this corresponds to a point source of light of unit intensity. Thus, $\Theta\delta(x-\alpha,y-\beta)$ tells us what this bright point of light looks like in our blurred image, which is why $h(x-\alpha,y-\beta)$ is called the point spread function — it is how a point of light is spread out after blurring.

Returning to what we had earlier,

$$g(x,y) = \int_{-\infty}^{\infty}\int_{-\infty}^{\infty} f(\alpha,\beta)h(x-\alpha,y-\beta)\,d\alpha\,d\beta. \quad (3.56)$$

This is a very important result. It tells us that the intensity of pixels in our blurred image can be computed by performing a *convolution* of the original image with the point spread function. This is now where Fourier transforms and the convolution theorem become very useful. The latter tells us that if

$$g = f * h \quad (3.57)$$

then

$$G = F \times H \quad (3.58)$$

where ∗ denotes convolution, and F, G and H are the Fourier transforms of f, g and h respectively. The product of the transforms gives the transform of the convolution.

Thus, if we assume we know g, which is our degraded image, and that we can model h in some way, it is possible to get back the original image f by the following sequence:

1. Find the Fourier transform G of the degraded image g.

2. Find the Fourier transform H of the point spread function h.

3. Compute $F(u,v) = G(u,v)/H(u,v)$.

4. Take the inverse Fourier transform of F to give the reconstruction of the original image, f.

How well this process works depends on how well we can model h, and also on how valid our assumptions about Θ being a linear, position independent operator are.

In practice, there are one or two further problems. Firstly, if $H(u,v)$ becomes zero or very small at any values of (u,v), this will cause $F(u,v)$ to become infinite, which will adversely affect the resulting image when the inverse transform is computed. The simplest solution in this case is just to ignore the hopefully few points where $H(u,v) = 0$.

A second problem is noise in the image. If we include the effects of noise, referring back to Eqn. 3.48, we now have

$$G(u,v) = H(u,v)F(u,v) + N(u,v) \qquad (3.59)$$

where N is the Fourier transform of n, or

$$\frac{G(u,v)}{H(u,v)} = F(u,v) + \frac{N(u,v)}{H(u,v)}. \qquad (3.60)$$

Note that we cannot just subtract N/H, as we assume the noise to be random, and so G/H is the best estimate we can construct of the Fourier transform of the original image. Now, this means that when H is small, the second term on the right tends to swamp the first term, and so the noise term can greatly exceed the information from the original image. This is a serious problem, as in practice, $H(u,v)$ often tends rapidly to zero as we go away from $(0,0)$ in the u-v plane.

The solution to this problem is to carry out restoration in a limited region of the u-v plane only, to avoid excessively small values of H.

Let us return to modelling the degradation. If we know the cause of the degradation, then it may be possible to model h theoretically, as in the case where the camera is moving at a known velocity while the image is captured, for example. Details of this calculation and various other techniques for restoring degraded images can be found in the book by Gonzalez and Wintz[105].

However, one generally applicable idea which is of use even when the exact source of the degradation is not known is to use the fact that h is the point spread function. If there are some point sources of light in the image, or points whose brightness varies appreciably from the background, we can look at the degraded image, and see how they have been spread, or blurred, in the degraded image. This directly gives the h we want without the need for a theoretical model. Obviously, such a method can be used directly as part of the camera calibration procedure if required.

3.6 Image Compression

3.6.1 Introduction

Images contain large amounts of data. Even a modest monochrome image of 512 × 512 resolution, with 8 bits stored per pixel, contains 1/4Mb of information. However, an image is not a random collection of pixel values, but usually represents highly structured data. Thus, adjacent pixels often have the same or similar values — they may both belong to the background, or both lie on the same surface of the object, with almost the same position and orientation with respect to the light, and hence have very similar shades. This property, referred to as *image coherence*, means that we are often able to compress image data so that it occupies a much smaller amount of memory.

Image compression is of use for two main purposes. The first and perhaps most obvious of these is for the storage of images, primarily so that they will take up less room on disk or tapes. The second application is in the transmission of images over computer networks, to reduce transmission times and hence costs, or to increase overall throughput of a limited capacity link.

A third and perhaps less widely used application of image compression is the actual processing of compressed images still in compressed form. Less random access memory will be required by the program while working on the image, and in general the algorithms will have less data to consider, at the price of increased complexity of the algorithms themselves. This results in algorithms which are expected to run faster than the standard ones operating on uncompressed data. Such algorithms are described in [70, 123] for example.

We can conveniently divide image compression techniques into two classes: those which preserve all the information in the image, so that it can be restored exactly (*lossless* methods), and those methods which actually throw away some information (*lossy* methods). The latter aim to do so in such a way that the human brain can still extract useful information, and indeed, at least at low compression ratios these methods give an image upon reconstruction which is substantially unaffected as far as the human eye is concerned. Basically, we are prepared to tolerate some loss of data in such cases for increased efficiency of compaction.

As far as computer vision is concerned, it is usually the case that the more initial data we have, the better, so information preserving compression methods are preferable. Image encoding techniques which reduce the information content of images are more applicable to transmission of television pictures over satellite links, for example.

The real space techniques we shall examine below are information preserving, while the transform and colour space methods are not.

3.6.2 *Real Space Compression*

Both techniques discussed here only use one dimensional image coherence along horizontal rows of pixels, and do not take into account any correlation between vertically adjacent rows.

Probably the simplest technique is *run length coding*. Rather than store the value of each pixel along each row, we store pairs (p, c), where p is the pixel value, and c is a count of successive pixels with that value. Normally, we make the storage location for the count large enough to

3.6. IMAGE COMPRESSION

allow for every pixel in the row having the same colour.

If we have black and white images, a simplification of this method can be used, as we know that successive runs of pixels must have opposite colours. In this case, all we need to store is the colour of the first run, followed by the lengths of each of the runs on the line.

While run length encoding works quite well for images generated by computer graphics, where whole areas may be of a uniform colour, in real images such areas often tend to vary slowly in shade, or are not quite uniform because of noise, which limits the usefulness of this method.

A second possibility is to use *differential coding*, which works well in such cases. Let us suppose that pixel values range from 0 to 255. Going across a row of the image, there will usually be fairly small changes in value from one pixel to the next, except near edges in the image. Indeed, the differences in a large proportion of the cases will probably lie in the range 0 to ± 7, so although the pixel values themselves need 8-bit storage locations, we can often store the pixel differences in 4-bit locations.

Thus, the differential coding technique stores the first pixel value on each line, and 4-bit differences as two's complement numbers for successive pixels. The only problem arises when we come to an edge or other large change in pixel values. There are several ways of handling this. One simple method is to store a value of -8, which does not indicate a difference, but is an escape code meaning that the following 8 bits give the next pixel value directly as a new starting value.

Obviously, in favourable cases, the differential coding method as described can give storage savings of almost 50%.

3.6.3 Transform Encoding

We shall now consider how transform methods can be of use in image compression. We start off by subdividing our original $N \times N$ image into subimages each of size $M \times M$, where $M < N$, and M divides N. Each subimage is then coded independently of each of the others.

Within each subimage, it is likely that there will be a high degree of correlation between the pixel values. Let us suppose for a moment that we are trying to encode one-dimensional images, and that we have split up our original image into subimages each containing just two pixels, with intensity values I_1 and I_2. Then, because of the correlation, I_1 and I_2 are likely to be fairly similar. If we replace I_1 and I_2 by a new pair of values, $J_1 = (I_1 + I_2)/2$ and $J_2 = (I_1 - I_2)/2$, we can see that we can recover I_1 and I_2 from them. While J_1 is the average of I_1 and I_2, and can thus be stored in a similar sized memory location, J_2 is half of the difference of I_1 and I_2, and is likely to be rather smaller. (Note that this is very similar to the principle of the differential coding method so far.)

In general, we take our original $M \times M$ set of pixel values, and from them produce a new set of values, where a subset of them (like I_1) are expected to be large, and others are likely to be smaller (like I_2). We then either store the second subset in reduced size memory locations (when we ignore any numbers too large to fit — giving an important difference from the differential coding method), or indeed, we discard them altogether. In the example given, if we keep I_1 and discard I_2 for each pair, the net result is that on reconstruction, each pair of pixels would be replaced by the average value for the pair, giving a still recognisable image and a 50% saving in storage.

In general, the new set of values is produced from the old ones by means of linear combinations. Thus, in one dimension we have

$$J_k = \sum_{i=0}^{N-1} a_{ki} I_i, \qquad (3.61)$$

where the a_{ki} are constants. Note that in particular, the discrete Fourier transform is of this form

$$F(u) = \sum_{x=0}^{N-1} \left(\frac{1}{N} e^{-2\pi i x u / N} \right) f(x), \qquad (3.62)$$

and indeed, there are many other transforms where the a_{ki} have other values, such as the *Walsh-Hadamard* transform.

Choosing the optimum transform to use depends upon the image in quite a complex way, and in fact it turns out that using the Fourier transform for this purpose is quite a reasonable choice in general.

In practice, we might subdivide an original 256×256 image into 16×16 subimages, take the Fourier transform of each, and throw away the

3.6. IMAGE COMPRESSION

128 highest frequency Fourier transform coefficients of the 256 produced for each subimage as we expect the low frequency components to be much larger than the high frequency components. We may get a little blurring of edges in each of the subimages on reconstruction, but still quite a useful and recognisable image.

Indeed, if we calculate the approximation error as

$$\frac{1}{N}\sqrt{\sum_{x=0}^{N-1}\sum_{y=0}^{N-1}(g(x,y)-f(x,y))^2}, \qquad (3.63)$$

where $f(x,y)$ are the original pixel values, and $g(x,y)$ are the reconstructed ones, it may well be less than 1% of the average pixel value even though we have reduced the data by 50%.

Much more detail on transform coding of images can be found in the book by Clarke [57].

3.6.4 Colour Space Compression

A rather different view of compression is sometimes used in conjunction with colour images. Often when colour images are captured, 24 bits are stored for each pixel, 8 bits for each of the red, green and blue components. However, even for a moderately large image of size 1024×1024 pixels, there can clearly be only 2^{20} different colours in the image, as there are only that many pixels. Using 24 bits to specify the colour of each pixel is obviously wasteful. In practice, colour workstations often allow an 8-bit value to be stored in each pixel, and these values are used as an index into a *colour look-up table* which stores 24-bit values. Thus, at any one time we can have up to 2^8 different colours on the screen chosen from the full set of 2^{24} possible values.

This type of compression is referred to as colour space compression. We still use one storage location per pixel, but with a decreased number of bits to store the colour information for that pixel. The problem of colour space compression is thus as follows. Given an original colour image represented by 24 bits per pixel, it is required to reduce it a new image with 8 bits per pixel, plus a description of which 24-bit value each of the 8-bit index values represents.

There are two parts to the problem. The first is to decide which set of 2^8 colours should be chosen — for best results, we should obviously choose different sets for a picture of a red teapot, and a picture of an outdoor scene containing mainly sky and grass. The second issue is then how to map each of the original 24-bit colour values to the appropriate 8-bit value which is its closest match. Note that in general this transformation is not information preserving, except in the unlikely case that there are only 256 or less colours in the original image, when the colour look-up table is large enough to allow each 24-bit value to be replaced by an exact match.

The simplest method, which ignores the first part of the problem, and hence does not generally give very good results, is to discard the low order bits of the colour information for each pixel, and to just keep the 3 most significant bits for red and green, and 2 most significant bits for blue, rounding the values appropriately. (The human eye is less sensitive to blue.) Thus, we always put the same set of colour values into the look-up table. These values are spaced a fixed distance apart in terms of colour, and reducing a 24-bit colour value to its equivalent 8-bit value is a simple calculation as described above.

More sophisticated methods have been described by Heckbert [115]. The *popularity algorithm* chooses the 256 most frequently occurring colours in the image to be the members of the look-up table. Matching each pixel to its corresponding 8-bit value is now rather more difficult. A second and better method is the *median cut algorithm* which tries to select the 256 colours in such a way that approximately an equal number of pixels is represented by each of the colours in the look-up table. A similar approach to the latter, but more efficient, based on octrees and hence called *octree quantization* has been described by Gervautz and Purgathofer [97].

3.7 Exercises

Exercise 3.1 *Evaluate the Fourier transform of $e^{-|x|}$.*

Exercise 3.2 *Evaluate the Fourier transform of $e^{-|x|-|y|}$.*

Exercise 3.3 *Find the convolution of the two functions $f(x)$ and $g(x)$, where $f(x) = 1$ for $0 \le x \le 1$ and $f(x) = 0$ otherwise, and $g(x) = x$ for $0 \le x \le 1$ and $g(x) = 0$ otherwise.*

3.7. EXERCISES

Exercise 3.4 *Prove that the Fourier transform of the convolution of $f(x)$ and $g(x)$ is equal to the product of the Fourier transforms of $f(x)$ and $g(x)$.*

Exercise 3.5 *Prove the Fourier theorems given in Section 3.2.5.*

Exercise 3.6 *Write an implementation in one dimension of the fast Fourier transform algorithm. It should perform forward and inverse transforms.*

Exercise 3.7 *Write a procedure which performs a fast Fourier transform on a two dimensional image array which works by repeatedly calling the one dimensional routine. Again, it should perform forward and inverse transforms.*

Exercise 3.8 *Write a procedure which performs histogram equalisation on an image passed to it as a rectangular array of integer intensity values in the range 0 to 255.*

Exercise 3.9 *Write a procedure which performs neighbourhood averaging, neighbourhood averaging with thresholding, or median filtering on an image passed to it as a rectangular array of integer intensity values in the range 0 to 255. What problems arise at the boundaries of the image, and how do you propose to cope with them?*

Exercise 3.10 *Write a procedure which performs image smoothing by taking the Fourier transform of the image, applying a low-pass filter or a Butterworth filter, and taking the inverse transform to give back the new image.*

Exercise 3.11 *Write an implementation of the Fourier transform encoding method for compressing 256×256 pixel images (with 8 bits per pixel) to half of their original size.*

Exercise 3.12 *Discuss the limitations of the popularity algorithm for colour space compression, and describe what types of input data you would expect it to not work particularly well for.*

Chapter 4

Segmentation and Feature Representation

4.1 Introduction

Having obtained either a two- or three-dimensional image of a scene using some vision acquisition system, the next stage for many vision applications is to recover useful information at a higher level from the large amount of low level data present in the image. This Chapter discusses the types of high level information that may be extracted from images, particularly depth maps, and the associated problems in doing so. The methods employed to extract and represent such features are also discussed.

Since any image contains a large number of pixels it is difficult and time consuming to perform any realistic computer vision tasks utilising this raw form of the data. In particular it is difficult to

- directly reason intelligently about the nature of objects in the scene;
- make any direct comparison of an object contained in the scene with a corresponding object model stored by the computer.

It is often convenient to partition images into regions of pixels that have similar properties. As we shall shortly see, most properties used to partition regions are based on measuring *gradients*. For greyscale images, the gradient corresponds to the rate of change in pixel intensity along some line or across an area of pixels. For depth images (where the x, y and z values are now directly related to physical dimensions) the gradients correspond to changes in depth with distance and thus directly correspond to geometric notions of gradient, such as the slope of a plane. Thus, one useful method of partitioning an image is to group together regions of similar gradient, or gradually changing gradient, which corresponds to pixels lying in the same plane or surface, while edges between them occur at sharp changes in gradient.

Clearly, if we can reliably extract some description of each region or, equivalently, the edges which bound that region, it is possible to work with this more concise form of information to obtain more efficient image processing methods. These *higher order* descriptions (than the pixel level descriptions) can be more easily processed to provide understanding of how parts of the image interact to form objects in the scene. Also, comparisons with a stored model of the object can be more easily carried out. The problems of reasoning with scene descriptions and object models are dealt with later in Chapters 5 and 8.

The term *segmentation* is often used to refer to such processing of the pixel data to yield some higher order description of the scene. Such a region extracted from the scene, of interest or importance to any subsequent vision processes, is called a *feature* or, alternatively, a *primitive*.

This rest of this Chapter is organised as follows. Firstly various goals for segmentation methods are outlined. Secondly, most features extracted from the scene are geometric in nature. The types of geometric data that can be found (both two- and three-dimensional in nature) are discussed in Section 4.3. Problems that arise in the extraction and use of such features are also addressed. Thirdly, we then discuss further the extraction of three-dimensional features since these are heavily used in later parts of this book. Planes are shown to be the features that can be extracted most reliably and details of our chosen method for extracting planes from a depth map are given. Higher level surfaces are also considered. Finally, we also consider the extraction of statistical measures for describing regions of pixels in an image.

4.2 Segmentation Goals

The problems of choosing suitable geometric representations of the scene are addressed in this Section. This choice is based on our practical experience of segmentation methods. An analysis of our segmentation algorithm in practice is given in Section 9.6.

In order that any subsequent processing of an image can occur successfully and efficiently it is necessary to achieve *reliable* segmentation of the image. Amongst other things, this means that we obtain the same basic description of features present in the image *irrespective* firstly, of the viewpoint we have chosen for the scene (assuming, of course, that we can see the same things in the scene), and secondly, of the position within the image where we have chosen to start the segmentation routines from.

Many computer vision systems require a single object in the scene to be recognised and perhaps accurately located (both in terms of position and orientation) within the scene. These requirements are addressed in some detail later in the book. Here we can say, however, that any representation of the scene must contain enough information to permit these tasks to be carried out relatively easily. Furthermore, for efficient recognition algorithms, the representation must be easily understood and simple to use. We do not wish to have ambiguous representations of the scene where effort may be wasted in deciding which representations are valid and correct when reasoning with this data.

The following requirements are desirable properties of a segmentation of an image. Most of these requirements are necessary for solving the object recognition, location and inspection tasks discussed later in the book. However, in general, these requirements are equally valid for any segmentation technique, although the emphasis is on segmentation for model based matching purposes. These requirements are:

- The features of the representation must be fairly stable with respect to partial occlusion. Partial occlusion occurs when a feature is partially obscured from view by another feature or even by another part of itself. Thus the underlying feature description must remain the same whether the feature is wholly visible or partially occluded.

- The feature representation must be stable with respect to the position of the viewpoint. For example, a plane is still a plane when visible from any viewpoint (unless it is parallel to our line of sight).

- The features must be easily extracted from the scene *i.e.*, not require excessive computational effort.

- The features extracted must be accurate and reliable. This affects the efficiency, as well as the reliability and accuracy of subsequent recognition algorithms, as discussed in Chapters 8 and 9

- The features must be easily understood, simple to use and not ambiguous — again so that subsequent algorithms using this data can be made efficient.

- There must be many fewer segmented features than the number of pixels in the original image.

- The features must retain enough information to drive subsequent vision processes. In particular, with reference to object recognition or model-based matching the features must carry enough information to recover position and orientation of the object.

- The features must correspond to those which are easily derivable from the stored model of the object.

4.3 Extracting and Describing Features in an Image

Various different types of features that can be extracted from a scene are discussed in the following Sections together with suitable (geometric) representations of such features. Problems associated with extracting these features are discussed as well as ones of using such features. The extraction of both two and three dimensional features is addressed. A scene representation (segmentation) may be made up of one or more of these types of features that are discussed below.

4.4 Point Extraction

4.4.1 Oriented Surface Points

We may distinguish between various types of points which can be extracted from images. Firstly, we may mean points of special interest in the scene, such as vertices where several surfaces meet, or centres of symmetry. Secondly and more generally, we may be interested in obtaining samples of points over surfaces present in the scene. Let us consider the latter a little further.

Let us initially assume that our image is a depth map. For points lying on a surface, a direction vector can be associated with each point which gives the direction of the surface normal at that point, as shown in Fig 4.1. This is useful later in resolving ambiguities when matching points to surfaces of the stored model, as the recognition strategies can be constrained a little by using the fact that the direction of the surface normal always points away from the object (see Chapter 8).

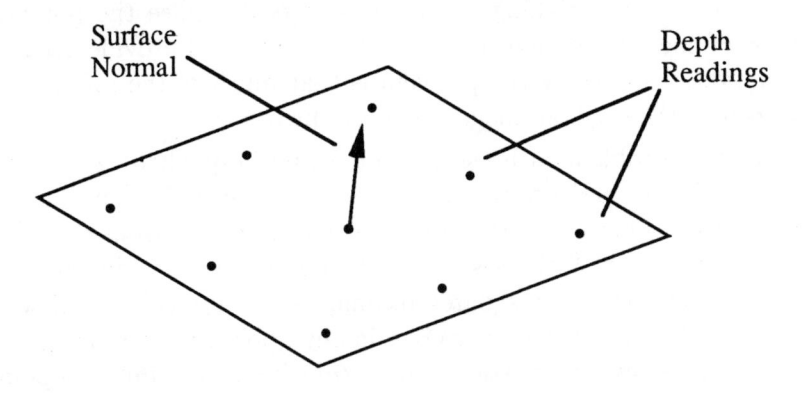

Figure 4.1: Example of an oriented surface point

Estimating the surface normal at a point is relatively simple. A region of variable size around the point is considered and a plane is fitted to all points contained within the region [94, 95, 113, 218] (for fitting see Appendix B). The resultant normal to this plane is then taken to be the associated direction vector at the point. A point with an associated direction vector is usually referred to as an *oriented surface point*. A similar process is also used in the first stages when segmenting planes as will be described in Section 4.6. Typically, the size of the region used is 3 × 3 pixels.

With two-dimensional images an analogous process can be performed (substituting grey scale intensities for depth values) where the vector associated with the point now gives an estimate of the rate of change of intensity over the region.

4.4.2 Problems with Points as Primitives

One problem with using point primitives is that the huge number of points that are present in an image makes it necessary to choose candidate points to restrict the massive amount of data available. One way of choosing candidate points is to select points of particular interest such as edge points and centres of symmetry as already mentioned. In general, however, it is difficult to choose appropriate candidate points to pass on to the matching processes that both utilise the potential richness of the information present in the image and provide accurate information. Essentially the problem is that much of the information contained in the original image is simply thrown away.

The main problem with using points (especially when we desire to accurately locate and position an object in a scene) is that they are highly prone to noise. Even if we take sample points from a surface, *accurate* estimation of the associated direction vectors is difficult. We shall see later that least squares techniques are typically used when segmenting lines or planes. The error in any approximation using least squares techniques is proportional to $1/\sqrt{n}$, where n is number of points used in the approximation. Since the number of points employed in approximating the normal vector associated with a point is much less than the number of points used when extracting either edges or planes, the corresponding error for a point approximation is much worse.

Sometimes the sensing methods do not yield reliable information of the kind required for this type of representation. For instance, active sensing methods can give poor information at edges due to specular reflections. This means that points which are naturally of greatest interest such as vertices or edge points cannot be accurately located.

4.5 Edge Extraction

An edge may be regarded as a boundary between two dissimilar regions in an image. These may be different surfaces of the object, or perhaps a boundary between light and shadow falling on a single surface. In principle an edge is easy to find since differences in pixel values between regions are relatively easy to calculate by considering gradients. The discussion here mainly considers edges represented as straight lines or chains of straight lines. Our justification is that:

- Many objects can be considered to consist of mainly planar features with straight line edges.

- Straight line edges are relatively easy to reliably extract from a scene.

- Reasoning and matching with non-linear quantities (curved edges in this case) is not desirable. It is difficult to perform and the results may well not be very reliable or accurate (see [80] and Chapters 5 and 8).

4.5.1 Representing Lines

The representation usually used for a line in two dimensions is of the form

$$y = mx + c \tag{4.1}$$

where m is the gradient of the line and c is the intercept of the line with the y axis (Fig 4.2). An alternative representation of a line is

$$r = x\cos\theta + y\sin\theta \tag{4.2}$$

where r is the perpendicular distance from the line to the origin and θ is the angle the line makes with the x axis, as shown in Fig 4.2. The latter form has the advantage that the gradient m, with a range $-\infty \leq m \leq +\infty$, which is hard to deal with computationally, has been replaced by the range of angles $0 \leq \theta \leq \pi$ (see also Section 4.5.4).

An alternative representation of an edge or line (again, see Fig 4.2) is by the vector pair $(\mathbf{n}, \mathbf{d})$, where $\mathbf{n}$ is a direction vector (usually normalised) along the edge and $\mathbf{d}$ is a vector from the origin to the closest

(a) $y = mx + c$ equation of line

(b) Alternative form of line equation

(c) Vector representation of a line

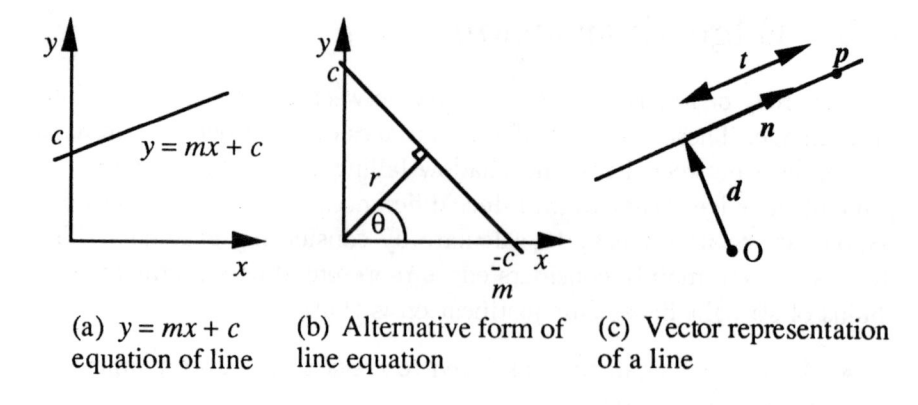

Figure 4.2: Line representation

point on the line. Thus, the length of **d** is the perpendicular distance of the line from the origin. This form of line representation is useful for both two- and three-dimensional lines, and indeed for three-dimensional lines this form is preferable. Another representation of a line in three dimensions is as the intersection of two planes. However more than one pair of planes can describe the same line, so this is an ambiguous type of line representation, which is not desirable when matching.

Another advantage of this form of line representation is that the line can be *parametrised*. Thus, we can specify the position of any point on the line, such as the end of an edge, by its distance t along the line. Therefore the coordinates of a point **p** ($\mathbf{p}(x,y)$ or $\mathbf{p}(x,y,z)$) are

$$\mathbf{p} = \mathbf{d} + t\mathbf{n}. \tag{4.3}$$

A good introduction to the representation of lines, and calculations using these representations, as well as many of the other geometric ideas discussed in this Chapter is to be found in the book by Bowyer and Woodwark [29].

4.5.2 Extracting Edges from Images

Many methods have been developed for extracting edges from images. Most use gradient based operators. We shall give a brief overview of these methods, discussing approaches which are common to many of them and highlighting details of specific techniques where appropriate.

4.5. EDGE EXTRACTION

Many edge extraction techniques can be broken up into two distinct phases:

- Finding pixels in the image where edges are likely to occur by looking for discontinuities in gradients. Candidate points for edges in the image are usually referred to as *edge points*, *edge pixels*, or *edgels*.

- Linking these edge points in some way to produce descriptions of edges in terms of lines, curves *etc*.

Each phase in turn will be discussed in the following Sections.

4.5.3 Detecting Edge Points

An edge point can be regarded as a point in an image where a discontinuity (in gradient) occurs across some line. A discontinuity may be classified as one of three types (see Fig 4.3):

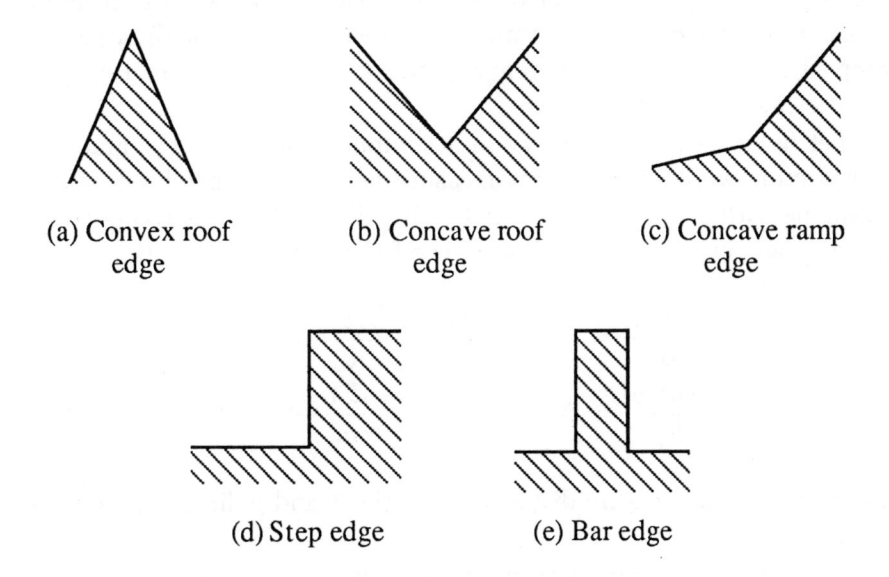

(a) Convex roof edge

(b) Concave roof edge

(c) Concave ramp edge

(d) Step edge

(e) Bar edge

Figure 4.3: Different types of edges

A Gradient Discontinuity — where the gradient of the pixel values changes across a line. This type of discontinuity is sometimes broken down into further classes such as *roof* and *ramp* edges or *convex* and *concave* edges by noting the sign of the component of the gradient perpendicular to the edge on either side of the edge. Ramp edges have the same signs in the gradient components on either side of the discontinuity, while roof edges have opposite signs in the gradient components.

A Jump or Step Discontinuity — where pixel values themselves change suddenly across some line.

A Bar Discontinuity — where pixel values rapidly increase then decrease again (or *vice versa*) across some line.

For example, if the pixel values are depth values, jump discontinuities occur where one object occludes another (or another part of itself), while gradient discontinuities usually occur between adjacent faces of the same object. If the pixel values are intensities, a bar discontinuity would represent cases like a thin black line on a white piece of paper. Step edges may separate different objects, or may occur where a shadow falls across an object.

Let us now consider how we may compute gradients. The gradient is a vector, whose components measure how rapidly pixel values are changing with distance in the x and y directions. Thus, the components of the gradient may be found using the following approximation:

$$\frac{\partial f(x,y)}{\partial x} = \Delta_x = \frac{f(x+d_x, y) - f(x,y)}{d_x},$$
$$\frac{\partial f(x,y)}{\partial y} = \Delta_y = \frac{f(x, y+d_y) - f(x,y)}{d_y}, \qquad (4.4)$$

where d_x and d_y measure distance along the x and y directions respectively.

Since we are dealing with discrete images we can consider d_x and d_y in terms of numbers of pixels between two points. Thus, when $d_x = d_y = 1$ (pixel spacing) and we are at the point whose pixel coordinates

4.5. EDGE EXTRACTION

are[1] (i,j) we have

$$\begin{aligned}\Delta_x &= f(i+1,j) - f(i,j),\\ \Delta_y &= f(i,j+1) - f(i,j).\end{aligned} \quad (4.5)$$

In order to detect the presence of a gradient discontinuity we must calculate the *change in gradient* at (i,j). We can do this by finding the following *gradient magnitude* measure,

$$M = \sqrt{\Delta_x^2 + \Delta_y^2}, \quad (4.6)$$

and the *gradient direction*, θ, given by

$$\theta = \tan^{-1}\left[\frac{\Delta_y}{\Delta_x}\right]. \quad (4.7)$$

We have described the basic theory of determining edges from an image above, and the implementation of these ideas is straightforward. The difference operators in Eqn. 4.5 correspond to convolving the image with the two masks in Fig. 4.4. In other words, the top left-hand corner of the appropriate mask is superimposed over each pixel of the image in turn, and a value is calculated for Δ_x or Δ_y by using the mask coefficients in a weighted sum of the value of pixel (i,j) and its neighbours. These masks are referred to as *convolution masks* or sometimes *convolution kernels*.

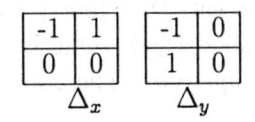

-1	1
0	0
Δ_x

-1	0
1	0
Δ_y

Figure 4.4: Edge operator convolution masks

It is now possible, instead of finding approximate gradient components along the x and y directions to approximate gradient components

---

[1] We assume i increases from left to right and j increases from top to bottom in the image, and hence in the masks shown

along directions at 45° and 135° to the axes respectively. In this case the following equations are used:

$$\Delta_1 = f(i+1, j+1) - f(i,j),$$
$$\Delta_2 = f(i, j+1) - f(i+1, j). \tag{4.8}$$

This form of operator is known as the *Roberts edge operator* and was one of the first operators used [203] to detect edges in images. The corresponding convolution masks are given by:

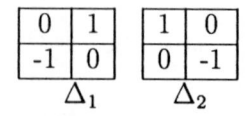

0	1		1	0
-1	0		0	-1
Δ_1			Δ_2	

Figure 4.5: Roberts edge operator convolution masks

Many edge detectors have been designed using convolution mask techniques, often using 3×3 mask sizes or even larger. The size of the mask is sometimes referred to as the *connectivity* of the mask. This is the maximum number of connected neighbouring pixels along any line contained within the mask. Thus a 3×3 mask is a three-neighbour connected mask.

An advantage of using a larger mask size is that errors due to the effects of noise are reduced by local averaging within the neighbourhood of the mask. An advantage of using a mask of odd size is that the operators are *centred* and can therefore provide an estimate that is biased towards a centre pixel (i, j). One important edge operator of this type is the *Sobel edge operator*. The Sobel edge operator masks are given in Fig 4.6.

All of the above edge detectors have approximated the first order derivatives of pixel values in an image. It is also possible to use *second order derivatives* to detect edges. Probably the most commonly used second order operator is the *Laplacian* operator. Mathematically, the Laplacian of a function $f(x,y)$, denoted by $\nabla^2 f(x,y)$, is defined by:

$$\nabla^2 f(x,y) = \frac{\partial^2 f(x,y)}{\partial x^2} + \frac{\partial^2 f(x,y)}{\partial y^2}. \tag{4.9}$$

4.5. EDGE EXTRACTION

-1	0	1
-2	0	2
-1	0	1

Δ_x

1	2	1
0	0	0
-1	-2	-1

Δ_y

Figure 4.6: Sobel edge operator convolution masks

Once more we can use discrete difference approximations to estimate the derivatives and represent the Laplacian operator with the 3 × 3 convolution mask shown in Fig 4.7.

0	1	0
1	-4	1
0	1	0

Figure 4.7: Laplacian operator convolution mask

However there are disadvantages to the use of second order derivatives. We should note that first derivative operators exaggerate the effects of noise. The very operation of taking second derivatives will lead to any noise present in the image being exaggerated twice. Another disadvantage of an operator such as the Laplacian is that it does not provide any directional information about the edge, which may be a severe handicap when linking edge points together (see Section 4.5.4).

The problems that the presence of noise causes when using the above edge detectors has led many researchers to look at ways of reducing the noise in an image prior to or in conjunction with the edge detection process, and indeed most of the other segmentation processes described later in this Chapter.

We have already discussed some methods of reducing or smoothing noise in Chapter 3, and some of these methods may be of use here. Alternatively we may use *Gaussian smoothing* techniques to achieve this result. Gaussian smoothing is performed by convolving an image with a Gaussian operator which is defined below. By using Gaussian smoothing in conjunction with the Laplacian operator, or another Gaussian

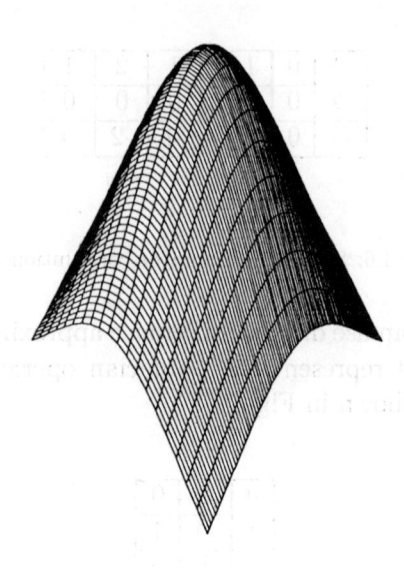

Figure 4.8: The Gaussian distribution in two variables

operator, it is possible to detect edges. Firstly the approach to Gaussian smoothing is briefly outlined.

The *Gaussian distribution* function in two variables, $g(x,y)$, is illustrated in Fig. 4.8 and is defined by

$$g(x,y) = \frac{1}{2\pi\sigma^2} e^{-(x^2+y^2)/2\sigma^2} \qquad (4.10)$$

where σ is the standard deviation representing the width of the Gaussian distribution. Thus the shape of the distribution and hence the amount of smoothing can be controlled by varying σ. In order to smooth an image $f(x,y)$, we convolve it with $g(x,y)$ to produce a smoothed image $s(x,y)$ i.e. $s(x,y) = f(x,y) * g(x,y)$.

Having smoothed the image with a Gaussian operator we can now take the Laplacian of the smoothed image. Therefore the total operation of edge detection after smoothing on the original image is $\nabla^2 (f(x,y) * g(x,y))$. It is simple to show that this operation can be reduced to convolving the original image $f(x,y)$ with a "Laplacian of a Gaussian" (LOG) operator $\nabla^2 g(x,y)$, which is shown in Fig. 4.9. Thus the edge pixels in an image are determined by a single convolution operation with a mask representing the composite LOG operator whose

4.5. EDGE EXTRACTION

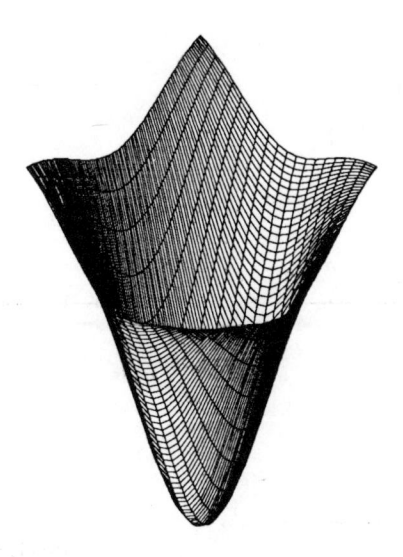

Figure 4.9: The LOG operator

form can easily be determined. Further properties of the LOG operator are addressed in the exercises at the end of this Chapter.

This method of edge detection was first proposed by Marr and Hildreth [157] who introduced the principle of the *zero-crossing* method. The basic principle of this method is to find the position in an image where the second derivatives become zero. These positions correspond to edge positions as shown in Fig. 4.10. The Gaussian function firstly smooths or blurs any step edges. Next, the second derivative of the blurred image is taken; it has a zero-crossing at the edge.

Although the LOG operator is still susceptible to noise, the effects of noise can be reduced by ignoring zero-crossings produced by small changes in image intensity. Also note that the LOG operator gives edge direction information as well as edge points. This can be determined from the direction of the zero-crossing.

A related method of edge detection is that of applying the "Difference of Gaussian" (DOG) operator to an image. This is formed by applying two Gaussian operators with different values of σ to an image and forming the difference of the resulting two smoothed images. It can be shown (see Exercise 4.6) that the DOG operator approximates

(a) f(x)

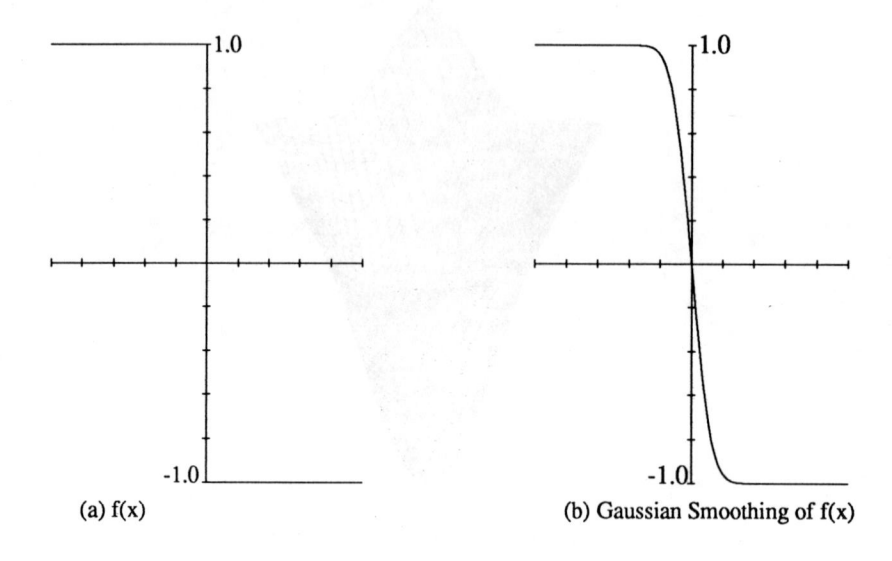

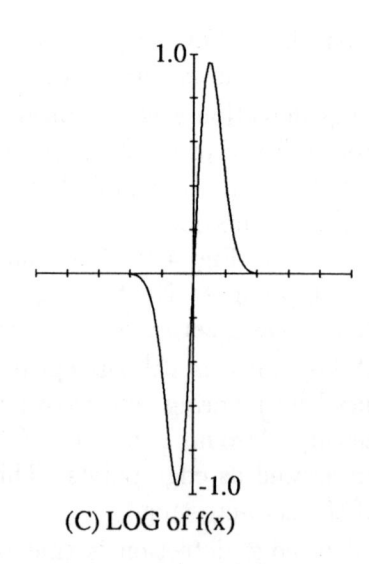

(b) Gaussian Smoothing of f(x)

(C) LOG of f(x)

Figure 4.10: Steps of the LOG operator

4.5. EDGE EXTRACTION

the LOG operator and indeed, evidence exists that the human visual system uses a similar method [157, 156, 236].

Another important edge detection method which has been used in many recent computer vision systems is the *Canny* edge detector [48]. Canny's approach is based on optimising the trade-off between two performance criteria, namely:

- Good edge detection — there should be low probabilities of failing to mark real edge points and marking false edge points.

- Good edge localisation — the positions of edge points marked by the edge detector should be as close as possible to the real edge.

Canny showed that the optimisation can be formulated by maximising a function that is expressed in terms of the signal-to-noise ratio of the image, the localisation of the edges and a probability that the edge detector only produces a single response to each actual edge in an image.

One further promising approach to edge detection has been described by Morrone, Owens and their co-workers [169, 170, 183]. A common failure of both the LOG and Canny's methods is that they suffer from "false positives", in other words some points of maximum gradient will be marked as lying on edges when a human observer would not consider an edge to be present. Morrone and Owens show that the human visual system perceives an edge to occur at points where the phase of all the components of the Fourier transform of the image are zero simultaneously. Similarly, a bar in an image appears at a point where all the spectral components have 90° phase congruency.

The Morrone and Owens edge detector thus finds edges by algorithmically seeking points of strong phase congruency. A point of strong phase congruency in turn implies a local maximum in the local energy function for the image. To find such maxima, the pixel values are again convolved with a pair of masks, one of which is the Hilbert transform of the other. The local energy function is the sum of the squares of the values output from these two convolution processes, and local maxima are sought in its value.

The resulting edge detector has the advantages that it is very robust to noise, it does not mark false positives, and can distinguish between step edges and bars. Furthermore, unlike the LOG operator, repeated application of this type of method to the output produced does not proliferate edges, *i.e.* edges of edges and so on are not output.

Many other methods of edge detection have been proposed as any survey of computer vision literature quickly verifies. In this Section we have attempted to give a broad introduction to the theory of edge detection and highlight important results in this area — there are far too many edge detection methods to discuss all of their relative merits or even to refer to more than just a few of them.

4.5.4 Edge Linking

The previous Section considered how to detect which pixels in an image lie on edges. The next step is to try to collect these pixels together into a set of edges. Thus, our aim is to replace many points on edges with a few edges themselves. The practical problem may be much more difficult than the idealised case. Small pieces of edges may be missing, small edge segments may appear to be present due to noise where there is no real edge, and so on. Some edge linking methods are able to deal with these problems better than others.

In general, edge linking methods can be classified into two categories:

Local Edge Linkers — where edge points are grouped to form edges by considering each point's relationship to any neighbouring edge points.

Global Edge Linkers — where all edge points in the image plane are considered at the same time and sets of edge points are sought according to some similarity constraint, such as points which share the same edge equation.

Firstly we shall consider methods employed by local edge linkers. Such techniques are similar to those that will be discussed in other contexts, particularly that of region growing, later in the book (see Section 4.6.4) and so these methods are only dealt with briefly here. Secondly, we shall also consider a global edge linking technique called the Hough transform.

4.5. EDGE EXTRACTION

Local Edge Linking Methods

We have seen that most edge detectors yield information about the magnitude of the gradient at an edge point and, more importantly, the direction of the edge in the locality of the point. The latter information is obviously useful when deciding which edge points to link together since edge points in a neighbourhood which have similar gradients directions (to within some threshold) are likely to lie on the same edge. It should be noted that this type of method lends itself to linking both straight and curved edges.

Local edge linking methods usually start at some arbitrary edge point and consider points in a local neighbourhood for similarity of edge direction as shown in Fig. 4.11. If the points satisfy the similarity constraint then the points are added to the current edge set. The neighbourhoods based around the recently added edge points are then considered in turn and so on. If the points do not satisfy the constraint then we conclude we are at the end of the edge, and so the process stops. A new starting edge point is found which does not belong to any edge set found so far, and the process is repeated. The algorithm terminates when all edge points have been linked to one edge or at least have been considered for linking once.

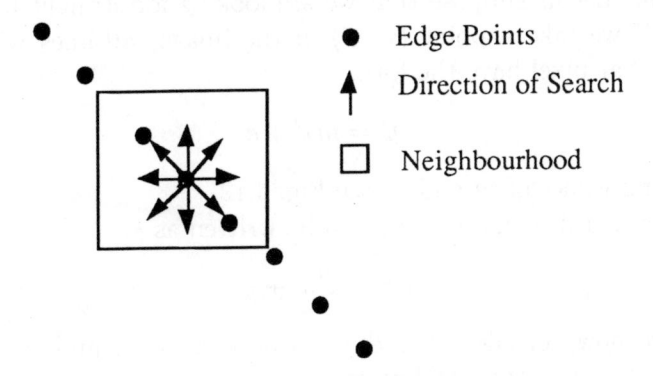

Figure 4.11: Edge linking

Thus the basic process used by local edge linkers is that of tracking a sequence of edge points. An advantage of such methods is that they can readily be used to find arbitrary curves. Furthermore, they can also use the information already obtained to help control the process. For instance, the equation of the current line being linked can be approximated and constantly updated when new points are added to the line. Points can be accepted or rejected as lying on the line according to their distance away from the line.

Many strategies have been adopted to control the search and selection processes used for edge linking. One such possibility is to use relaxation labelling techniques (see Section 5.3) to find sets of edges [246]. Here the assignment of each edge point to an edge is based on a probability estimate that the particular edge point and its local neighbours lie on the same edge. Other methods have posed the edge linking problem as a graph or tree search problem, or have employed dynamic programming techniques, where functions measuring the error in the fitting of an edge to a set of points are minimised to find the best fitting edges in the image. Many other edge linking techniques exist, and a good survey as well as further details of those described above can be found in the book by Ballard and Brown [6].

Hough Transforms

One powerful global method for detecting edges is called the *Hough transform*. Let us suppose that we are looking for straight lines in an image. If we take a point (x', y') in the image, *all* lines which pass through that pixel have the form

$$y' = mx' + c \qquad (4.11)$$

for varying values of m and c. See Fig. 4.12.

However, this equation can also be written as

$$c = -x'm + y' \qquad (4.12)$$

where we now consider x' and y' to be constants, and m and c as varying. This is a straight line on a graph of c against m as shown in Fig. 4.13.

Each different *line* through the point (x', y') corresponds to one of the *points* on the line in (m, c) space.

Now consider two pixels p and q in (x, y) space which lie on the same line. For each pixel, all of the possible lines through it are represented

4.5. EDGE EXTRACTION

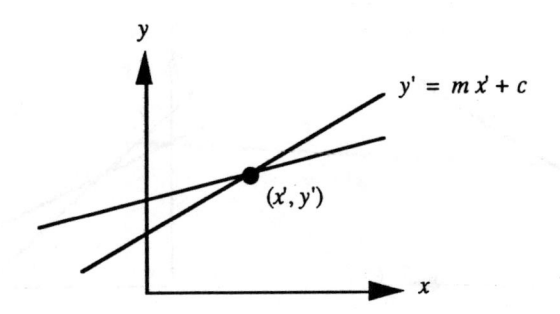

Figure 4.12: Lines through a point

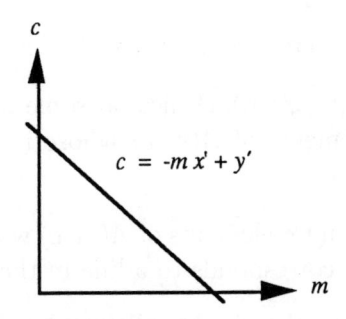

Figure 4.13: Lines though a point in terms of m and c

by a single line in (m, c) space. Thus the single line in (x, y) space which goes through both pixels lies on the intersection of the two lines representing p and q in (m, c) space, as illustrated by Fig. 4.14.

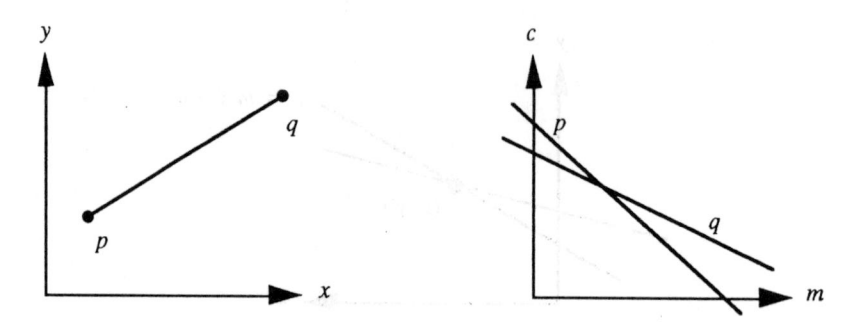

Figure 4.14: Points on the same line

Taking this one step further, all pixels which lie on the same line in (x, y) space are represented by lines which all pass through a single point in (m, c) space. The single point through which they all pass gives the values of m and c in the equation of the line $y = mx + c$.

To detect straight lines in an image, we thus proceed as follows:

1. We quantise (m, c) space into a two-dimensional array A for appropriate steps of m and c.

2. We initialise all elements of $A(m, c)$ to zero.

3. For each pixel (x', y') which lies on some edge in the image, we add 1 to all elements of $A(m, c)$ whose indices m and c satisfy $y' = mx' + c$.

4. Finally, we search for elements of $A(m, c)$ which have large values. Each one found corresponds to a line in the original image.

In the last step it is usual to ignore all (m, c) pairs with values of less than, say five, which will reject short lines with four or less pixels on them as spurious or unimportant.

One useful property of the Hough transform is that the pixels which lie on the line need not all be contiguous. For example, all of the pixels lying on the two dotted lines in Fig. 4.15 will be recognised as lying

4.5. EDGE EXTRACTION

on the same straight line. This can be very useful when trying to detect lines with short breaks in them due to noise, or when objects are partially occluded as shown.

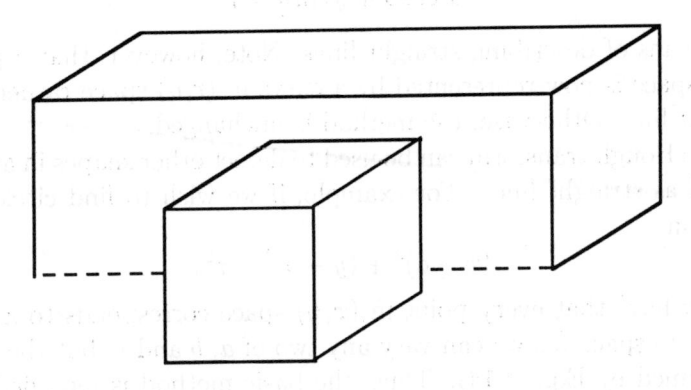

Figure 4.15: Non-contiguous pixels on the same line

On the other hand, it can also give misleading results when objects happen to be aligned by chance, as shown by the two dotted lines in Fig. 4.16.

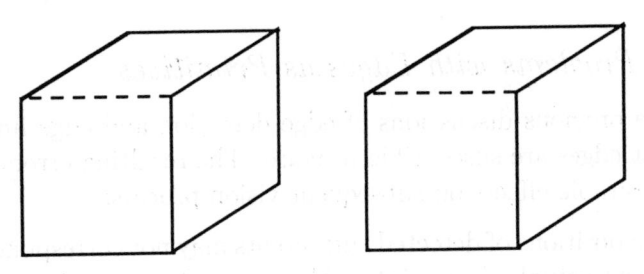

Figure 4.16: Aligned objects

Here, the Hough transform method still only returns one line for what a human observer would quite clearly recognise as two separate line segments. Indeed, this clearly shows that one disadvantage of the Hough transform method is that it gives an *infinite line* as expressed by the pair of m and c values, rather than a finite *line segment* with two well-defined endpoints.

One practical detail is that the $y = mx + c$ form for representing a straight line breaks down for vertical lines, when m becomes infinite.

To avoid this problem, it is better to use the alternative formulation given in Section 4.5.1,

$$x \cos \theta + y \sin \theta = r, \tag{4.13}$$

as a means of describing straight lines. Note, however, that a point in (x, y) space is now represented by a curve in (r, θ) space rather than a straight line. Otherwise, the method is unchanged.

The Hough transform can be used to detect other shapes in an image as well as straight lines. For example, if we wish to find circles, with equation

$$(x - a)^2 + (y - b)^2 = r^2, \tag{4.14}$$

we now have that every point in (x, y) space corresponds to a surface in (a, b, r) space (as we can vary any two of a, b and r, but the third is determined by Eqn. 4.14). Thus, the basic method is modified to use a three-dimensional array $A(a, b, r)$, and all points in it which satisfy the equation for a circle are incremented. In practice, the technique takes rapidly increasing amounts of time for more complicated curves as the number of variables (and hence the number of dimensions of A) increases, and so the method is really only of use for simple curves.

4.5.5 Problems with Edges as Primitives

From the previous discussions of edge detection and edge linking it is clear that edges are susceptible to noise. The resulting errors can have two undesirable effects on subsequent vision process:

- The positions of detected edge points may not correspond to those of the actual edge points. Many object recognition and image reasoning methods (see Chapter 5) produce a description of an object's orientation and position (pose) in a scene. These methods depend on receiving accurate segmentation information to produce accurate object pose descriptions. Clearly, if edges are used and the edge detector produces inaccurate output then the resultant pose description will also be in error.

- Errors due to noise can increase the complexity of subsequent reasoning and recognition stages. Edge detectors in general are liable to

4.5. EDGE EXTRACTION

- detect more than one edge for a single actual edge in the scene,
- detect edges that do not actually exist in the scene, and
- not detect some actual edges at all.

Clearly if any of the above occur often then the strategies that deal with higher level reasoning, such as which edges bound which faces of solid objects (see Chapter 5), or recognising objects from stored geometric models (see Chapter 8), will require more sophisticated processing to tolerate such errors.

Edges are also unsuited to many subsequent vision tasks in comparison to surfaces for several other reasons not necessarily associated with noise:

- Edges only represent locally important data in an image and the global richness of the whole data set is not exploited, a large amount of the information in the image being totally disregarded. As a simple example, given a planar face of an object, there are far fewer pixels contained along its edges than are contained within its interior. All of the data provided by these interior pixels will be wasted.

- Also related to the above item and as we discussed when considering extracting points (Section 4.4), it can be reasonably expected that the estimate of the parameters of a surface by least squares fitting to a set of pixel values will be much more accurate than any estimate of edge parameters due to the much larger number of points used in the fit.

- Certain types of objects bounded by curved surfaces which smoothly blend into one another may have no edges. The only edges which can be extracted for such objects are their silhouette edges, where the object occludes the background. Information provided by such edges is obviously limited.

In closing this Section, we would like to state that whilst we may have appeared to give edges "a bad press" in the above discussion, these criticisms are of most relevance when edges are used as the *sole* source of information or used as a basis for methods that require very accurate data. Edges do provide strong clues to object shape. For example, in depth images it is simple to recognise a cylinder from its outline of two concentric circles, with at least one partially occluded, flanked by two parallel lines. This is possible even if the edges are fragmented or not complete. Furthermore, if edge information is used in conjunction with other methods (see Chapter 5) then it is possible to develop some very powerful vision applications. Indeed, as remarked earlier, evidence exists [156, 236] that various low level human visual processes are based on edge detection, and it seems that this information is integrated with many other types of visual information in human vision. Finally, one advantage that edge detection methods have over other methods to be discussed shortly is that they are relatively cheap to compute.

4.6 Surface Extraction

Another way of extracting and representing information from an image is to group pixels together into regions of similarity. In a greyscale or a colour image we would group pixels together according to the rate of change of their intensity over a region. In three-dimensional images we group together pixels according to the rate of change of depth in the image, corresponding to pixels lying on the same surface such as a plane, cylinder, sphere etc.

There are many types of three-dimensional surface description, some of which are easier to extract from images than others. In the following Section we will discuss various ways in which surfaces can be described. We will then discuss methods of extracting similar regions of pixels from both two and three-dimensional images.

4.6.1 *Representing Surfaces*

The simplest form of surface is the plane, described by a *first order* equation. It is a relatively easy task to fit a plane to a set of points using least squares methods (see Appendix B) with low errors in the resultant approximation, and indeed, as we will see later, planes can

4.6. SURFACE EXTRACTION

be extracted more reliably than other surfaces. It is also possible to describe more complicated surfaces in terms of planar approximations as will be discussed in Chapter 7.

A plane is usually represented by vector **n** and a scalar d, where $\mathbf{n} = (a, b, c)$ is the unit normal vector to the plane (assumed to point outwards when the plane is part of the surface of an object) and d is the signed perpendicular distance of the plane from the origin (see Fig. 4.17).

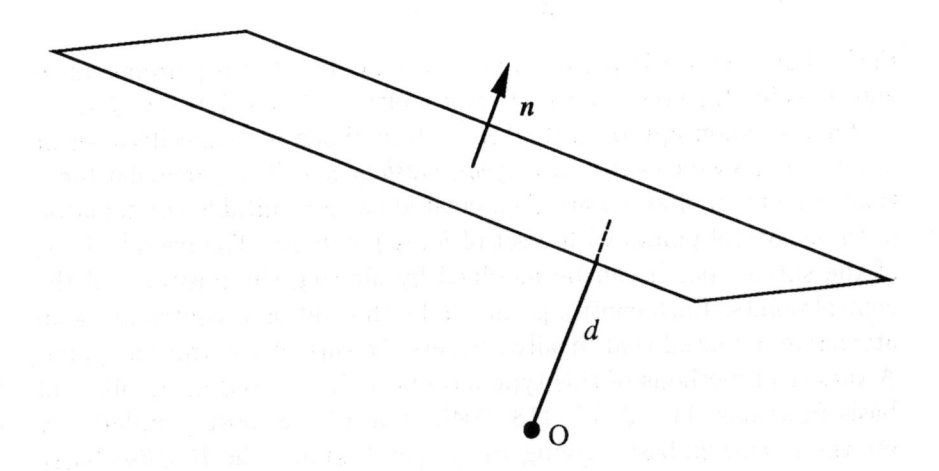

Figure 4.17: Representation of a plane

The equation of the plane is then

$$ax + by + cz + d = 0. \tag{4.15}$$

For use with depth maps this equation may also be usefully written in the form

$$z = a'x + b'y + c' \tag{4.16}$$

where $a' = -a/c$, $b' = -b/c$ and $c' = -d/c$, as any planes with $c = 0$ in Eqn. 4.15 would not be visible in the scene because they would be parallel to the line of sight of the master camera.

112   CHAPTER 4. SEGMENTATION AND FEATURE REPRESENTATION

The use of higher order surfaces than planes is necessary if real world objects are to be processed by a computer vision system. There are many ways of representing higher order surfaces.

Some representations model surfaces by means of surface patches. The simplest surfaces to use are planar [76] but many methods exist [11, 12, 74, 118, 204] to model non-planar surface patches. These are usually based on a *parametric* representation for the surface, *i.e.*

$$\begin{aligned} x &= x(u,v), \\ y &= y(u,v), \\ z &= z(u,v). \end{aligned} \qquad (4.17)$$

Each of x, y and z is represented by a function of two parameters, u and v, which take on a range of values, often $0 \leq u \leq 1, 0 \leq v \leq 1$.

One common approach used in such methods is to specify a set of *control points* which define a *polygon mesh* or *net*. The particular functions used for x, y and z are then created using a suitable combination of these control points with a set of *basis functions*. The overall shape of the surface patch can be modified by altering the positions of the control points. Each control point can be thought of as contributing an attractive potential that locally attracts the surface towards the point. A variety of methods of this type have been formulated using different basis functions [11, 12, 74, 118, 204]. Two of the most popular ones are the *Bernstein* basis (giving *Bézier* patches) and the *B-spline* basis. An example of a curved surface formed using Bézier patches is shown in Fig. 4.18.

However, this type of surface representation is not easy to use for computer vision strategies, as will be explained in Chapter 5, and consequently not many vision systems have adopted this type of representation. Among the problems are difficulties in uniquely representing a surface, specifically with control points. It is possible to describe the same surface with totally different control points. This will lead to difficulties with any subsequent recognition or other strategies that need to match a segmented surface with a stored one.

Another way of representing surfaces is to describe them as *implicit surfaces*. An implicit surface is of the form

$$f(x,y,z) = 0 \qquad (4.18)$$

4.6. SURFACE EXTRACTION

• Control point

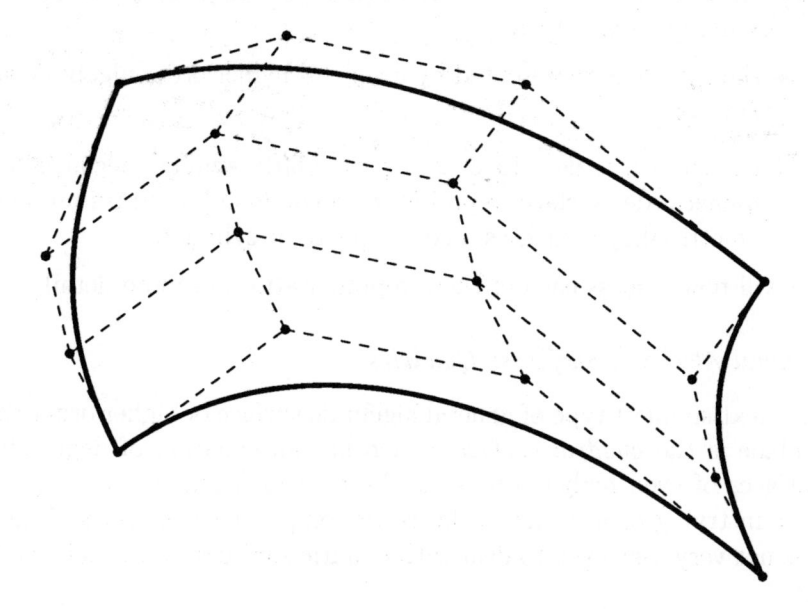

Figure 4.18: Bézier patch surface representation

where $f(x,y,z)$ is usually a polynomial function of x, y and z, although it may take some more complicated form. When $f(x,y,z)$ *is* a polynomial function, the surface generated is called an *algebraic surface*. As a simple example, the points on the surface of a unit sphere are those which satisfy

$$x^2 + y^2 + z^2 - 1 = 0. \qquad (4.19)$$

Further discussion of algebraic surfaces can be found in the book by Hoffmann [118].

Because of problems with parametric representations, in vision applications, two alternative representations of higher order surfaces are usually used instead. Either

- the surface is represented by a general higher order algebraic surface, or

- the surface is described as a particularly simple special case of higher order surface, representing a commonly used primitive geometric shape, such as a cone, sphere or cylinder.

We will now discuss each of these representations in more detail.

General Algebraic Surfaces: Quadrics

The next simplest type of general algebraic surface of higher order than a plane is the *quadric surface*, which has an equation of degree two. Surfaces of even higher orders can be described but many difficulties arise in trying to use such surfaces for computer vision tasks. Indeed, it is not very easy even to deal with quadric surfaces as we shall shortly see.

A quadric surface has the general form:

$$ax^2 + by^2 + cz^2 + 2dxy + 2exz + 2fyz + 2gx + 2hy + 2iz + j = 0 \quad (4.20)$$

Eqn. 4.20 can also be expressed in matrix form as:

$$\mathbf{x}\mathbf{A}\mathbf{x}^t + \mathbf{v}\mathbf{x} + j = 0 \qquad (4.21)$$

where

$$\mathbf{A} = \begin{pmatrix} a & d & e \\ d & b & f \\ e & f & c \end{pmatrix}, \quad \mathbf{v} = (2g\ 2h\ 2z) \quad \text{and} \quad \mathbf{x} = (x\ y\ z)$$

4.6. SURFACE EXTRACTION

Using this form of equation Faugeras has developed methods for segmenting quadric surfaces from image data [81, 79] (see also Appendix B); other vision workers [72] have also worked with quadric surfaces. One important aspect of this representation which also falls in line with our segmentation criteria is that we can now represent a quadric in terms of lower order geometric features derived from the above. This substantially simplifies subsequent vision tasks such as recognition of quadric surfaces. These are namely:

- Principal axes — the physical meaning of which will be described shortly. The direction vectors of the principle axes correspond to the eigenvectors of $\mathbf{A}$ (see Appendix A for definitions of various matrix operations).

- The centre $\mathbf{c}$ of the quadric if it exists:

$$\mathbf{c} = -\frac{1}{2}\mathbf{A}^{-1}\mathbf{v}. \tag{4.22}$$

Note that $\mathbf{c}$ only exists if the matrix $\mathbf{A}$ is non-singular and not all quadric surfaces satisfy this requirement (see Table 4.1).

Quadric surfaces can be classified into many types of surface — for example spheres, cylinders, ellipsoids and paraboloids. We can determine the type of surface by considering the sign of the eigenvalues of $\mathbf{A}$ in Eqn. 4.21 . However the surface classification is only reliable if the eigenvalues are large which can not be guaranteed.

A more detailed classification can be achieved by expressing the equation of a quadric as follows

$$\mathbf{x}\mathbf{Q}\mathbf{x}^t = 0 \tag{4.23}$$

where

$$\mathbf{Q} = \begin{pmatrix} a & d & e & g \\ d & b & f & h \\ e & f & c & i \\ g & h & i & j \end{pmatrix} \quad \text{and} \quad \mathbf{x} = (x\,y\,z\,1).$$

For ease of notation we will rewrite the elements of $\mathbf{Q}$ as

$$\mathbf{Q} = \begin{pmatrix} q_{11} & q_{12} & q_{13} & q_{14} \\ q_{21} & q_{22} & q_{23} & q_{24} \\ q_{31} & q_{32} & q_{33} & q_{34} \\ q_{41} & q_{42} & q_{43} & q_{44} \end{pmatrix}.$$

Also let **A** be defined as in Eqn. 4.21 and let q_r and a_r be the ranks of the matrices **Q** and **A** respectively. Let a_s be the modulus of the signature of **A**.

Define T_a to be the trace of matrix **A**, and T_q, D_2 and D_3 as follows:

$$T_a = \sum_{i=1}^{3} q_{ii}, \qquad (4.24)$$

$$T_q = \sum_{i=1}^{2}\sum_{j=i+1}^{3} \begin{vmatrix} q_{ii} & q_{ij} \\ q_{ji} & q_{jj} \end{vmatrix}, \qquad (4.25)$$

$$D_2 = \sum_{i=1}^{3}\sum_{j=i+1}^{4} \begin{vmatrix} q_{ii} & q_{ij} \\ q_{ji} & q_{jj} \end{vmatrix}, \qquad (4.26)$$

$$D_3 = \sum_{i=1}^{2}\sum_{j=i+1}^{3}\sum_{k=j+1}^{4} \begin{vmatrix} q_{ii} & q_{ij} & q_{ik} \\ q_{ji} & q_{jj} & q_{jk} \\ q_{ki} & q_{kj} & q_{kk} \end{vmatrix}. \qquad (4.27)$$

Let

$$C_1 = \begin{cases} 1 \text{ (boolean true)} & \text{if } T_q > 0 \text{ and } |\mathbf{A}|T_a \leq 0 \\ & \text{or } T_q \leq 0 \\ 0 \text{ (false)} & \text{otherwise} \end{cases} \qquad (4.28)$$

and

$$C_2 = \begin{cases} 1 \text{ (true)} & \text{if } T_q > 0 \text{ and } |\mathbf{A}|T_a > 0 \\ 0 \text{ (false)} & \text{otherwise} \end{cases} \qquad (4.29)$$

The quantities defined above can now be used to classify a given quadric form as follows. We will only list *valid* or *physical* geometric forms of quadrics that are relevant to vision problems. A more complete treatment of this theory can be found in Levin [149].

Clearly for the matrix **Q** to be non-singular q_r must be equal to 4 and for **A** to be non-singular $a_r = 3$. If **Q** is singular we can tell the type of quadric by looking at the values of q_r, a_r and a_s and using the other quantities defined above where appropriate. A classification of quadric surfaces when **Q** is singular is provided by Table 4.1 which is based on Table 1 in Levin's paper [149].

If **Q** is non-singular than we can classify the quadric surface by looking at the value of the determinants of **A** and **Q** as shown in Table 4.2 which is also based on Table 1 in Levin's paper [149].

4.6. SURFACE EXTRACTION

q_r	a_r	a_s	Additional Condition	Surface Type
1	1	1		2 coincident planes
2	0	0		single plane
2	1	1	$D_2 > 0$	2 parallel planes
2	2	0	$T_q < 0$	2 intersecting planes
2	2	2	$T_q > 0$	line
3	1	1		parabolic cylinder
3	2	0	$T_q < 0$	hyperbolic cylinder
3	2	2	$T_q > 0, T_1 D_3 < 0$	elliptic cylinder
3	3	1	$C_1 = 1$	cone
3	3	3	$C_2 = 1$	point
all other cases are invalid physical surfaces				

Table 4.1: Classification of singular quadric surfaces

$|A|$	$|Q|$	a_s	Additional Condition	Surface Type
0	> 0	0	$T_q < 0$	hyperbolic paraboloid
0	< 0	2	$T_q > 0$	elliptic paraboloid
$\neq 0$	> 0	1	$C_1 = 1$	hyperboloid of one sheet
$\neq 0$	< 0	1	$C_1 = 1$	hyperboloid of two sheets
$\neq 0$	< 0	3	$C_2 = 1$	ellipsoid
all other cases are invalid physical surfaces				

Table 4.2: Classification of non-singular quadric surfaces

118   CHAPTER 4. SEGMENTATION AND FEATURE REPRESENTATION

In general it is difficult to fit data to a quadric surface for two reasons which are discussed further in Section 4.6.8:

- It is difficult to express the requirements for fitting in a form that can be used easily by approximation algorithms (see Appendix B). For example, for least squares error fitting it is necessary to express the error of fit of points to a surface. This is usually a measure of distance from the point to the surface. However this is not a simple quantity to compute for a quadric surface except in special cases (see Examples 4.10 and 4.11).

- It is difficult to cope with the number of free parameters associated with a general quadric surface (see Section 4.6.8 and Examples 4.10 and 4.11). In general, a plane has three free parameters (after normalisation) while a quadric has nine, so many more unknowns need to be estimated from the data.

Particular Geometric Primitives

Instead of trying to deal with a general class of algebraic surfaces such as quadrics it is often simpler to restrict our attention to a few special classes, which we will call geometric primitives. Such classes might be cylinders, spheres, cones *etc.* and sometimes special generalised cylinders and cones. Such surfaces will be discussed in this Section. This type of approach has been popular in vision research for many years [7, 20, 26, 40, 85, 148, 158, 178]. Many objects can be approximated by or thought of as mainly consisting of these simple shapes.

Although many of these shapes are special cases of quadric surfaces, and could be represented as described in the previous section, in practice treating each class separately leads to alternative representations which are easier to use.

We will discuss the representation of the circular cylinder and sphere first:

Cylinders:

A cylindrical surface can be represented as follows (see Fig 4.19) :

- A unit vector, **n**, along its axis of symmetry,

4.6. SURFACE EXTRACTION

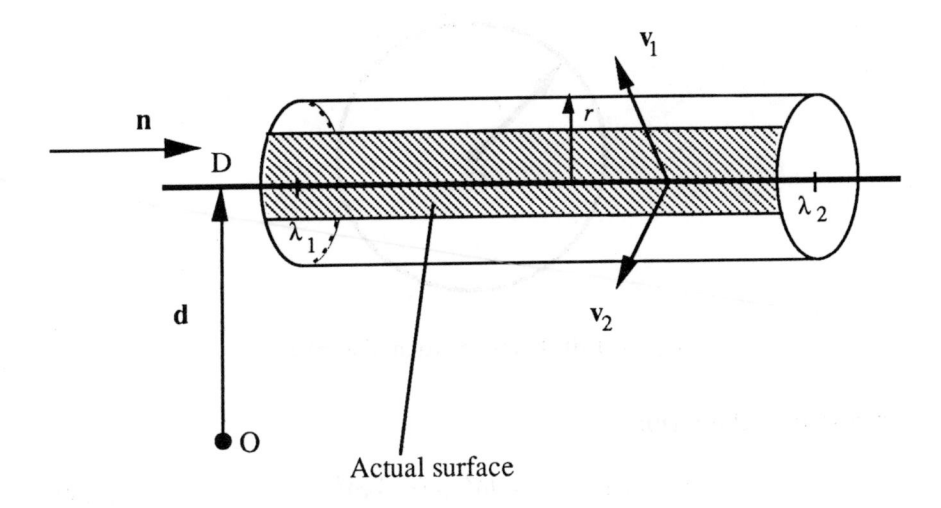

Figure 4.19: Representation of a cylinder

- A vector, **d**, from the origin to the closest point on the axis of symmetry,

- The radius of the cylinder, r,

- Two scalars, λ_1 and λ_2, denoting the start and end points of the cylinder along the axis of symmetry, as exemplified by

$$start\_point = \mathbf{d} + \lambda_1 \mathbf{n}.$$

- Two vectors, $\mathbf{v}_1$ and $\mathbf{v}_2$, denoting the sector of the cylinder which is of interest. The cylindrical surface is that part of the surface of the cylinder bounded by the two edges formed by taking a line parallel to the axis of symmetry and displacing it a distance r along $\mathbf{v}_1$ and $\mathbf{v}_2$. The vectors $\mathbf{v}_1$ and $\mathbf{v}_2$ must be perpendicular to the axis of symmetry. The bounded surface is taken to be the surface defined turning clockwise, from $\mathbf{v}_1$ to $\mathbf{v}_2$, when looking along **n**.

Spheres:
A complete sphere may be represented see (Fig 4.20) by its centre, $\mathbf{C} = (a\ b\ c)$, and its radius, r. The equation of the sphere may be

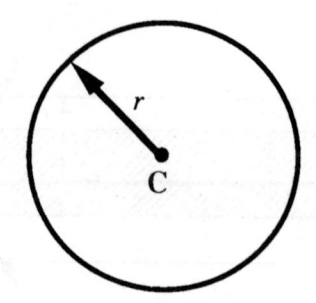

Figure 4.20: Representation of a sphere

expressed in the form:

$$(x - a)^2 + (y - b)^2 + (z - c)^2 = r^2. \qquad (4.30)$$

We may also represent incomplete spherical surfaces by choosing vectors that specify the visible portions of the surface similar to the method used for cylinders.

This approach has been extended further to include generalised cylinders and cones [20, 148, 178]. We will not pursue this topic here, except to note that a cylinder may be generalised in one or more of the following ways:

- The axis may be a curve rather than a straight line.
- The radius of the cylinder may vary along its axis.
- The cross-section may be some planar shape other than a circle.
- The cross-section may be held at some angle other than a right angle to the axis.

4.6.2 Extracting Surfaces from Images

There are two main approaches that can be employed to extract planes and other surfaces from greyscale or depth images. These are:

Region Splitting — Here the basic approach is to initially consider the scene as a whole and to recursively split it into smaller regions

4.6. SURFACE EXTRACTION

until the subdivision produces regions which meet some terminating criteria for belonging to a single surface. The algorithms used to perform these tasks fall into the *divide and conquer* or *top down* category.

Region Growing — This is basically a *bottom up* approach where each pixel is grouped with neighbouring pixels to form larger regions while the composite neighbourhood continues to satisfy some similarity constraint(*e.g.* all pixels lie on the same plane, or all pixels have similar intensity gradients or texture). Such methods commence at some arbitrarily chosen *seed* points from which regions start to grow.

We will consider each of these methods in more detail in the following sections. We will start by describing each method in a form that can be applied to most types of images, simply referring to similarity constraints in general. Subsequently we will discuss particular similarity constraints that can be applied to different types of images. Finally, we will describe in more detail particular methods that use these constraints.

4.6.3 Region Splitting

As previously stated, the basic idea of region splitting is to break the image into a set of disjoint regions which are coherent within themselves. Initially (in the simplest case) we take the image as a whole to be the area of interest. We then look at the area of interest and decide if all pixels contained in the region satisfy the chosen similarity constraint. If they do then we know that the area of interest corresponds to a region in the image. If they do not we split the area of interest (usually into four equal sub-areas) and consider each of the sub-areas as the area of interest in turn. We continue this process until no further splitting occurs. In the worst case this happens when the areas are just one pixel in size.

If only a splitting schedule as above is used then the final segmentation would probably contain many neighbouring regions that have identical or similar properties. Thus, a *merging* process is used after each split which compares adjacent regions and merges them if necessary. Algorithms of this nature are called *split and merge* algorithms.

Most region splitting methods which adopt the split and merge approach start with the whole image as the initial area of interest [126, 187]. Although some [185] have adopted a medium resolution approach which start by considering areas of some arbitrary size, merging similar areas first and then splitting and merging the remaining areas.

To illustrate the basic principle of these methods let us consider an imaginary image. Let **I** denote the whole image shown in Fig 4.21(a). Not all the pixels in **I** are similar so the region is split as in Fig 4.21(b). Assume that all pixels within regions $\mathbf{I}_1$, $\mathbf{I}_2$ and $\mathbf{I}_3$ respectively are similar but those in $\mathbf{I}_4$ are not. Therefore $\mathbf{I}_4$ is split next as in Fig 4.21(c). Now assume that all pixels within each region are similar with respect to that region, and that after comparing the split regions, regions $\mathbf{I}_{43}$ and $\mathbf{I}_{44}$ are found to be identical. These are thus merged together as in Fig 4.21(d).

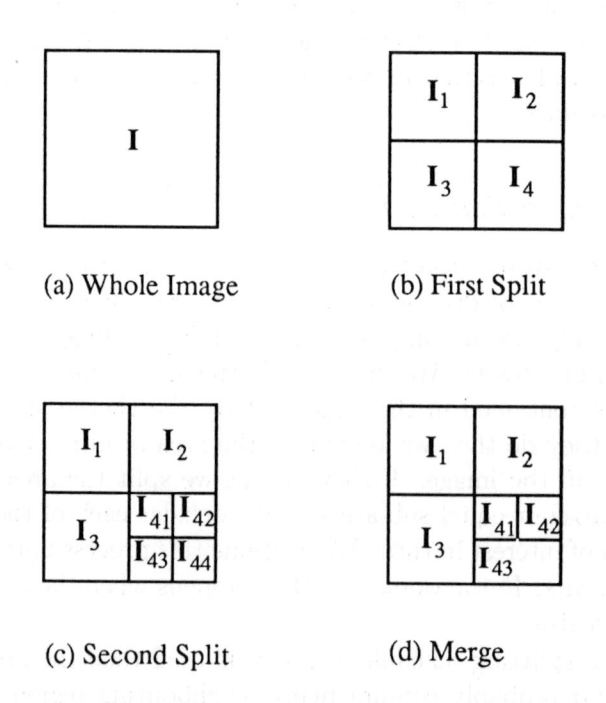

(a) Whole Image

(b) First Split

(c) Second Split

(d) Merge

Figure 4.21: Example of region splitting and merging

4.6. SURFACE EXTRACTION

We can describe the splitting of the image using a tree structure, using a modified *quadtree*. Each non-terminal node in the tree has at most four descendants, although it may have less due to merging. See Fig. 4.22. If we start with three-dimensional depth map data, in a

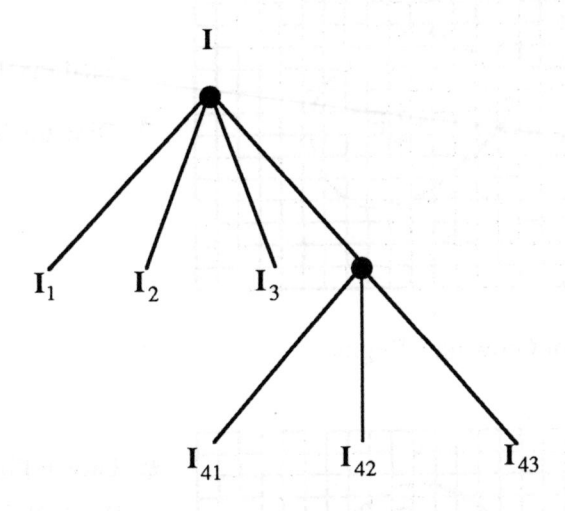

Figure 4.22: Region splitting and merging tree

similar way, octrees can be used.

In practice, the tests for equality described above have to be relaxed, and less strict tests are used to group pixels into regions. Because of this, the resulting regions often end up with more rectangular edges than they really should have, as boundary pixels are occasionally grouped into incorrect regions.

4.6.4 Region Growing

The region growing approach is the opposite of the split and merge approach in that an initial set of small areas are iteratively merged according to similarity constraints while these are still satisfied. Initially it is usual to start by choosing an arbitrary *seed pixel* and to compare it with neighbouring pixels (see Fig 4.23). The region is *grown* from the seed pixel by adding in neighbouring pixels that are similar, increasing the size of the region. When the growth of one region stops we simply choose another seed pixel which does not yet belong to any region and start again. This whole process is continued until all pixels belong to some region.

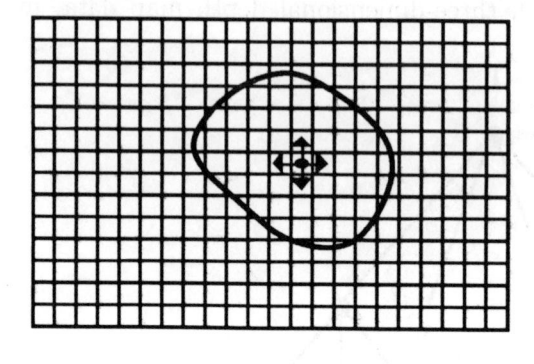

(a) Start of Growing a Region

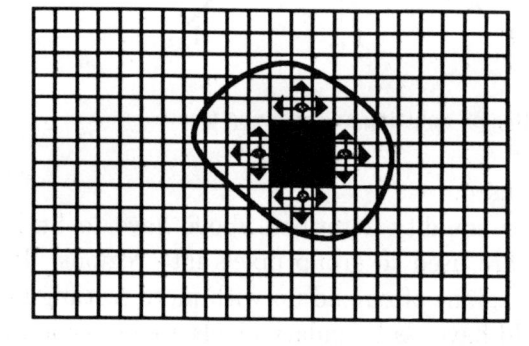

• Seed Pixel

↑ Direction of Growth

(b) Growing Process After a Few Iterations

■ Grown Pixels

● Pixels Being Considered

Figure 4.23: Example of region growing

Region growing methods often give very good segmentations that correspond well to the observed edges. However one problem is that starting with a particular seed pixel and letting this region grow completely before trying other seeds biases the segmentation in favour of the regions which are segmented first. This can have several undesirable effects:

- Since the region currently being considered dominates the growth

4.6. SURFACE EXTRACTION

process any ambiguity around edges of adjacent regions that have similar properties may not be resolved correctly. In particular, a wrong decision may be made as for pixels close to the edge to which side of the edge they lie. *Edge leaking* occurs in this case and the region currently being considered overflows into the pixels that ought to be identified as belonging to an adjacent region.

- The segmentation may not be robust. Different choices of seeds may give different segmentation results, again as a result of allowing one region to dominate the process at any given time. This is particularly a problem for region growing methods which adaptively change their similarity constraints as growth proceeds.

- Problems can occur in arbitrarily choosing the seed point, particularly if the seed point lies on an edge.

To counter the above problems, *simultaneous region growing* techniques have been developed [14, 95, 133, 139, 159, 184] where similarities of neighbouring regions are taken into account in the growing process. Here no single region is allowed to completely dominate the proceedings, but rather a number of regions is allowed to grow at the same time, and indeed, initially all pixels may be allowed to grow. In these circumstances similar regions will gradually coalesce into expanding regions and the problems associated with the order of segmentation are avoided. The processing to control these methods may be quite complicated [14, 139] but efficient methods have been developed [95]. Another advantage of this approach is that it is very easy and efficient to implement on parallel computers [14].

Many researchers have looked at methods to inhibit regions from leaking. The simultaneous region growing process partly solves the problem since individual regions can no longer dominate. One popular method that has been adopted to limit the problem further is to use edge information obtained from an edge detector (Section 4.5.3) in conjunction with the region growing method. The basic idea of such a hybrid approach is to prevent a region from growing over an edge *unless* there is sufficient information present to safely allow this to happen. Many variations on this idea have been proposed.

Geman and Geman [96] devised a probabilistic approach to achieve region segmentation in which they also reduce image noise at the same time. They express the likelihoods of a pixel lying within a local region and of lying on an edge as independent probabilities. They then use statistical reasoning techniques to only allow a region to cross an edge boundary if the probability that regions on either side of the boundary are similar is high and the probability of an edge occurring is weak. Their original method was applied to greyscale values but this approach has been extended to segment images using texture [58, 62], colour [244] and three-dimensional information [1, 18, 119]. This type of method will be described in more detail in Section 5.3.

Other methods [14, 18, 37, 95, 139, 159, 226] have modelled the region and edge information as *attractive* and *repulsive forces* respectively, acting to group or separate the pixels into regions. This formulation is similar to the probability-based approaches described above except that here the *magnitude* of the forces is determined empirically. We will discuss such ideas in more detail in the next Section on similarity constraints and methods.

One further approach to region growing is to apply histogram filtering techniques [241] to either the greyscale values or range data. The basic approach of these methods is to construct a histogram of intensity values in the image and then to form regions by grouping together pixels whose intensity values lie between a pair of thresholds. For example, this technique can often readily separate an object from the background. However, this approach tends to be too simple for complex scenes [184]. When depth values are used, it does not work well on long objects that occupy a large range of depth values in the image [16].

4.6.5 Similarity Constraints

In the previous two Sections we have considered general methods for segmenting an image into regions. We have referred to the criteria used to decide if pixels belong to the same region as *similarity constraints*. In this Section we will discuss how to define and use such constraints for different types of source image.

For greyscale images it is usual to use the pixel values directly in the

4.6. SURFACE EXTRACTION

segmentation process and to calculate intensity gradients as part of the segmentation process. However in probability-based methods such as the Geman and Geman method [96] discussed in the previous Section we must first calculate a probability that a pixel is similar to its neighbours. Usually a 3 × 3 mask is passed over the image which calculates the similarity between the centre pixel and its neighbours, expressed as a probability that they belong to the same region. Similar approaches are used when these methods are used for other image segmentation tasks.

For colour images there are various possibilities upon which to base a segmentation. The red, green and blue components of the image can be treated as separate bands. Each is dealt with as an independent greyscale image and the information is brought together later in the processing [244]. Thus, for example, a red region would only be detected in one band whereas a magenta region would be detected in two bands (red and blue). An alternative is to treat the three colour bands as a three dimensional (vector) space where the red, green and blue bands each contribute components to a vector, $\mathbf{v} = (rgb)$. Methods employing this approach can then use similar methods to those used for three-dimensional surface extraction to be described later. Many researchers working with colour images prefer to represent colour in terms of the three alternative components of image *brightness* and *chromaticity* (also referred to as *hue*, *saturation* and *intensity*). The segmentation methods proceed in the same way, however.

For depth maps it is not normal to attempt to segment the raw depth data. Instead, in practice, we usually require some information about changes in local gradients to help us determine surface shape. This can be provided by initially fitting planes to small regions by passing a mask (typically 3 × 3 pixels) over the depth map (using the method described for oriented points given in Section 4.4). This information is then used, in particular the direction of the plane normal, to drive the segmentation.

We will now briefly discuss one such segmentation method for extracting planar regions from depth maps. As we will shortly find out it is based on previous work on colour and greyscale segmentation techniques. The method is based on work by Gay [95] and Page [184]. Details of its application to segment three-dimensional images can be found in [132, 133, 159]. This segmentation method is the one used in our vision system as described in various parts of the book. Results of the method in practical use are given in Chapter 9.

4.6.6 A Method for Segmenting Planes

The ideas behind this segmentation algorithm arose from the involvement of British Aerospace's Sowerby Research Centre in many vision research projects over recent years, and in particular the Alvey project "Object Identification in 2-D Images" for which new segmentation methods for greyscale images were developed [95, 184]. It has been shown that the technique can be used to segment either monochrome or colour images [95, 184], and indeed, as might be expected, it was demonstrated that segmentation using colour images gives better results than that using monochrome images due to the increase in information available to constrain the process [184]. Since segmentation of colour images can be achieved by treating the red, green and blue colour space as a three-dimensional space and using gradient based information in this space to segment the images. Thus it is reasonable to adopt a similar approach to segment depth maps which use three-dimensional space explicitly. The depth map segmentation is performed in the three-dimensional gradient space of the approximated planes (*i.e.* (x, y, a', b', c') where $z = a'x + b'y + c'$ is the plane equation) whereas colour segmentation is performed in (x, y, r, g, b) space. The only real differences are in the preprocessing of the depth data and modifications in the similarity constraints used to determine regions.

The method also makes use of edge information as discussed in the previous Section. Edge information is obtained using a variant of Canny's edge detector. The principles of this detector were described in Section 4.5.3.

The algorithm proceeds through four stages:

Initial Stage — a mask of variable dimension (usually 3×3) is passed over the entire image and small planar approximations are made at each pixel by estimating the local surface normal. Each oriented point is thus initially considered as a region.

Coalescing Stage — this quickly reduces the initially large number of

4.6. SURFACE EXTRACTION

regions to approximately 3,000 regions which are more rigorously tested by subsequent stages. The method is based on the physical phenomenon of condensation; regions are iteratively grouped together under the condition that the difference between regions is less than some allowable amount. This amount is increased on each iteration until a specified limit is reached.

Forcing Stage — here a more rigorous merging procedure is used. The basic idea is to examine pairs of regions and consider *forces of attraction* and *repulsion* between them. The attractive forces are based on similarity (gradient, distance of the plane from the origin, area, goodness of fit *etc.* can all be used) of the two regions. The repulsive forces are based on the probability that an edge (boundary) lies between the regions and on the strength of the edge — the strength of the edge being based on the gradient magnitude of a detected edge as described in Section 4.5. Only regions with positive resultant forces after finding the difference of the attractive and repulsive forces are merged together. This merging process is based on simultaneous region growing.

Featuring Stage — various features of each region such as its plane equation, area, errors in plane fit, centre of gravity and number of connected regions are derived from the segmentation.

Throughout the segmentation use is made of a region map (see Fig 4.24). A region map is an array of numbers, one for each pixel, which indicates the unique region which that pixel belongs to, as determined by the segmentation process.

4.6.7 Segmenting Higher Order Surfaces

The segmentation of higher order surfaces can be performed in a similar manner to the segmentation of planar surfaces. Region growing approaches generally provide the best results.

The basic region growing approach for extracting higher order surfaces may be described as follows:

1. Small neighbourhoods of points are grouped together by approximating small planar regions.

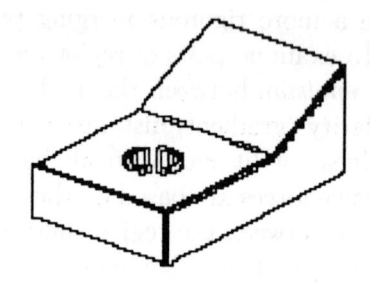

130   CHAPTER 4. SEGMENTATION AND FEATURE REPRESENTATION

(a) Artificially created Widget with cylindrical hole

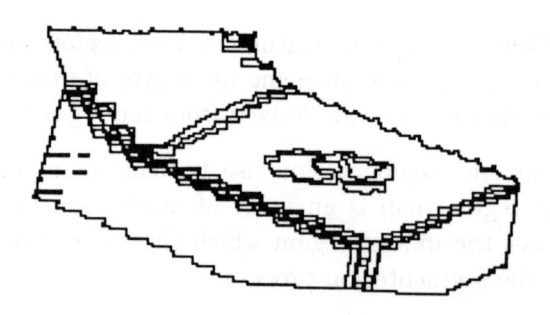

(b) Widget with Coin on a face

Figure 4.24: Regions determined by region maps

4.6. SURFACE EXTRACTION

2. These groups are merged together into regions that are restricted to a small area. This is similar to the case for planes although less strict similarity constraints are required.

3. These regions are classified as planar, curved (quadric, sphere, cylinder *etc.* depending on requirements — see below) or unclassified. This may be achieved by examining the errors in the initial plane fit, or by examining curvatures obtained from a sample of points contained in the region, or both. If the region is determined to be non-planar then the region may be approximated according to its new class.

4. Similar regions are merged until approximation errors become too large. Errors are measured in the least-squares sense as described in Appendix B

Many methods based on the above scheme have been developed. Good reviews of these can be found in Besl's book [13] and other papers [16, 25, 139]. Faugeras *et al.* [81] also use similar ideas to find planes and general quadric surfaces. Oshima and Shirai [182, 216] have developed methods that group surface elements as belonging to spherical, cylindrical or conical regions after initial planar approximations. These elementary curved surfaces are then merged into consistent larger regions which are fitted with quadric surfaces.

Besl and Jain [14, 18, 139] have developed a generic segmentation algorithm based on fitting surfaces of variable order to three-dimensional range data or intensity images. Surfaces are classified according to the order of their parametric defining equations. These are, in ascending order, planar (order 1), biquadratic, bicubic and biquartic (usually the highest order considered). Data for each surface is recorded as a list of surface coefficients, error of surface fit, surface and region labels, and so on. The basic approach initially attempts to fit small planar regions to the image points (as described above) and the error of the fit is noted. As the regions are grown, errors in the fit and other acceptance tests are considered which allow a surface to be described by a higher order surface if a lower order one is no longer satisfactory.

Besl and Jain's method applies both *differential geometry* techniques and classification of edges (as described in Section 4.5) to growing and labelling the surfaces. Many other segmentation methods have applied techniques from differential geometry [33, 73, 168, 234]. Below, we briefly present some fundamental concepts of differential geometry that are useful to the present problem of classifying surfaces for segmentation. Further details of these concepts can be found in many books, including [145, 179]. Details of using image data to estimate the values of the quantities that we define can be found in [33, 73, 168, 234].

Through each point on a surface there is a unique *surface normal*, a vector locally perpendicular to the surface. If we choose any plane containing this normal vector, it intersects the surface in a plane curve. The curvature of this curve at the given point gives the local *normal curvature* of the surface in the direction of the curve.

If we take all possible planes containing the surface normal at a given point, we find that (except for special points called umbilics) the normal curvature varies with the direction of the plane. In particular, we find that there are two orthogonal *principal directions*, in one of which the normal curvature is a maximum and in the other is a minimum. The normal curvatures in these principal directions are called the *principal curvatures*.

An alternative way of defining the principal directions is to consider what happens to the normal vector. If we erect a small normal vector at a given point on the surface, and then move it across the surface, generally speaking the tip of the normal vector will not only move in the direction the base of the normal vector has moved, but will tend to tip over to one side or the other. This is due to *geodesic torsion* of the surface. In certain special directions this does not happen. These special directions are the principal directions as defined earlier. Thus the geodesic torsion is zero in a principal direction.

Surface curvature at a point is often defined not in terms of the principal curvatures, but two other quantities. Let the two principle curvatures be denoted by n_1 and n_2. Then we define the *Gaussian curvature*, K, and the *mean curvature*, H, by

$$K = n_1 n_2, \qquad (4.31)$$

$$H = \frac{n_1 + n_2}{2}. \qquad (4.32)$$

Obviously, a specification of H and K at a point is equivalent to

4.6. SURFACE EXTRACTION

specifying n_1 and n_2. In practice, however, Gaussian and mean curvatures are usually preferred in computational vision work since:

- The sign of the Gaussian curvature immediately determines the convexity of a surface.

- Mean curvature is less sensitive to noise in numerical computations since it is the average of the principal curvatures. Gaussian curvature is more sensitive to noise however.

- Surfaces can be classified into eight types using Gaussian and mean curvature but only six in terms of principal curvatures [18].

- Gaussian curvature is invariant to viewpoint direction.

- Less computations are required to calculate Gaussian and mean curvatures from image data as shown by Besl [18].

The eight surface types that can be determined from the Gaussian and mean curvatures are, using the terminology of Besl and Jain [18]:

1. *Planar Surface* — $K = 0$, $H = 0$.

2. *Peak Surface* — $K > 0$, $H < 0$.

3. *Ridge Surface* — $K = 0$, $H < 0$.

4. *Saddle Ridge Surface* — $K < 0$, $H < 0$.

5. *Saddle Valley Surface* — $K < 0$, $H > 0$.

6. *Valley Surface* — $K = 0$, $H > 0$.

7. *Cupped Surface* — $K > 0$, $H > 0$.

8. *Minimal Surface* — $K < 0$, $H = 0$.

Fig. 4.25 illustrates these surface types.

(a) Planar surface

(b) Peak surface

(c) Ridge surface

(d) Saddle ridge surface

(e) Saddle valley surface

(f) Valley surface

(g) Cupped surface

(h) Minimal surface

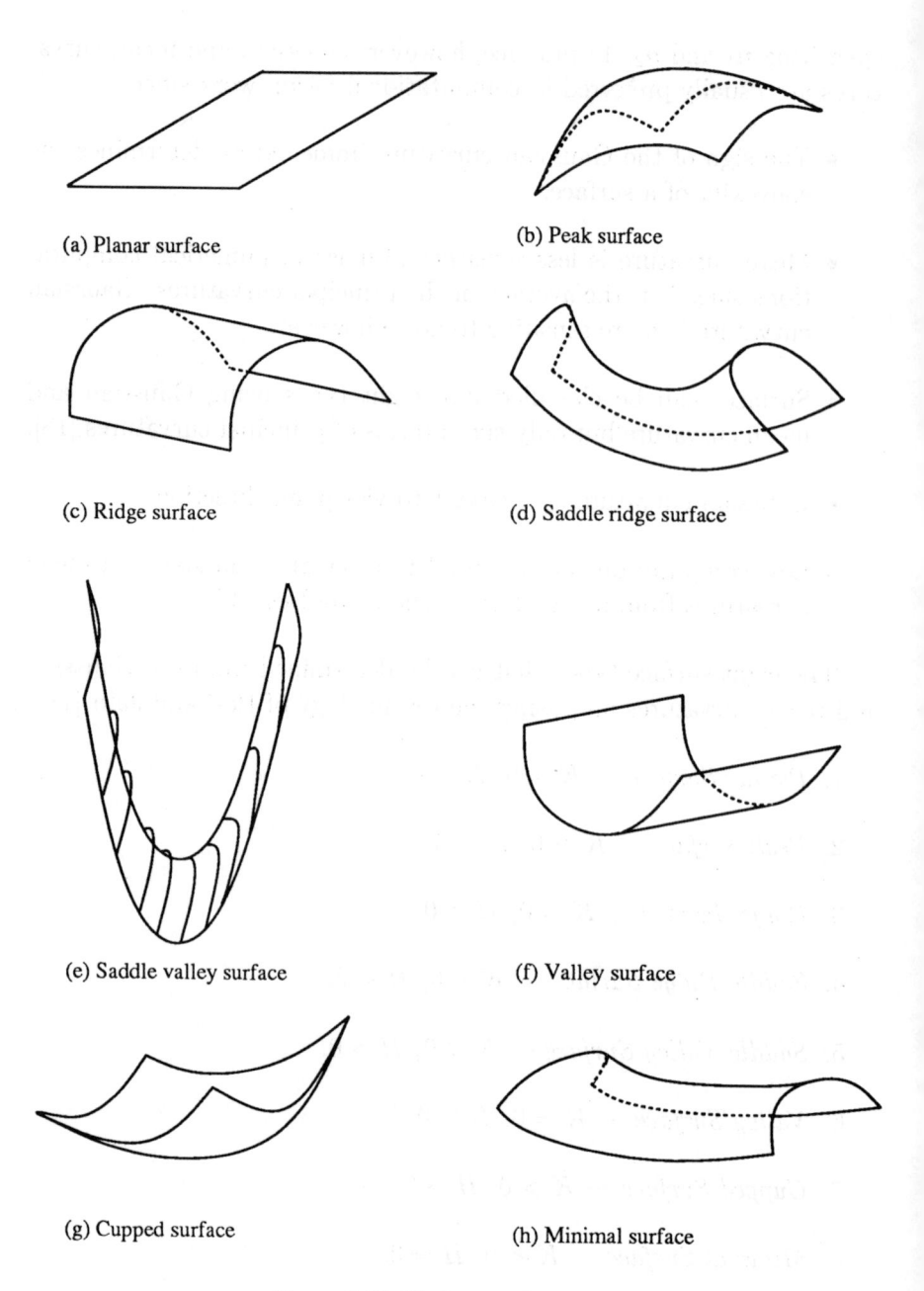

Figure 4.25: Various surface types

4.6. SURFACE EXTRACTION

4.6.8 Problems with Surface Segmentation

We have described various ways of extracting surface information from images. In this Section we address the problems arising at this stage and during the subsequent tasks that utilise the segmentation results. In the course of previous Sections we have already considered some of these problems when using surface *patches*. We will now consider in particular various surface types in turn.

Planes seem an ideal choice, when appropriate, for many subsequent vision tasks such as accurate recognition, localisation and inspection, since all of the requirements set out in Section 4.2 are easily satisfied:

- A plane is robust to occlusion — the equation of a plane may still be found even when it is partially hidden.

- A plane is robust to variations in the position it is viewed from, except in the unlikely case that it is viewed sideways on, *i.e.* along a direction which lies in the plane.

- The full richness of the depth data is exploited by the approximation of planes to the data (*i.e.* all depth data is used, not just some as when edges and points are segmented).

- Planes are easily extracted from the scene with very few problems in their accuracy and reliability, as discussed in this Chapter. This is not the case with other surface types.

- Planes can be easily and efficiently used to drive subsequent reasoning and recognition processes as we shall see in Chapters 5 and 8.

The extraction of planes seems to suffer no serious drawbacks and planes would seem to be the most useful feature that can be extracted from three-dimensional images. On the other hand, attempting to segment scenes into regions which are described by general higher order surfaces is often very difficult and time consuming to perform reliably, as shown by Hebert and Kanade [114] and Faugeras *et al.* [79, 80] for quadric surfaces. Since these are the "next simplest" type of general surfaces after planes, it can safely be assumed that any approach that uses *general* higher order surfaces will be unreliable. Unfortunately, some real world scenes do not contain or cannot be satisfactorily described by planes, which currently is a serious problem for vision systems.

The main problem in fitting general higher order algebraic surfaces to a set of points is that there is a large number of free parameters that need determining. Thus, a plane (Eqn. 4.16) has three free parameters, namely a', b' and c', whereas a general quadric (Eqn. 4.20) has nine free parameters, $a/j, \ldots, i/j$. In the case of planes we can employ natural geometric interpretations of the plane's parameters to derive an expression that can be minimised using least squares methods (Appendix B). When a plane is expressed as $ax + by + cz + d = 0$ we can interpret (a, b, c) as its normal vector, and d as its perpendicular distance from the origin. Altering any of these values has an obvious geometric effect. However in the case of quadric surfaces the parameters $a, \ldots, j$ have no such simple geometric meaning. In particular, calculating the distance from a point to the quadric is much more difficult. In practice, for least squares fitting of data to general quadrics, distances of points from the quadric are not used [80], but rather some other artificial measure of error, as shown in Appendix B.

In least squares techniques, it is necessary to use more points than the number of free parameters present so that an overdetermined system of equations is formed. In the case of quadrics there are nine parameters while for a plane there are only three. This leads to two problems. Firstly, the time required to solve the system of equations is increased. Consequently the time to segment the scene increases greatly since very many such systems of equations have to be solved. Secondly, larger errors in segmentation will result due to the inherent numerical instability of the formulation of the problem.

Therefore an approach is required that can reduce the number of free parameters to be determined. One method is to use the curvature of a surface to make an initial estimate of surface type. By looking for curved surfaces of *specific* types such as cones, spheres, and cylinders (rather than general quadrics), the number of free parameters is reduced. Secondly, the remaining parameters often can be given simple geometric interpretations, and in turn, distances can be simply calculated with reference to their centres (in the case of spheres) or axis of symmetry (for cylinders and cones), as discussed in Exercises 4.10

and 4.11. Thus, both fitting and inspection tasks are simplified if only particular types of quadrics are employed, as both rely on distance calculations.

Another approach which may prove useful, particularly for tasks like inspection when we have a good idea of what what the vision system is looking at, is to use the stored model of the object to drive the segmentation process. Using model information it is possible to restrict the number of free parameters considerably, perhaps only looking for cylinders with a certain radius or quadrics with certain parameters fixed. One particular variant of this idea is to match the model on planes segmented from the scene first, and then to fit the remaining points to the known surfaces in the model. Such a method is currently being investigated for inclusion in the segmentation procedures [94] used our inspection system. Results from Chapter 12 imply that segmentations of cylinders and spheres should be possible with this approach.

This type of approach to segmentation has overtones of data fusion (see Section 5.3.2) where data from many sources is gathered together with the aim of obtaining better results than those which could be achieved by using a single source. Another place where we have already seen such ideas is in the context of using edges with regions for segmentation, as described in Sections 4.5.5 and 4.6.6.

4.7 Statistical Region Description

4.7.1 Statistical Descriptions

The features we have described so far have been geometric in nature. Instead of trying to extract such higher level geometric features, another approach is to work directly with regions of pixels in the image, and to describe them by various statistical measures. Such measures are usually represented by a single value. These can be calculated as a simple by-product of the segmentation procedures previously described.

Such *statistical descriptions* may be divided into two distinct classes. Examples of each class are given below:

- *Geometric descriptions*: area, length, perimeter, elongation, compactness, moments of inertia.

- *Topological descriptions*: connectivity and Euler number.

Some of the above measures have already been defined or have obvious meanings. The others will be described briefly below in the context of two-dimensional images.

Elongation — sometimes called *eccentricity*. This is the ratio of the maximum length of line or *chord* that spans the region to the minimum length chord. We can also define this in terms of moments as we will see shortly.

Compactness — this is the ratio of the square of the perimeter to the area of the region.

Moments of Inertia — the ijth discrete central moment m_{ij} of a region is defined by

$$m_{ij} = \sum (x - \bar{x})^i (y - \bar{y})^j \tag{4.33}$$

where the sums are taken over all points (x, y) contained within the region and $(\bar{x}, \bar{y})$ is the centre of gravity of the region:

$$\bar{x} = \frac{1}{n} \sum_x x \quad \text{and} \quad \bar{y} = \frac{1}{n} \sum_y y.$$

Note that, n, the total number of points contained in the region, is a measure of its area.

We can form seven new moments from the central moments that are invariant [127, 236] to changes of position, scale and orientation of the object represented by the region (invariants will be discussed below shortly), although these new moments are *not* invariant under perspective projection. For moments of order up to three, these are:

$$M_1 = m_{20} + m_{02}$$
$$M_2 = (m_{20} - m_{02})^2 + 4m_{11}^2$$
$$M_3 = (m_{30} - 3m_{12})^2 + (3m_{21} - m_{03})^2$$
$$M_4 = (m_{30} + m_{12})^2 + (m_{21} + m_{03})^2$$

4.7. STATISTICAL REGION DESCRIPTION

$$M_5 = (m_{30} - 3m_{12})(m_{30} + m_{12})\left[(m_{30} + m_{12})^2 - 3(m_{21} + m_{03})^2\right]$$
$$+ (3m_{21} - m_{03})(m_{21} + m_{03})\left[3(m_{30} + m_{12})^2 - (m_{21} + m_{03})^2\right]$$
$$M_6 = (m_{20} + m_{02})\left[(m_{30} + m_{12})^2 - 3(m_{21} + m_{03})^2\right]$$
$$+ 4m_{11}(m_{30} + m_{12})(m_{03} + m_{21})$$
$$M_7 = (3m_{21} - m_{03})(m_{12} + m_{30})\left[(m_{30} + m_{12}^2 - 3(m_{21} + m_{03})^2\right]$$
$$- (m_{30} - 3m_{12})(m_{12} + m_{03})\left[3(m_{30} + m_{12})^2 - (m_{21} + m_{03})^2\right] \quad (4.34)$$

We can also define eccentricity using moments as

$$\text{eccentricity} = \frac{m_{20} + m_{02} + \sqrt{(m_{20} - m_{02})^2 + 4m_{11}^2}}{m_{20} + m_{02} - \sqrt{(m_{20} - m_{02})^2 + 4m_{11}^2}}. \quad (4.35)$$

We can also find *principal axes of inertia* that define a natural coordinate system for a region. Let θ be given by

$$\theta = \frac{1}{2}\tan^{-1}\left[\frac{2m_{11}}{m_{20} - m_{02}}\right] \quad (4.36)$$

We will get two values for θ which are 90° apart. The pair of lines which make an angle θ with the x axis passing through the centre of gravity of the region define a pair of principal axes which are generally aligned along what would intuitively be called the length and width of the region.

Three-dimensional moments can be similarly calculated [210].

Connectivity — the number of neighbouring features adjoining the region.

Euler Number — for a single region, one minus the number of holes in that region. The Euler number for a set of connected regions can be calculated as the number of regions minus the number of holes.

Most of these statistical measures may be useful in helping to control image reasoning and recognition, even if they are not sufficient for reasoning by themselves. However, these measures are very cheap to calculate as byproducts of the segmentation strategies detailed, especially region growing. Consequently, when these measures are used in conjunction with other reliably extracted features, they can help to improve recognition techniques by constraining the search space for matches between image and object features (see Chapters 5 and 8).

4.7.2 Invariant Measures

Several of the statistical measures we have met are *invariant* measures, which is to say that the value of the measure does not vary with, for example, the position of the region in the image, or perhaps its orientation or scale. Thus, while the centre of gravity of a region obviously varies with its position, its area does not. While area *does* vary with scale (and thus closeness of the camera to the object, for example), compactness as defined above is invariant with respect to scale as well as position and orientation. Invariant measures can be quite useful in recognising objects [17].

Other invariant measures we have met before are:

- The *ranks* of the matrices **Q** and **A** of a quadric (see Section 4.6.1) are invariant with respect to rotation and translation. This means that they are viewpoint invariant to some degree.

- *Gaussian curvature* and *mean curvature* are invariant with respect to rotation and translation of a surface [14].

- *Moments of inertia* $M_1 \ldots M_7$ (Eqn. 4.34) are invariant with respect to scaling, rotation and translation.

- *Eccentricity* is invariant with respect to scaling, rotation and translation.

- *Fourier Transforms* are rotation invariant as described in Section 3.2. We can use the Fourier transform to compute *Fourier descriptors* of an object which are invariant [105, 236] with respect to position and orientation.

Chapter 3 of Wechsler's book [236] provides a thorough treatment of the whole topic of invariance.

4.8 Exercises

Exercise 4.1 *Write down the difference equations for the Sobel edge operator described in Section 4.5.3.*

Exercise 4.2 *Give the 5×5 convolution masks for the Sobel edge operator described in Section 4.5.3.*

Exercise 4.3 *By approximating second order derivatives by finite difference equations or otherwise verify that the Laplacian operator can be represented by the convolution mask given in Fig. 4.7, assuming that we are considering four neighbours only. Derive the appropriate convolution mask assuming we use all eight nearest neighbours.*

Exercise 4.4 *Calculate the Fourier transform of the Laplacian operator.*

Exercise 4.5 *Show that the Fourier transform of a Gaussian function (defined in Section 4.5.3 is also a Gaussian function but of different standard deviation and height.*

Exercise 4.6 *Show that as the difference in the standard deviations used in the difference of Gaussians (DOG) operator approaches zero that the DOG operator approximates the LOG operator.*

Exercise 4.7 *We can express the Gaussian operator in polar representation as $g(r)$ where*

$$g(r) = \frac{1}{2\pi\sigma^2} e^{-r^2/2\sigma^2}$$

and $r = \sqrt{x^2 + y^2}$. Calculate the polar form of the LOG operator $\nabla^2 g(r)$. Calculate the Fourier transform of $\nabla^2 g(r)$.

Exercise 4.8 *How could the Hough transform method be used to detect circles of radius 10 units? Show that pixels in (x, y) space would map to circles in this case if Eqn. 4.14 is used to define circles.*

CHAPTER 4. SEGMENTATION AND FEATURE REPRESENTATION

Exercise 4.9 *Suppose the Hough transform method is to be used to detect circles of arbitrary radius. What would pixels in (x,y) space now map to? How much less efficient would it be to search for arbitrary circles than circles of a fixed radius?*

Exercise 4.10 *Take the equation of a sphere (Eqn. 4.30). By following least squares approximation techniques given in Appendix B and applying distance constraints show how to find the best fit of a set of points to a spherical surface.*

Exercise 4.11 *Repeat Exercise 4.10 to find the best fit of a set of points to a cylindrical surface.*

Exercise 4.12 *A general point $\mathbf{P}(u,v)$ on a Bézier surface with a rectangular array of $(n+1) \times (n+1)$ control points $\mathbf{p}_{ij}$ is given by*

$$\mathbf{P}(u,v) = \sum_{i=0}^{n}\sum_{j=0}^{n} \mathbf{p}_{ij} u^i (1-u)^{n-i} v^j (1-v)^{n-j} \frac{n!}{i!(n-i)!} \frac{n!}{j!(n-j)!} \quad (4.37)$$

where $0 \le u \le 1$ and $0 \le v \le 1$. Show that the surface goes through each of the four corner control points when $(u,v) = (0,0), (0,1), (1,0)$ or $(1,1)$ respectively. Note that the surface does not pass through any of the other control points in general.

Chapter 5

Reasoning With Images

5.1 Introduction

So far we have only considered "reasoning" with images to a limited degree, in that we have been interested in extracting particular features from images. Other reasoning methods are also considered in this Chapter, which are concerned with both low level reasoning and higher level reasoning. In the latter we try to understand the relationships between and meanings of features extracted by various low level methods. Initially we often attempt to conceptually group features together which are related, such as a set of edges which form the boundary of a face, or a group of faces belong to the same object. The overall goal in many cases is to try to recognise the objects in the scene.

5.2 Line Labelling

5.2.1 Introduction

We saw in the last Chapter how to extract edges from an image. Although other methods extract planes or other surfaces directly, some reasoning techniques work from the edges, and use them to try to build up information about the structure of the whole object. The method we shall now describe is applicable when two-dimensional lines have been extracted from a two-dimensional image.

Initially, we shall restrict ourselves to the assumption that the objects in our scene are polyhedra, *i.e.* have flat faces bounded by straight edges. Under this assumption it becomes possible to interpret the edges, and in particular, to determine which edges bound which faces, and to deduce certain relationships between faces.

Consider the line drawing shown in Fig. 5.1. It can be seen that the

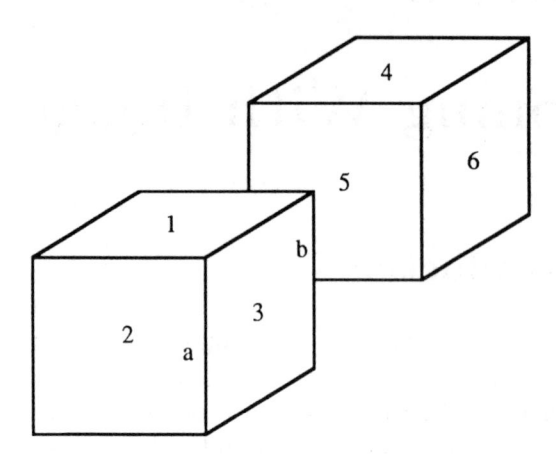

Figure 5.1: Line drawing of two cubes

edges bound the numbered faces, and that faces 1, 2 and 3 belong to one cube, while faces 4, 5 and 6 belong to another (assuming that the two cubes are not joined by a thin sliver that cannot be seen).

An important observation is that every three-dimensional edge is associated with exactly two faces, one on either side of it, except for a few rather ill-conditioned objects called *non-manifold objects* which we shall not consider further here. Sometimes both of these faces can be seen from a given viewpoint (as in the case of line a in Fig. 5.1), but if only one of the faces is visible, the edge is called an *occluding* edge. An example of such an edge is line b in Fig. 5.1. On the other side of an occluding edge, the background or a non-touching face is visible.

The basis of the line labelling method is to attempt to classify each edge in the image as being a concave edge, a convex edge, or an occluding edge. Edges of each type are shown in Fig. 5.2. Where edges meet in the image, only certain possible interpretations of the meeting lines are mutually consistent. This fact is used to reduce the number

5.2. LINE LABELLING

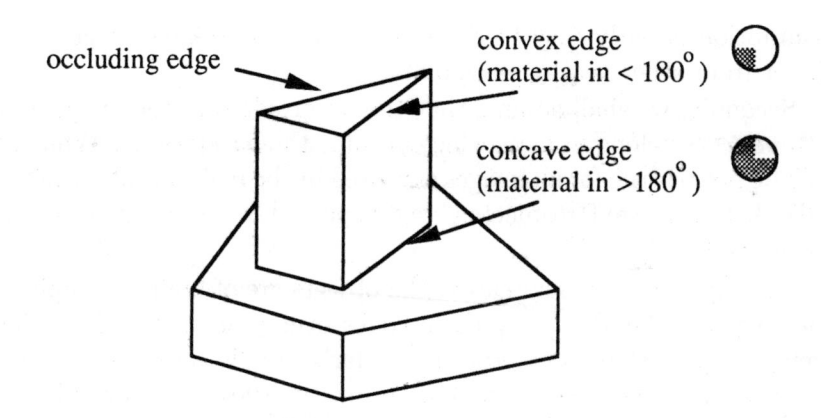

Figure 5.2: Types of edge in a line drawing

of possibilities, hopefully until we have a single possible classification for each edge. Even if we are not able to arrive at a unique solution, it is still very likely that we shall have obtained some useful information concerning the edges and objects in the scene.

If a line corresponds to an occluding edge, it is marked with an arrow. The orientation of the arrow gives the side of the edge on which the visible face lies, using the convention that the visible face lies on the right hand side of the edge, looking along the direction of the arrow. Lines corresponding to convex edges are marked with a '+' symbol, and lines corresponding to concave edges are marked with a '−' symbol. Each edge in the image thus has one of four possible labellings.

5.2.2 Assumptions

Before proceeding, we shall make some some simplifying assumptions concerning the view of the object and its complexity. Firstly, we shall assume that the object is in *general position*, by which we mean that no two things are accidentally lined up in the image because of a special viewing position. Thus, if the viewpoint is slightly changed, the same edges will be seen in the same relative positions to each other, and new edges will not appear, or old ones disappear. In other words, if we look at the object from some arbitrary viewpoint, we expect this not to be a special case, which is what we mean by general position. This assumption is fairly easy to check in practice by taking images of the object from other nearby viewpoints.

Secondly, we shall assume that the outer edges in the image — its *silhouette* — belong to faces which occlude the background. While this may be not be true for all images captured in the real world, it is not too difficult to arrange for objects being inspected in an industrial setting, for example.

Thirdly, we shall assume that the objects are of limited complexity, and in particular that only three edges ever meet at a vertex. This limitation is just to provide a particularly simple universe of objects for this explanation, as the number of possibilities which must be considered increases with the number of edges allowed to meet at a vertex. Real systems which do not have this limitation can readily be created.

5.2.3 Junction Types

It is now possible to say that all line junctions will belong to one of four types, as shown in Fig. 5.3. At an *L-junction* only two edges meet, while

Y-junction

Arrow-junction

L-junction

T-junction

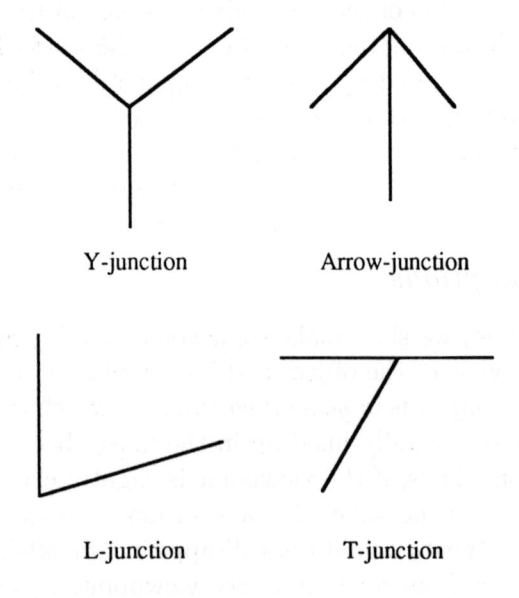

Figure 5.3: Types of junction in a line drawing

5.2. LINE LABELLING

three edges meet at the other junction types. The distinction between an *Arrow-junction* and a *Y-junction* is whether the three edges subtend a total angle of less or more than 180°. The dividing case between them is specifically identified as a *T-junction*.

Consideration of what such junctions can represent in a view of a polyhedral model leads to the conclusion that, out of all these which are possible, only certain labellings of the edges at a junction can arise in images of a real scene. These are illustrated in Fig. 5.4. There are

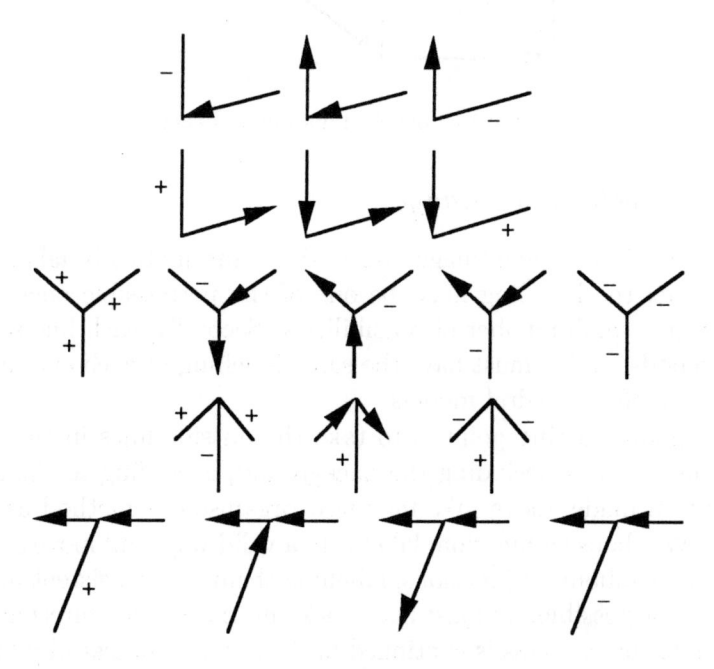

Figure 5.4: Permissible labellings at a junction

far fewer of these than the total number of all possible labellings. Some of the permissible labellings are shown occurring in a scene in Fig. 5.5.

One immediate deduction that can be made is that T-junctions only occur where one object occludes another, or perhaps an object occludes another part of itself.

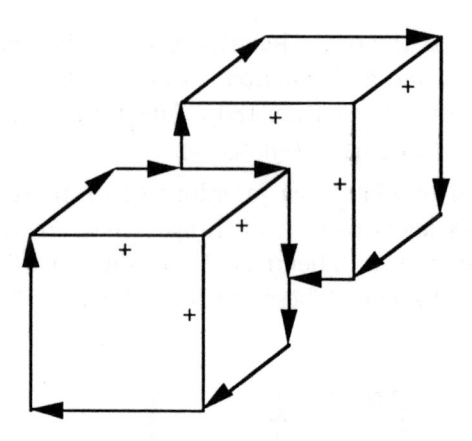

Figure 5.5: Labelled lines in an image

5.2.4 Labelling an Image

Suppose we have a line image and wish to automatically label it. At each vertex the labelling must be one of the permissible ones — one of a relative small number of possibilities. Secondly, each line segment has two ends, and it must have the same labelling at each end, at least in the case of polyhedral models.

One good starting point is to take the outside lines in the image, and label them as occluding the background, according to the second assumption made above. We may now organise our method as a tree search. We choose a junction, label it in a valid way, and move along its edges to neighbouring junctions, labelling them in a consistent manner. If this is not possible, we must backtrack and try another alternative for a previous choice. This is continued until we have successfully labelled the drawing, or we have no possibilities left. In the latter case the line image does not represent a valid physical object.

Note that more than one valid interpretation may exist, so we may wish to make our tree search exhaustive to find other permissible labellings, rather than just stopping after after finding one consistent labelling for the whole image.

Much work has been done to take the basic method described here further. For example, Waltz[235] has shown how shadow edges can also be allowed for. On the one hand, the number of different line types and

5.3. RELAXATION LABELLING

possible junctions is increased, but on the other, the extra information gained from the shadows also tends to help resolve ambiguities. He has also shown how to use a *constraint propagation* method to quickly reduce the number of possibilities before a full tree search is used. Further details of line labelling may also be found in the books by Ballard and Brown[6], Horn[123], and Winston[239].

5.3 Relaxation Labelling

5.3.1 Introduction

Relaxation labelling techniques can be applied to many areas of computer vision. We have already mentioned several applications to image segmentation and edge detection in Chapter 4. Relaxation techniques have been applied to many other areas of computation in general, particularly to the solution of simultaneous nonlinear equations. We shall first describe the basic principles behind relaxation labelling methods and then discuss various applications.

The basic elements of the relaxation labelling method are a set of features belonging to an object and a set of labels. In the context of vision, these features are usually points, edges and surfaces, as we have already described. Normally, the labelling schemes used are *probabilistic* in that for each feature, weights or probabilities are assigned to each label in the set giving an estimate of the likelihood that the particular label is the correct one for that feature. Probabilistic approaches are then used to maximise (or minimise) the probabilities by iterative adjustment, taking into account the probabilities associated with neighbouring features. It should be noted that relaxation strategies do not necessarily guarantee convergence, and thus, we may not arrive at a final labelling solution with a unique label having probability one for each feature. However if the strategies can be constrained so that probabilities are always changing in the same direction, by rejecting inconsistent labels for example, then convergence *can* be guaranteed. Even if the end result is not a single possible label for each feature, the probabilities of the labels can be used to decide which labellings are more likely to be correct.

Let us now consider the labelling approach in more detail. Let us assume:

- O is the set $\{o_1, \ldots, o_n\}$ of n object features to be labelled.
- L is the set $\{l_1, \ldots, l_m\}$ of m possible labels for the features.

Let $P_i(l_k)$ be the probability that the label l_k is the correct label for object feature o_i. More generally, weights rather than probabilities can be used, which only give some indication of the relative likelihood of a given label occurring. Nevertheless, if probabilities are used then more reliable and robust methods can be developed since more constraints can be applied to the reasoning. Thus, the usual probability axioms can be applied that:

- Each probability satisfies $0 \leq P_i(l_k) \leq 1$ where $P_i(l_k) = 0$ implies that label l_k is impossible for feature o_i and $P_i(l_k) = 1$ implies that this labelling is certain.
- The set of labels are *mutually exclusive* and *exhaustive*. Thus we may write for each i:

$$\sum_L P_i(l_k) = 1. \tag{5.1}$$

Thus each feature is correctly described by *exactly one* label from the set of labels.

The labelling process starts with an initial, and perhaps arbitrary, assignment of probabilities for each label for each feature. The basic algorithm then transforms these probabilities into to a new set according to some relaxation schedule. This process is repeated until the labelling method converges or stabilises. This occurs when little or no change occurs between successive sets of probability values.

Many transformation methods have been designed. Popular methods [96, 128, 246] often take *stochastic* approaches to update the probability functions. Here an operator considers the *compatibility* of label probabilities as constraints in the labelling algorithm. The compatibility $C_{ij}(l_k, l_l)$ is a correlation between labels defined as the conditional probability that feature o_i has a label l_k given that feature o_j has a

5.3. RELAXATION LABELLING

label l_l, i.e. $C_{ij}(l_k, l_l) = P(l_k|l_l)$. Thus, updating the probabilities of labels is done by considering the probabilities of labels for neighbouring features.

In detail, let us assume that we have changed all probabilities up to some step, S, and we now seek an updated probability for the next step $S+1$. We can estimate the change in confidence of $P_i(l_k)$ by:

$$\delta P_i(l_k) = \sum_{j \in N} w_{ij} \left[\sum_{l \in L} C_{ij}(l_k, l_l) P_j(l_l) \right] \quad (5.2)$$

where N is the set of features neighbouring o_i, and w_{ij} is a factor that weights the labellings of these neighbours, defined in such a way that

$$\sum_{j \in N} w_{ij} = 1. \quad (5.3)$$

The new probability for label $P_i(l_k)$ in generation $S+1$ can be computed from the values from generation S using

$$P_i(l_k) = \frac{P_i(l_k)\left[1 + \delta P_i(l_k)\right]}{\sum_l P_i(l_l)\left[1 + \delta P_i(l_l)\right]} \quad (5.4)$$

5.3.2 Statistical Relaxation Techniques

As already mentioned, relaxation methods have been applied to many problems, particularly optimisation problems. Such problems are very common in computer vision and so relaxation methods have been applied in a wide variety of ways to computer vision. We will briefly mention some applications here and also highlight uses of other statistical methods.

Relaxation methods have been applied to edge linking problems [246]. The probabilities of edge points lying on particular edges is determined by considering neighbouring edge points. Different labels are used for each edge, and a relaxation schedule is then used to find the appropriate label for each edge point.

Line labelling techniques (see Section 5.2) can be expressed a relaxation problem (see Exercise 5.5). Here we seek to label lines as belonging to a certain class of edge (occluding, concave, convex *etc.*). Probabilities can be assigned to each type of labelling fairly easily.

As previously noted, only certain sets of edge labellings are mutually compatible at line junctions. These restrictions can be expressed as constraints when the conditional probabilities for the particular labels are estimated.

Relaxation methods have proved to be very popular for segmentation of images. Segmentation can be interpreted in two slightly different ways here:

- The process of grouping pixels into regions of similarity. Here the relaxation processes amount to self-organisation of the image. The regions are simply labelled as $region_1, \ldots, region_n$.

- The process of labelling regions of image as belonging to recognised physical entities such as sky, grass, trees, car and road.

We shall briefly consider problems of the first type above. Problems of the second type require higher-order reasoning semantics to label the scene (*e.g.* cars cannot fly, so they should be connected to the ground, trees grow from the ground, and so on).

Relaxation techniques have been applied to many segmentation problems using greyscale and colour information [96, 244], texture [58, 62], and three-dimensional images [1, 18, 119]. The basic approaches used in each of these cases can be expressed in terms of the usual relaxation labelling method described above. We shall now describe one further method which adopts a slightly different approach and introduces a few new concepts.

Geman and Geman [96] describe how to simultaneously reduce noise and segment a scene. The conditional probability that a pixel belongs to a current region given the intensities of pixels in a finite neighbourhood is estimated, and these conditional probabilities are stored in a two-dimensional array, one element for each pixel. The array is called a *Markov Random Field* (MRF). Geman and Geman show that the probability that the MRF is in a given state at a given time is uniquely determined by these conditional probabilities and moreover can be expressed by a Gibbs or Boltzmann distribution. We can therefore calculate the probability $P(\mathbf{M})$ that the MRF is in a particular state by

$$P(\mathbf{M}) = \frac{1}{Z} e^{U(\mathbf{M})/T} \tag{5.5}$$

5.3. RELAXATION LABELLING

where $\mathbf{M} = \{m_i\}$ is the state or configuration of the MRF at some given instance, and the m_i are the neighbourhoods which completely determine the array. Z is used to normalise the distribution, and T is a parameter which controls the shape (sharpness) of the probability distribution. Finally, $U(\mathbf{M})$ is a sum over all neighbourhoods of a *potential* calculated for each neighbourhood [96]. These potentials measure how neighbouring pixels influence the probability that a given pixel has a certain value, and are calculated using the principles that

- the intensity values of neighbouring pixels are expected to be similar, and

- when edges are present between pixels, the potentials inhibit pixels on either side of the edge from influencing one another (see Section 4.6.4).

It can be seen that noise reduction is achieved by noting that the state of the MRF of maximum probability is the state that most likely corresponds to the restored image. Segmentation is achieved at the same time by considering the probability that each pixel is similar to its neighbours. These probabilities are maximised by a process known as *simulated annealing*.

Simulated annealing is a technique that is suited to the solution of very complex optimisation problems. The techniques are derived from physical systems that model the slow cooling (annealing) of metallurgical processes. Let us consider some energy function U which corresponds to the energy state of excited atoms in a hot metal. We now reduce the temperature of the metal. If we reduce the temperature too quickly, the metal becomes brittle. This corresponds to the energy states for each atom getting stuck in local minima — the metal has cooled too quickly for the atoms to reach their most stable configuration. Cooling the metal more slowly avoids this problem, and allows the atoms to reach more stable configurations. The technique of simulated annealing has been applied to many vision problems [51, 144].

Returning to our segmentation problem, as given in Eqn. 5.5, the function $U(\mathbf{M})$ corresponds to the energy measure mentioned above. The state at any instant (or temperature) T is defined by the Gibbs (or Boltzmann) distribution.

We start with a random configuration of the system (MRF). We can then calculate for a small change in the system Δm_i the change in energy $\Delta U = U(m_i + \Delta m_i) - U(m_i)$. We decide to accept this change of state with a probability of $e^{-\Delta U/T}$. Our annealing schedule starts with a high value (temperature) of T and we slowly *cool* our system down to $T = 0$. Thus, initially, we are quite happy to accept large changes in state, allowing us to search widely for an optimum, but as the system cools down, we search more carefully in a local region of the state space. We must be careful how we cool our system down to avoid falling into local minima traps. Many approaches to reducing the temperature have been considered [144]. Some adopt an exponential decrease in temperature while others use a logarithmic decrease in temperature — $T(t)/T_0 = 1/\log(1+t)$ where $T(t)$ is the temperature at some time t and T_0 is the initial temperature. Recently Szu and Hartley [223] have introduced a fast annealing schedule where they reduce the temperature inversely proportional to time, $T(t)/T_0 = 1/(1+t)$.

Most relaxation methods can be easily parallelised (since they are local and iterative) which has led to much interest in recent years. Simulated annealing is closely related [60] to genetic algorithms [61, 120] which are also currently receiving a lot of research effort from many varying research disciplines including computer vision.

Another area of application of statistical techniques is to the problems of *data fusion*. Data fusion involves the integration of information obtained either from different sources, or from different image processing techniques. The advantages of data fusion are potentially very great. This book discusses many sensing and processing strategies all of which have their relative merits and drawbacks. If we could somehow harness all of the useful information from a selection of methods then the resultant hybrid approach should provide a better understanding of the scene. One example where we have already seen such ideas was when discussing segmentation in Chapter 4, where edge information was used to improve the region growing process. An advantage of the Geman and Geman approach above is that it can readily be extended for general data fusion methods [1, 244]. One important application area for data fusion is in controlling *autonomous guided vehicles* [32, 34, 68, 139]. A good introduction to data fusion may be found in [68].

5.3. RELAXATION LABELLING

The difficulty of data fusion is that there is potentially an enormous amount of data present and we have to decide which is useful and which is not. There may be some uncertainty in this decision and clearly probabilistic methods provide a way of analysing this uncertainty.

All of the the probabilistic methods discussed so far in this section have used the basic Bayesian approach to probabilistic reasoning. Thus, some evidence E is assumed and a set of alternative hypotheses $H = \{H_1, \ldots, H_n\}$ is considered. Then the probability $P(H_i|E)$ that the particular hypothesis H_i is correct given E is:

$$P(H_i|E) = \frac{P(E|H_i)P(H_i)}{\sum_{k=1}^{n} P(E|H_k)P(H_k)} \qquad (5.6)$$

However this approach is open to several criticisms:

- The Bayesian model requires all outcomes to be disjoint or mutually exclusive. This is often not the case for many vision tasks.

- The model does not allow exceptions to the classification, such as "none of the above", arising through lack of information.

- It is difficult to add new information, in the form of new outcomes, should they arise, to the reasoning procedure.

- The approach is computationally expensive. All probabilities have to be updated when the information changes.

However, Pearl in his book [186] argues that many of these problems can be avoided or minimised using his approach to Bayesian probabilistic reasoning. He argues that many of the problems discussed above are related to the Markov approach to the formulation. The problems arise in that in reality certain probabilities may depend on only a subset of the other probabilities and furthermore, may only operate in a single direction. Thus he proposes an approach based on *directed graphs* (see Chapter 8 for further discussion of graph structures in general) where the directions of the links of the graph permit probability dependencies to be represented more realistically and efficiently. These graphs are called *Bayesian networks*. Pearl shows that this approach can be used to permit new information to be added and the probabilities to be updated on a more local basis.

Dickson [64] has reported preliminary work using Pearl's ideas to group edges in an image. The aim of this technique is to group lines in an image formed by a perspective projection which were originally parallel in three dimensions. The purpose of this grouping is find particular structures in an image which can be used to aid the recognition process. Dickson's work uses this information about groupings in a matching strategy similar to that of the *SCERPO* system which we shall describe in Section 8.10.

Various other methods than Bayesian probability have also been developed for handling uncertain information:

Belief Models and Certainty Factors have been suggested by Shortliffe and Buchanan [217]. They define measures of belief B and disbelief D of hypothesis H_i given evidence E as follows:

$$B(H_i|E) = \begin{cases} 1 & \text{if } P(H_i) = 1 \\ \frac{\max[P(H_i|E), P(H_i)] - P(H_i)}{1 - P(H_i)} & \text{otherwise} \end{cases}$$

$$D(H_i|E) = \begin{cases} 1 & \text{if } P(H_i) = 0 \\ \frac{P(H_i) - \min[P(H_i|E), P(H_i)]}{P(H_i)} & \text{otherwise} \end{cases} \quad (5.7)$$

In the above, $P(H_i)$ is the standard probability. The certainty factor C of some hypothesis H_i given evidence E is defined as:

$$C(H_i|E) = B(H_i|E) - D(H_i|E). \quad (5.8)$$

Dempster-Shafer Models set up a confidence interval — an interval of probabilities within which the true probability lies with a certain confidence — based on the support S and plausibility PL provided by some evidence E for a proposition P [214]. The support brings together all the evidence that would lead us to believe in P with some certainty. The plausibility brings together the evidence that is compatible with P and is not inconsistent with it. This method allows for further additions to the set of knowledge and does not assume disjoint outcomes. If Ω is the set of possible outcomes, then a *mass probability*, M, is defined for each member of the set 2^Ω and takes values in the range $[0, 1]$.

We can then define the confidence interval as $[S(E), PL(E)]$ where

$$S(E) = \sum_A M \qquad (5.9)$$

where $A \subseteq E$ i.e. all the evidence that makes us believe in the correctness of P, and

$$\begin{aligned} PL(E) &= 1 - S(\bar{E}) \\ &= 1 - \sum_{\bar{A}} M \end{aligned} \qquad (5.10)$$

where $\bar{A} \subseteq \bar{E}$ i.e. all the evidence that contradicts P.

Statistical reasoning methods can be developed from the above alternative approaches to probability for application to vision problems. A good collection of important papers [215] in the field of probabilistic and uncertain reasoning contains many strategies based on the above approaches with applications to robot control, medicine, psychology and computer vision.

5.4 Shape from Shading

5.4.1 Introduction

Imagine that an image is captured of an object illuminated from a given direction. In general, different parts of the object will give rise to different pixel intensities in the image, even if the object is made from a single material with a single surface finish. This intensity will depend on the characteristics of the material itself, and on the local orientation of the surface with respect to the direction of the light source and the viewing direction. In this Section we shall consider how such intensity variations in an image can be used to tell us something about the shape of the object. Note however, that such methods do not directly give a description of the surfaces bounding an object, which is why we have not included them in the discussion of segmentation in the previous Chapter.

The treatment here follows closely that in Horn's book [123]. Firstly, we must see what dictates the intensity of pixels in an image.

5.4.2 Radiance and Irradiance

The intensity of a pixel representing part of the surface of a solid object depends on various things: the reflectance properties of that part of the object's surface, the orientation of that part of the surface relative to the camera, and the positions and intensities of the light sources. The amount of light falling on a piece of surface is called the *irradiance*, and is measured in units of power per unit area, Wm^{-2}. The amount of light radiated, or reflected, by a piece of surface is called the *radiance*, and is measured in units of power per unit area per unit solid angle, $Wm^{-2}sr^{-1}$. Note that the amount of light radiated in different directions may be different.

One useful result in the current context (for details, see Horn [123]) is that the image irradiance, and thus the intensity of the pixels in the image, is proportional to the scene radiance in the direction of the lens, and hence all we have to consider is how much light each part of the object is emitting.

In turn, the object radiance depends upon the amount of light reaching the object, the fraction of the light which is reflected, and the direction in which the surface is viewed. An ideal mirror reflects light from an incoming beam only in a single direction. A less than perfect mirror also tends to reflect some energy into other directions close to the ideal direction. At the opposite extreme, certain real world surfaces may reflect light more or less equally in all directions.

5.4.3 Bidirectional Reflectance Distribution Function

The information about the reflectance properties of a surface is provided by a function called the *bidirectional reflectance distribution function* (BRDF). It gives the ratio of the radiance of the surface to its irradiance per unit solid angle for given illumination and viewing conditions, or in other words, what fraction of the incident light in a given direction is reflected in a second given direction per unit solid angle.

Let us take some point on the surface. At that point there is a unique outwards surface normal vector, **n**. If we choose an arbitrary reference direction in the tangent plane to the surface as $\phi = 0$, we can specify the direction of any other vector **v** at that point in terms of ϕ

5.4. SHAPE FROM SHADING

(azimuth) and θ (polar angle), where ϕ measures the angle between the reference direction and the projection of **v** onto the tangent plane, and θ measures the angle between **n** and **v**, as shown in Fig. 5.6.

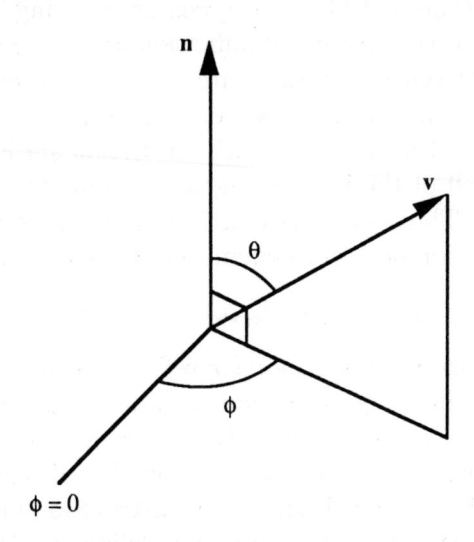

Figure 5.6: Azimuth and polar angle

In particular, we can now specify the direction from which incident light falls on some point of the surface by the pair of angles (θ_i, ϕ_i), and the direction in which light is emitted towards the viewer as the pair (θ_e, ϕ_e). The bidirectional reflectance distribution function depends in general on all four of these angles.

However, the laws of physics determine that if a given piece of surface is illuminated from one direction and viewed from another, and then we interchange the positions of the light source and the observer, exactly the same fraction of the incident light must be observed each time. Thus, if B is the bidirectional reflectance distribution function, it is always the case that

$$B(\theta_i, \phi_i, \theta_e, \phi_e) = B(\theta_e, \phi_e, \theta_i, \phi_i). \tag{5.11}$$

Secondly, for most surfaces, the direction of the reference direction chosen for $\phi = 0$ is unimportant. The bidirectional reflectance distribution function does not change if we rotate the surface about its surface normal vector n, and B depends simply on the difference $\phi_i - \phi_e$ rather than on ϕ_i and ϕ_e separately. One example where this is *not* true can be seen in the wings of some tropical butterflies, where the colour ranges between shades of blue and green as the wing is rotated about its normal, even with constant illumination and viewing conditions.

Two idealised types of surface are of special interest, as to some extent they represent limiting cases of real surfaces. The first of these is a *matt* surface, also called an *ideal Lambertian surface*. This type of surface reflects all of the incident light, and reflects an equal amount in all directions. Thus, for a matt surface, the bidirectional reflectance distribution function does not depend on any of θ_i, ϕ_i, θ_e or ϕ_e, and

$$B_{matt}(\theta_i, \phi_i, \theta_e, \phi_e) = \text{constant}. \tag{5.12}$$

At the other extreme is a *specular reflector* or *perfect mirror* which only reflects light in a single direction: light arriving from a direction θ_i, ϕ_i is reflected in the direction $\theta_e = \theta_i$, $\phi_e = \pi + \phi_i$. (This is the simple "angle of incidence equals angle of reflection" law of geometrical optics.) In this case the bidirectional reflectance distribution function takes the form (where δ is the Dirac delta function defined in Section 3.2.4)

$$B_{specular}(\theta_i, \phi_i, \theta_e, \phi_e) = \text{constant} \times \delta(\theta_e - \theta_i)\,\delta(\phi_e - \phi_i - \pi). \tag{5.13}$$

For other real surfaces the bidirectional reflectance distribution function can be measured experimentally, or even calculated theoretically from a physical model of the properties of the surface.

5.4.4 Radiance of a Surface

Let us now consider how the bidirectional reflectance distribution function is used to find the radiance of a small piece of surface, which in turn determines the intensity of pixels in our image. Let us start with a simple case: the radiance R of a matt surface illuminated by light uniformly from all directions (such as would occur under a cloudy sky), with total irradiance I. In this case

$$R = I, \tag{5.14}$$

because as much light is being reflected in each direction as is received in that direction.

5.4. SHAPE FROM SHADING

In general to find the radiance of a small piece of surface we must consider the spatial distribution of all light sources illuminating the surface. Noting that these light sources may be extended sources, not just point sources, we must then take the integral of the radiance of each one times the value of the bidirectional reflectance distribution function corresponding to the appropriate direction.

In detail then, this is done as follows: vectors giving the direction of part of an extended light source lying between θ_i and $\theta_i + \delta\theta_i$, and ϕ_i and $\phi_i + \delta\phi_i$ subtend a solid angle of $\delta\omega$, where

$$\delta\omega = \sin\theta_i\, \delta\theta_i\, \delta\phi_i. \tag{5.15}$$

Let $R(\theta_i, \phi_i)$ be the radiance of the light source in the direction given by (θ_i, ϕ_i). When a narrow beam of light falls obliquely at an angle θ_i onto a surface, the larger θ_i is, the wider the area of the surface this beam is spread out over, as shown in Fig. 5.7. The reduction in surface

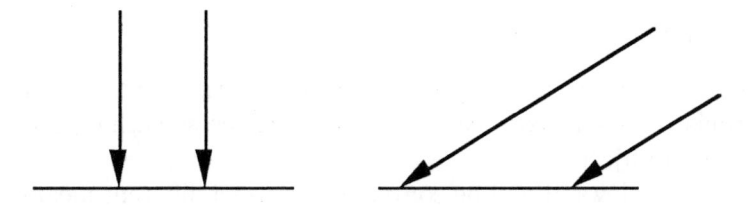

Figure 5.7: An oblique beam covers a larger area

irradiance as a result is by a factor $\cos\theta_i$. Overall, then the surface radiance $R(\theta_e, \phi_e)$ in the direction given by θ_e, ϕ_e is

$$R(\theta_e, \phi_e) = \int_{-\pi}^{\pi} \int_0^{\pi/2} B(\theta_i, \phi_i, \theta_e, \phi_e) R(\theta_i, \phi_i) \cos\theta_i \sin\theta_i\, d\theta_i\, d\phi_i. \tag{5.16}$$

As an example, let us consider the particular case of a matt surface illuminated by a point source of light. For a matt surface we know that $B(\theta_i, \phi_i, \theta_e, \phi_e)$ is a constant; this constant can readily be shown to be $1/\pi$ by noting that the total amount of light radiated in all directions must be equal to the total irradiance of the surface. Given a single point source lying in a direction θ_p, ϕ_p, of radiance I, the distribution of radiance is given by

$$R(\theta_i, \phi_i) = \frac{I\,\delta(\theta_p - \theta_i)\,\delta(\phi_p - \phi_i)}{\sin\theta_p}, \tag{5.17}$$

where the $\sin\theta_p$ term is again a constant to make sure that on integrating over all angles that the total radiance comes out as I. Thus, the radiance of a matt surface illuminated by this point source of light is given by

$$\begin{aligned} R(\theta_e,\phi_e) &= \int_{-\pi}^{\pi}\int_0^{\pi/2} \frac{1}{\pi}\frac{I\,\delta(\theta_p-\theta_i)\,\delta(\phi_p-\phi_i)}{\sin\theta_p} \cos\theta_i \sin\theta_i\, d\theta_i\, d\phi_i \\ &= \frac{I\cos\theta_p}{\pi}. \end{aligned} \quad (5.18)$$

This relation is known as *Lambert's law*. The $\cos\theta_p$ term arises, as remarked above, because a beam hitting a surface at an oblique angle is spread over a wider area of surface.

Further discussion of radiance and irradiance can be found in Horn's book [123].

5.4.5 Surface Orientation

We shall now return to our original aim, which was to see how intensity variations between pixels can be used to tell us something about the shape of an object.

In what follows, it will be convenient to use a different means of describing the orientation of a piece of surface than the polar and azimuth angles used earlier. If we consider a small piece of surface, then its surface normal **n** must point partly towards the camera if it is to be visible (if **n** points partly away from the camera, the piece of surface must belong to the back of the object). We shall choose the z-axis to point directly towards the camera. Let us suppose that as we go across the piece of surface by a distance δx in the x direction, the surface comes a distance $p\,\delta x$ closer to the camera. Remembering that the surface normal is perpendicular to the surface, the effect on the surface normal vector is that it tips over in the $-x$ direction by a similar amount: if the length of the normal vector in the z direction is 1 unit, its component in the $-x$ direction is p units. Similarly, if as we go across the piece of surface by a distance δy in the y direction, the surface comes a distance $q\,\delta y$ closer towards the camera, the component of the normal in the $-y$ direction is q units. See Fig. 5.8. Thus, the normal to the surface can be written as $(-p,-q,1)$. The value of 1 just tells us that the surface

5.4. SHAPE FROM SHADING

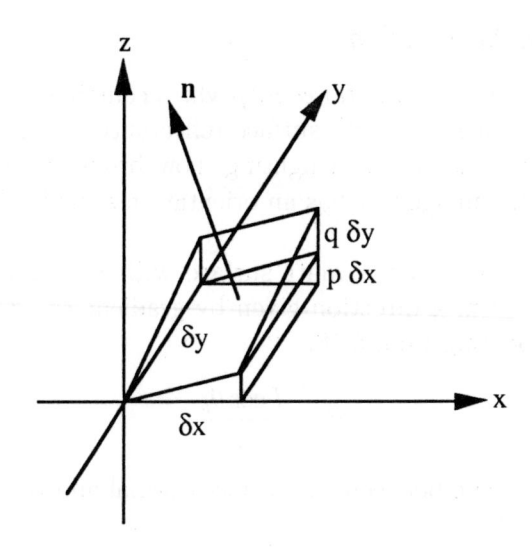

Figure 5.8: Orientation of a piece of surface

points towards us and not away from us, while p and q are the slopes of the surface in the x and y directions. The pair of numbers (p, q) is called the *gradient* of the surface at that point, which we have already discussed in connection with edge detection in Section 4.5.3. From this, we can derive the *unit* surface normal as

$$\hat{n} = \frac{(-p, -q, 1)}{\sqrt{1 + p^2 + q^2}}. \tag{5.19}$$

We now need to describe the angles between the surface normal, and the viewing direction and the light sources. The former angle is θ_e. As the z-axis points directly towards the camera, the viewing direction is $(0, 0, 1)$, and so

$$\cos \theta_e = \hat{n}.(0, 0, 1) = \frac{1}{\sqrt{1 + p^2 + q^2}}. \tag{5.20}$$

As for the light sources, their direction can be specified in the same way as the surface normal. Thus, a point light source lies in the direction $(-p_p, -q_p, 1)$ if a piece of surface with gradient (p_p, q_p) directly faces that light source.

5.4.6 Reflectance Map

We can now compute a *reflectance map* which combines the information of light source placements and surface reflectance properties. What it tells us is, under the specified lighting, how bright a piece of surface will appear in the image if it has an orientation specified by a gradient (p, q).

For example, consider a matt surface with a single point source of illumination from a direction given by gradient (p_p, q_p). The scene radiance is, repeating Eqn. 5.18,

$$R = \frac{I \cos \theta_p}{\pi} \tag{5.21}$$

where θ_p is the angle between the surface normal and the point source. Now

$$\begin{aligned} \cos \theta_p &= \hat{n} . \frac{(-p_p, -q_p, 1)}{\sqrt{1 + p_p^2 + q_p^2}} \\ &= \frac{1 + p_p p + q_p q}{\sqrt{1 + p^2 + q^2} \sqrt{1 + p_p^2 + q_p^2}} \end{aligned} \tag{5.22}$$

and hence

$$R(p, q) = \frac{I(1 + p_p p + q_p q)}{\pi(\sqrt{1 + p^2 + q^2} \sqrt{1 + p_p^2 + q_p^2})} \tag{5.23}$$

which is the reflectance map we require.

The easiest way to visualise such a reflectance map is to draw it as a contour map in the p-q plane, where all points with gradients (p, q) with the same value of R are joined by a curve as shown in Fig. 5.9.

As an aside, such a reflectance map can be used in computer graphics for producing shaded images — if we know the value of the gradient (p, q) for a piece of surface, we can then use $R(p, q)$ to decide how bright to make the pixel which represents that piece of surface.

However, in this case, we wish to do the opposite, and from the brightness of a pixel we wish to work out its orientation. We can immediately see that there is a problem. A pixel with a given level of brightness corresponds to an infinite number of possible orientations. For example, a pixel intensity of 0.8 corresponds to all orientations

5.4. SHAPE FROM SHADING

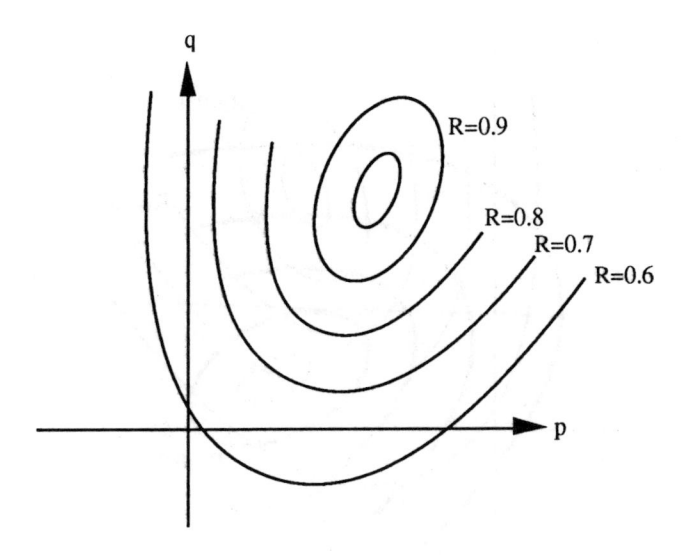

Figure 5.9: A reflectance map

given by the 0.8 contour in the diagram of $R(p,q)$. The problem is basically that we are trying to find the values of two unknowns, p and q, from a single known quantity, the pixel intensity.

5.4.7 Photometric Stereo

Thus, a method called *photometric stereo* is used. We take two separate images of the object with the camera and object in exactly the same positions, but under different lighting conditions. (This can usually be achieved readily in a manufacturing plant, where we can control the lighting. In a natural scene, a similar result can be obtained by taking two images at different times of day.) In simple terms, we now have two values of R, R_1 and R_2, for each pixel, which we can use to find our two unknowns p and q.

In practice, things are not quite as simple as this. Generally speaking, R will be a non-linear function of p and q, and so there may not be a unique solution to these equations. This is best understood by considering an example, as in Fig. 5.10. The thin contour lines correspond to the reflectance map for the first set of lighting conditions, and the

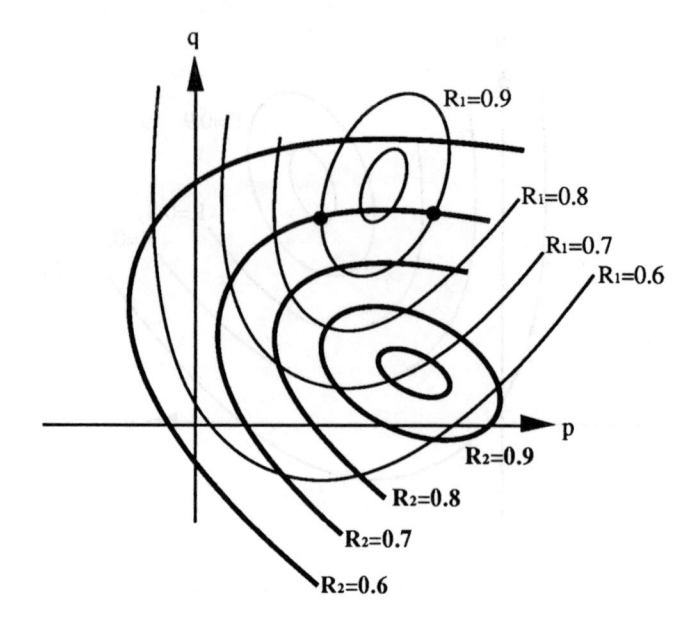

Figure 5.10: A pair of reflectance maps

thicker lines show the reflectance map for the second set of conditions. Now, consider a pixel with intensity given by $R_1 = 0.9$ in the first image, and intensity given by $R_2 = 0.7$ in the second image. It must have a gradient which lies on each of these contour lines respectively, which in the example shown intersect in two places as marked in Fig. 5.10. Thus, we deduce that the given pixel may have one of two possible pairs of (p, q) values.

One possible method of resolving this small final uncertainty is to use lighting from a third direction.

In practice, the reflectance maps required may be determined experimentally rather than theoretically, by viewing an object of known shape such as a sphere, made from the appropriate material, under the desired lighting conditions. The result can be stored in a two-dimensional array as a look-up table.

5.5 Object Recognition

5.5.1 Introduction

So far we have considered how various features can be grouped together to form higher level feature descriptions. These are often intermediate steps along a path starting with an initial image of some scene represented as an array of pixels and ending with the recognition of objects present in the scene. The recognition stage is the ultimate goal of many vision systems — having recognised objects in the scene, this information can be used to move around safely in the region shown by the scene while avoiding objects, to pick and place various objects, to inspect objects for compliance with predefined models, and to perform many other tasks.

The problems of recognising objects in an image has been the subject of much research for many years, and many approaches have been suggested.

Since we are usually trying to recognise instances of objects that correspond to preconceived notions of what the object should look like many approaches have tried to *match* scene descriptions (segmentations) with object models. The stored model descriptions may be based on geometric models (see Chapter 7), while other descriptions of the scene are built from one or more views of an object taken by a vision system.

Most types of features we have described so far in this book have been applied to the recognition problem. We shall discuss various aspects of model based matching and object recognition in Part 2 of this book. In the following Section we shall consider one particular type of higher-order description that may be of use in matching.

5.5.2 Extended Gaussian Images

The theory of *extended Gaussian images* (EGI) has been applied to attempt to solve the the problem of recognising objects in a three-dimensional scene by many researchers [42, 122, 124, 123, 136, 138]. The method may be thought of as providing a summary of the description of the shape of an object. We have already seen such summaries provided by statistical measures in Section 4.7; the summary provided by an extended Gaussian image provides more information, however.

Surface normal vector information for any object can be mapped onto a unit sphere, called the Gaussian sphere. This mapping is called the *Gaussian image* of the object. The mapping is formed as follows. Surface normals for each point of the object (for example, found as described in Chapter 4 on segmentation) are placed so that their tails lie at the centre of the Gaussian sphere and their heads lie on a point on the sphere appropriate to the particular surface orientation.

We can *extend* this process so that a weight is assigned to each point on the Gaussian sphere equal to the area of the surface having the given normal (for example, all points on a planar face of an object will have the same unit normal, and map to the same point on the Gaussian sphere). This mapping is called the *extended Gaussian image*. Sometimes the weights are represented by vectors parallel to the surface normals, with length equal to the weight. An example of such an extended Gaussian image is shown in Fig. 5.11.

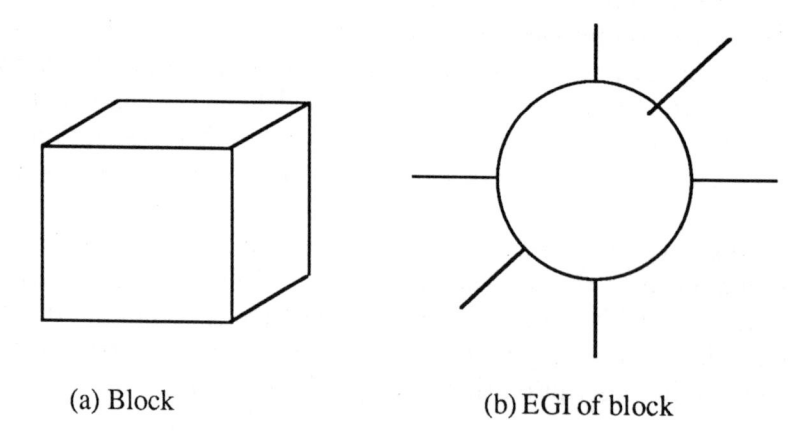

(a) Block   (b) EGI of block

Figure 5.11: The EGI of a block

Using three-dimensional solid models of objects (see Chapter 7), the corresponding EGIs for each stored object model can be computed and saved in the model database in the form of surface normal vector histograms. Depth maps corresponding to the observed data may be processed to give surface normal information either for individual

5.6. OPTICAL FLOW

depth points or for whole planes and some curved surfaces as described in Chapter 4. Thus, for any depth map an EGI histogram can be computed that will correspond to a visible half of one of the EGI histograms stored in the database. In principle, once a match is found (by comparing EGI histograms) both the identity and orientation of the object may be calculated.

The main drawback with EGI approaches is that EGIs uniquely define a solid only in the case of convex objects. There exist an infinite number of non-convex objects that possess the same EGI. Some examples of dissimilar objects which have the same EGI are illustrated in Fig. 5.12. Some approaches [122, 124] have attempted to resolve this problem by employing orientation histograms for every view belonging to a discrete set of views. This increases the complexity of the recognition problem, however. A good treatment of EGIs can be found in

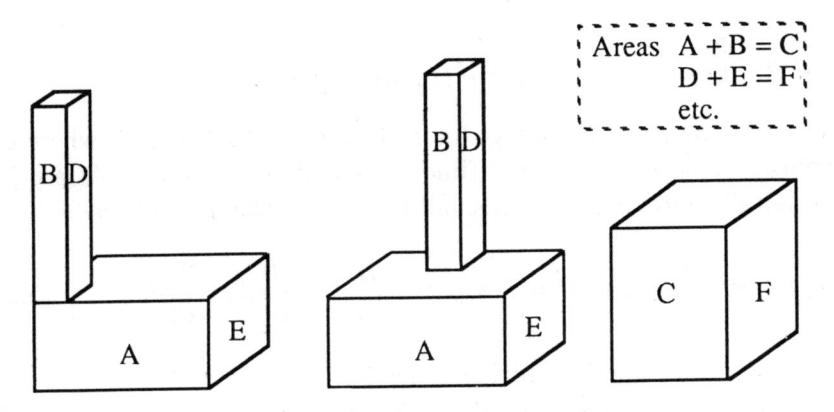

Figure 5.12: Examples of objects with the same EGI

Horn's book [123].

5.6 Optical Flow

5.6.1 Introduction

If we take a series of images in time, and there are moving objects in the scene, or perhaps the camera is itself on some moving vehicle, useful information about what the image contains can be obtained by analysing and understanding the difference between images caused by the motion. For example, given an image of a moving car, deciding which pixels in the image represent motion can help to decide which pixels belong to the car, and which to the static background. Going even further, and considering in detail the motion represented at each pixel, we can answer such questions as to how many moving objects there are, which directions they are moving in, whether they are undergoing linear or rotational motion, and how fast they are moving.

From an image sequence, we calculate a new function called the *optical flow*. For every pixel, a velocity vector $\mathbf{v} = (u, v)$ is found which says how quickly whatever is in that pixel is moving across the image, and in what direction. Drawings of optical flow can be made by drawing a small arrow in the direction of the velocity vector for a regular subset of the pixels, where the length of the arrow gives the speed.

5.6.2 Optical Flow Constraint Equation

Let us suppose that the image intensity is given by $I(x, y, t)$, where the intensity is now a function of time, t, as well as of x and y. At a point a small distance away, and a small time later, the intensity is

$$I(x+dx, y+dy, z+dz) = I(x,y,t) + \frac{\partial I}{\partial x}dx + \frac{\partial I}{\partial y}dy + \frac{\partial I}{\partial t}dt + \ldots, \quad (5.24)$$

where the dots stand for higher order terms.

Now, suppose that part of an object is at a position (x, y) in the image at a time t, and that by a time dt later it has moved through a distance (dx, dy) in the image. Furthermore, let us suppose that the intensity of that part of the object is just the same in our image before and afterwards. This is a reasonable assumption to make if, for example, the object is moving under the sky on a dull, cloudy day with uniform lighting conditions, but may well not be realistic if the object is moving or rotating significantly with respect to a nearby light source, where the lighting is highly directional. Provided that we are justified in making this assumption, we then have that

$$I(x + dx, y + dy, z + dz) = I(x, y, t), \quad (5.25)$$

5.6. OPTICAL FLOW

and so
$$\frac{\partial I}{\partial x}dx + \frac{\partial I}{\partial y}dy + \frac{\partial I}{\partial t}dt + \ldots = 0. \quad (5.26)$$

However, dividing through by dt, we have that
$$\frac{dx}{dt} = u, \qquad \frac{dy}{dt} = v, \quad (5.27)$$

as these are the speeds the object is moving in the x and y directions respectively. Thus, in the limit that dt tends to zero, we have
$$-\frac{\partial I}{\partial t} = \frac{\partial I}{\partial x}u + \frac{\partial I}{\partial y}v, \quad (5.28)$$

which is called the *optical flow constraint equation*.

Now, $\partial I/\partial t$ at a given pixel is just how fast the intensity is changing with time, while $\partial I/\partial x$ and $\partial I/\partial y$ are the spatial rates of change of intensity, *i.e.* how rapidly intensity changes on going across the picture, so all three of these quantities can be estimated for each pixel by considering the images.

5.6.3 Further Constraints

We wish to calculate u and v, but unfortunately the above constraint gives us only one equation per pixel for two unknowns, so this is not enough by itself. To see why, consider the images in Fig. 5.13. The lines are contours of equal intensity in the two images, which are taken at times Δt apart. The difficulty is in telling whether the part of the scene represented by point A in the first image has moved to point B, or B', or indeed any other point of the same intensity in the second image.

Thus further information is required allow determination of u and v. A key observation is that except near the edges of moving objects, the motion that is observed in adjacent pixels will be quite similar. This is clearly true for rigid objects, where the motion of one part of the object strongly constrains the motion of adjacent parts of the object, and indeed is also true for flexible objects which are bending rather than tearing. This idea can be used to formulate a further constraint on the optical flow. A measure of how much the optical flow deviates

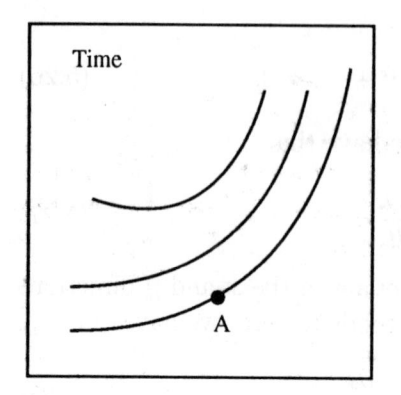

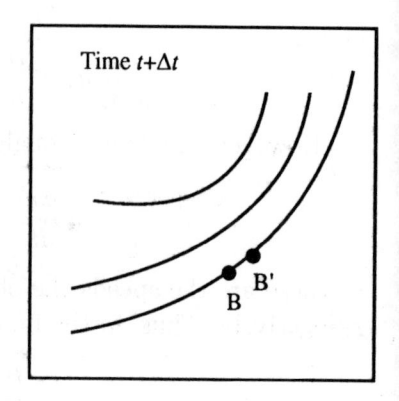

Figure 5.13: Ambiguity in determining optical flow

from this smoothly varying ideal can be calculated by evaluating the following integral

$$S = \iint_{\text{image}} \left(\frac{\partial u}{\partial x}\right)^2 + \left(\frac{\partial u}{\partial y}\right)^2 + \left(\frac{\partial v}{\partial x}\right)^2 + \left(\frac{\partial v}{\partial y}\right)^2 \, dx\, dy \quad (5.29)$$

over the whole image, where the derivatives like $\partial u/\partial x$ measure how rapidly the velocity is changing on going across the image.

5.6.4 Finding the Optical Flow

The above smoothness constraint is not necessarily entirely consistent with the optical flow constraint. We can express how much a solution for u and v deviates from the condition required by the optical flow constraint equation by evaluating

$$C = \iint_{\text{image}} \left(\frac{\partial I}{\partial x}u + \frac{\partial I}{\partial y}v + \frac{\partial I}{\partial t}\right)^2 \, dx\, dy. \quad (5.30)$$

To meaningfully combine these two constraints, we use the technique of *Lagrangian multipliers* [100, 123]. We attempt to find a solution for u and v which minimises $S + \lambda C$, where λ is a parameter

5.7. TEXTURE

giving the relative importance of these two error terms. Typically we would make λ large if the intensity measurements were accurate, when we should expect C to naturally be quite small. On the other hand, if the original data were noisy, λ would be made quite small. Interactive adjustment will in general be required to find the best value for λ.

Minimising the resulting integral can be done by using standard techniques from the calculus of variations, which show that the functions u and v which are required satisfy the following coupled pair of differential equations:

$$\frac{\partial^2 u}{\partial x^2} + \frac{\partial^2 u}{\partial y^2} = \lambda \left(\frac{\partial I}{\partial x} u + \frac{\partial I}{\partial y} v + \frac{\partial I}{\partial t} \right) \frac{\partial I}{\partial x}, \qquad (5.31)$$

$$\frac{\partial^2 v}{\partial x^2} + \frac{\partial^2 v}{\partial y^2} = \lambda \left(\frac{\partial I}{\partial x} u + \frac{\partial I}{\partial y} v + \frac{\partial I}{\partial t} \right) \frac{\partial I}{\partial y}.$$

The derivatives of I for each pixel are obtained from the original image, and λ is chosen as above. An iterative method can then be used to solve these equations for u and v at each pixel. Details are to be found in the book and papers by Horn and Schunck [123, 125, 212] as well as in the book by Murray and Buxton [174]. The heavy computations involved and the oversimplifying assumption of smoothly varying optical flow combine to make this method less than ideal in practice.

5.7 Texture

5.7.1 Introduction

Sometimes it is easier to tell what we are looking at by means of its texture, rather than its shape. For example, the pattern in Fig. 5.14 suggests a brick wall, irrespective of the overall outline or extent of the pattern.

The concept of texture is a rather difficult one to pin down exactly, but "something composed of one or more simple elements repeated in a periodic or random manner" comes close. A brick wall is an example of periodic texture (neglecting minor irregularities), while a pile of coins or the leaves on a tree represent a random texture.

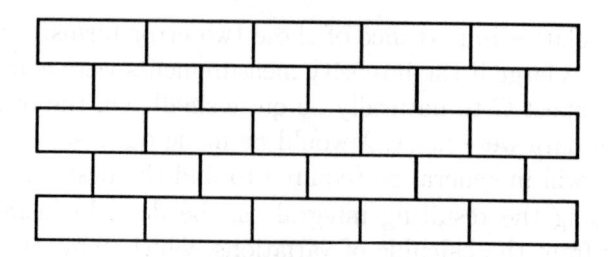

Figure 5.14: Texture suggesting a brick wall

Note that we may have to be fairly liberal in our interpretation of "simple" in this definition. Coins obviously can be quite detailed in appearance. Nevertheless, when we are thinking of them as making up a texture in an image, there is an inherent assumption that there are many of them in the image. Thus each one only occupies or contributes to a small number of pixels and so the level of detail seen for each coin is quite limited. On the other hand, each one must contribute to a sufficient number of pixels to be recognisable. For example, grains of sand on a beach are usually too small in most images to be detected. Generally, we are not interested in describing the elements as individual items, but only the entity which they identify.

Note also that there may be more than one type of simple element. Most types of coins have distinct appearances for heads and tails, for example. In a regular pattern there may be large and small tiles as shown in Fig. 5.15.

5.7.2 Texture Methods

Two main approaches are taken to recognising and classifying textures. The first method, the *structural* method, considers the texture to be composed of primitive elements arranged in a repeating pattern. Separate steps are used to find the primitives in the image, and then to decide if they are organised in an appropriate way.

The second method, the *statistical* method, describes a texture in terms of the distribution of pixel values in the image, for example in terms of the number of pixels of each intensity, or the spatial relationships between pixels of differing intensities, or both.

5.7. TEXTURE

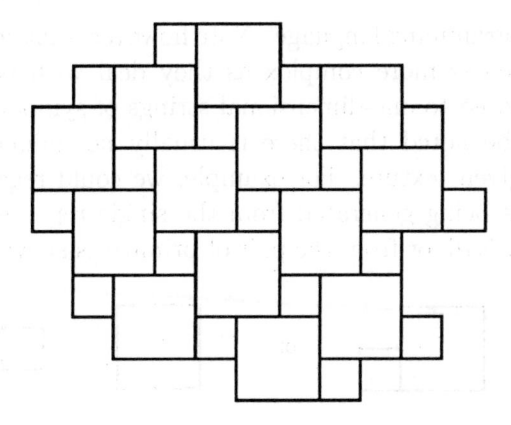

Figure 5.15: Texture containing more than one basic element

As will be fairly obvious, the former approach is better suited to textures which contain relatively large primitive elements which can be readily identified, and to textures where the pattern is regular, or almost so. The latter approach is more useful when considering random textures, or textures where the primitives are quite variable or otherwise difficult to identify.

As might be expected, many artificial textures, carefully arranged by man in exact patterns, fall into the former category, while many natural textures fall into the latter category. For this reason we shall concentrate on structural methods here. Further discussion of both approaches can be found in Ballard and Brown [6]. One particularly useful idea they discuss which we shall not pursue here is how information about the orientation of an object can be recovered from texture data. In contrast, Kanatani [141] shows how information about an object's shape can be recovered by analysing the appearance of texture in an image.

5.7.3 Shape Grammars

One structural method of texture recognition uses *shape grammars* to describe the primitive elements and they way in which they are laid out. Such grammars are analogous to the grammars used by compilers to recognise and understand the strings of symbols which form the programs in a programming language. Note however, that shape grammars are in some senses more complex as they deal with two-dimensional shapes as opposed to one-dimensional strings of symbols.

It should be noted that there is usually no unique grammar for describing a given texture. For example, we could regard the texture in Fig. 5.15 as being generated from the single repeated primitive on the left in Fig. 5.16, or from the pair of primitives shown on the right.

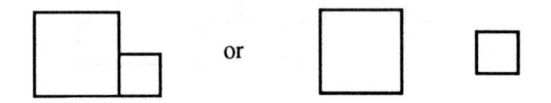

Figure 5.16: Alternative primitives

A shape grammar as defined by Stiny and Gips [222] consists of four parts:

- A set of one or more *terminal shapes*, $\{T\}$.

- A set of one or more *markers*, $\{M\}$.

- A set of one or more *shape rules*, $\{R\}$.

- An initial shape S.

The terminal shapes are the shapes which are to appear in the final texture, while the markers are special shapes whose presence is used to help the process along. No shape is allowed to be both a terminal shape and a marker. The function of the markers is to indicate which rules are applicable in which situations. The initial shape is normally a combination of one or more terminal shapes and one or more markers. The texture is built up by applying various rules in turn starting from the initial shape. Some terminal shapes, markers and an initial shape are shown in Fig. 5.17.

Each rule has a left-hand side and a right-hand side. The left-hand side of each rule is some combination of one or more terminal shapes and one or more markers, while the right-hand side of each rule is some combination of zero or more terminal shapes and zero or more markers. When a texture is being built up, if a combination of shapes is found in

5.7. TEXTURE

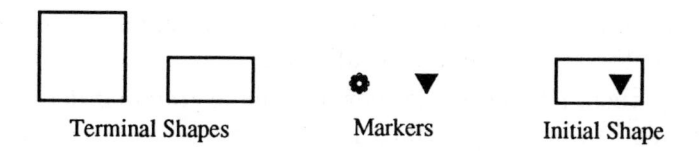

Figure 5.17: Grammar elements for texture generation

the texture which matches the left-hand side of a rule, that combination is allowed to be replaced by the right-hand side of the rule.

In one or more of the rules, there will be only terminal shapes on the right-hand side, which allows the markers to be removed. Once all of the markers have been removed, no more rules can be applied, and the process stops. Some rules using the previously given terminal shapes and markers are shown in Fig. 5.18.

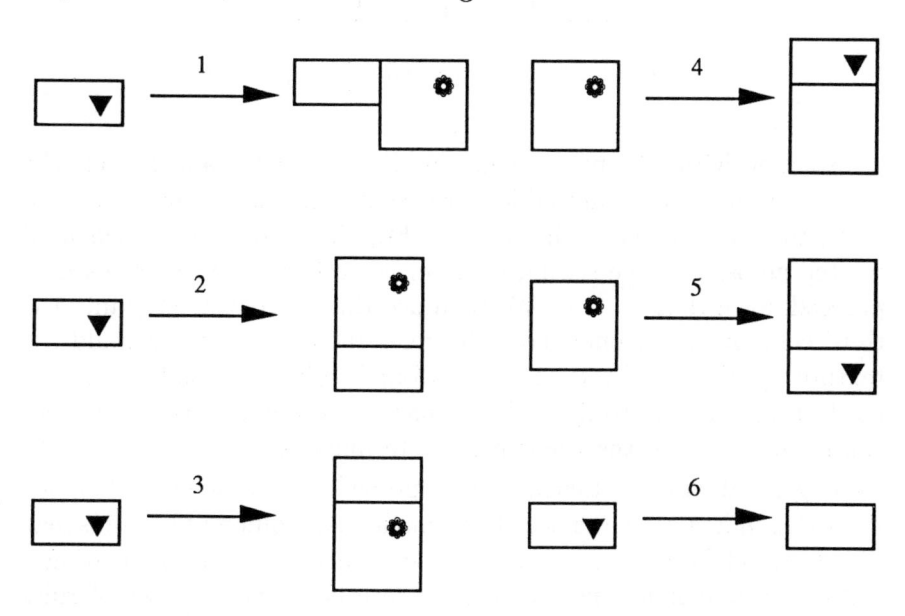

Figure 5.18: Rules for texture generation

When comparing part of the texture generated so far and the left-hand side of a rule, a match is considered to have been found even if the two are not identical, as long as they are geometrically related through a rotation, translation, scaling or mirroring. In this case, the same transformation is also applied to the right-hand side of the rule to find how the texture should be affected.

Finally, then, let us consider how the shape grammar defined above could be used to build up a tiling like the one shown in Fig. 5.19.

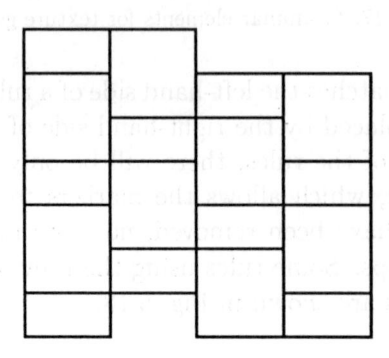

Figure 5.19: A texture generated by a grammar

Starting with the initial shape in the lower left hand corner, the other terminal shapes are added proceeding up the left hand column using the rules indicated in order in Fig. 5.20. Rule 1 is then used to step across into the second column, when further rules are used to proceed down that column. We then use Rule 1 again to step into the third column, where once more the rules shown allow us to build the texture upwards, and then down the fourth column. Finally, Rule 6 is used at the bottom to discard the marker, whereupon no more rules can be applied, and the whole process terminates.

In general, at each stage it may be possible to choose more than one rule. Depending on the choice of rule made, quite different textures may be generated from the same set of rules. Part of the art of creating a suitable grammar for describing a texture is to choose a set of rules which while generating the texture desired does not generate many other possible textures too. However, sometimes we *will* want to allow many possibilities when we are trying to detect a type of texture which may be very variable.

Note also that different extents of pattern may be created by deciding when to apply the rules which remove the markers.

5.7. TEXTURE

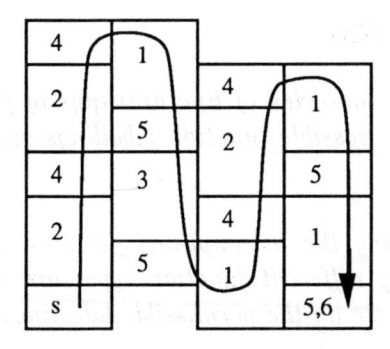

Figure 5.20: Rules applied at each stage of texture generation

5.7.4 Texture Recognition

Now, just as a compiler works, if we are given a complete texture, these rules can be used in reverse order, *i.e.* from right to left, to recognise a pattern rather than to build one up. Essentially, at each stage we chip away at part of the pattern, removing one or more terminal shapes, until we either get back to the starting shape, in which case we have recognised the texture, or we get stuck. In the latter case we must backtrack, and try possible alternative rules at an earlier step. If after trying all possible alternatives, we can not find a path back to the starting shape, we must deduce that the grammar does not describe the texture being observed.

One problem which remains is how to identify the terminal shapes themselves in the image. These are sometimes referred to as *texels* (texture elements, the smallest unit of a texture) by analogy with *pixels* (the smallest elements of a picture). Most approaches regard a texel as being a particular arrangement of pixels, with certain specified relationships between their pixel values. At the simplest level, it is possible to simply search for a fixed pattern of pixels. A more sophisticated method of identifying the texels is to again use a grammar to describe how each texel is made up from its component pixels. This leads to a two-level system which uses grammars both to seek out the texture elements, and then to find how they are grouped.

5.8 Exercises

Exercise 5.1 *Make a series of line drawings of polyhedral models to show how all of the possible junction labellings given in Fig. 5.4 may arise in practice.*

Exercise 5.2 *Making the assumptions given in the Section on line labelling, how many different junction types are potentially possible? What fraction of these do the permissible labellings correspond to?*

Exercise 5.3 *Draw a rectangular arch to show how T-junctions can arise when part of an object obscures another part of itself.*

Exercise 5.4 *Amend the line labelling descriptions to take into account shadow information. How many different junction types are potentially possible? What fraction of these do the permissible labellings correspond to? Compare your answer to that of Exercise 5.2.*

Exercise 5.5 *Design a relaxation labelling algorithm to implement the line labelling method discussed in Section 5.2.*

Exercise 5.6 *Instead of using gradient information to segment an image into regions, we have seen in Section 5.3 that we can use probabilities that pixels lie within certain regions to perform this task. Assuming that appropriate probabilities have been assigned initially for each pixel, design algorithms to segment the image using*

1. *Region growing techniques (see Section 4.6.4).*
2. *Relaxation labelling techniques (see Section 5.2).*

What are the basic differences between these approaches in this instance?

Exercise 5.7 *Show that for a matt surface where the bidirectional reflection distribution function is a constant, this constant has the value $1/\pi$. Hint: Integrate to find the total radiance of the surface in all directions and equate it to the irradiance of the surface.*

5.8. EXERCISES

Exercise 5.8 *Show that for a point source of light of radiance R, the distribution of radiance is given by (where θ_i, ϕ_i, θ_p and ϕ_p are as defined in Section 5.4.3)*

$$R(\theta_i, \phi_i) = \frac{R\,\delta(\theta_p - \theta_i)\,\delta(\phi_p - \phi_i)}{\sin\theta_p}, \quad (5.32)$$

by considering the integral of this expression over all directions.

Exercise 5.9 *An image of a matt sphere illuminated by a distant point source of light is produced. Firstly, by considering the form of $R(p,q)$, find the gradient of that point on the sphere which will appear brightest in the image. Justify the answer by referring back to the surface radiance. Secondly, what form does the contour $R(p,q) = 0$ take in the p-q plane?*

Exercise 5.10 *Draw some simple objects and work out their extended Gaussian images.*

Exercise 5.11 *Give some examples of sets of objects that have the same extended Gaussian image (see Fig. 5.12).*

Exercise 5.12 *A series of images is produced of a uniformly coloured sphere rotating about a vertical diameter under uniform lighting. How do successive images differ in appearance? What can be deduced from this series of images about the motion of the sphere?*

Exercise 5.13 *A patterned sphere is rotating such that its axis is inclined at 45° towards the plane of an image being made of it. Calculate the optical flow for a general point on the surface of the sphere. How would you locate in the image the point on the sphere through which the axis of rotation passes? Would you expect this to be a stable calculation?*

Exercise 5.14 *Generate further patterns from the shape grammar given in Figs. 5.17 and 5.18. What limitations are placed on the overall outline of the pattern by this set of rules? What extra rules could be added to overcome these limitations? How would they also affect the range of patterns which it is possible to generate?*

Part II

Model-Based Matching Techniques

Chapter 6

Introduction to Model Based Matching

6.1 Introduction

In the first Part of the book we have addressed many important issues involved in providing a computer with some sort of visual sensing capability. We have gone from the acquisition of various types of images to methods of understanding the scenes contained within these images. We have also already touched upon the major problem of recognising objects in images.

In this second Part of the book we shall describe in detail a range of approaches that have been taken by many researchers to solving the recognition problem. We shall only deal here with three-dimensional object recognition. As we shall see, there is still much work to be done before it will be possible to recognise relatively complex scenes containing many objects, or objects which have a complicated structure. Many of the methods we shall describe deal with highly constrained classes of objects. For example, some can only deal with polyhedra whilst even most others can only deal with objects that are bounded by a very restricted class of higher order surfaces such as quadrics or generalised cylinders. Some methods only use points extracted from the image as a basis for the matching method, others use edge data, others still use surface information, whilst a few use more than one of these types of data.

Clearly the segmentation of features is an important step in the recognition process and we shall build substantially on the work detailed in Chapter 4 throughout this Part of the book. In particular, in our system, we shall use the segmentation algorithm detailed in Section 4.6.6 to provide input to the matching algorithm we shall develop in Chapter 9. Also in that Chapter a full analysis of the practical performance of the segmentation algorithm will be provided.

In order to recognise an occurrence of an object in a scene, we must have some notion of what the object looks like. This information is usually provided by a geometric model of the object stored in the computer. The recognition process is then essentially a matter of trying to find a stored object in a database of such objects which agrees with the data we have extracted from the image containing the observed object. Thus, this is a matching problem where we seek to find the best correspondence between some set of object features for the observed object and the same types of features for known objects. For obvious reasons this type of recognition strategy is called *model based matching*.

Model based matching strategies have a wide range of applications. A few of the more obvious ones include:

Object Manipulation — Having recognised an object we can then use the known geometric information about it to decide how to pick it up, perhaps with a robot arm, so that it can be placed somewhere else in a desired position and orientation.

Navigation — In this case a vision system is attached to a mobile vehicle or robot. Recognising objects allows the vehicle to avoid collisions when it moving around its environment, and in more sophisticated uses, to plan a path to a destination.

Inspection — Here, a vision system is used to test if certain features of the object are present or missing, and to measure dimensions and other physical properties for conformance to expected values stored in an object model. This application will be considered in detail in Part 3 of the book.

As can be seen, nearly all of the above tasks involve determining the position and orientation of an object in the scene as well as its recognition . Fortunately, as a by-product of their means of operation, most of the model based matching methods *do* output a description of position and orientation.

6.2 Outline of Part 2

In this Part of the book we shall first discuss various methods of representing geometric information about objects. Computer based geometric models of objects have been in use for some time for computer aided design and, to a lesser extent, computer graphics. In Chapter 7 we shall survey a range of geometric modelling methods and discuss their suitability for use in a computer vision system intended to recognise objects.

In Chapter 8 we shall describe many of the currently popular model-based matching techniques developed in recent years.

Finally, in Chapter 9 we shall assess the performance of these matching techniques, on both real and artificial data, and provide further implementation details of many aspects of matching. We shall also provide details of a matching algorithm we have developed for use in our inspection system to be described in Part 3 of this book. A particular requirement of inspection is that the algorithm must provide a very accurate description of the position and orientation of the observed object. Thus, many of the details in Chapter 9 are concerned with meeting this goal. Nevertheless, we believe that the matching algorithm we shall describe is suitable for general recognition purposes also.

Chapter 7

Geometric Modelling for Computer Vision

7.1 Introduction

Geometric modelling's origins lie in the field of computer aided design. Early design systems were based on producing engineering drawings, and the computer was used merely as a means of speeding up the process of producing such drawings. Nevertheless, such drawings have limitations, and it fairly soon became obvious that what is really needed is a model of the shape of the object, not a model of a drawing of the object. We shall briefly consider a few of the issues that have lead to the use of *solid modellers*.

To start with, it should be noted that a drawing may not represent *any* valid object. Such an example is the well-known "Devil's Fork" shown in Fig. 7.1. Looking at the right hand end it appears as if the fork has three (round) prongs. However, looking at the left hand end, there appear to be two (square) prongs. Clearly, such an object does not exist, even though the illustration for this book was drawn using a computer draughting system!

Probably the first models to be used for three-dimensional objects were *wireframe* models, where an object is represented just by its set of edges. However, such models have inherent problems, as is shown by the example in Fig. 7.2. The wireframe at the top corresponds to either of the two different solid objects below it, which contain a front-to-back

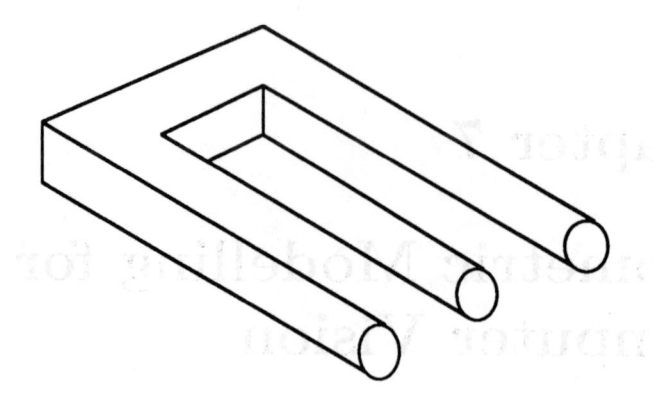

190   CHAPTER 7. GEOMETRIC MODELLING FOR COMPUTER VISION

Figure 7.1: The Devil's fork

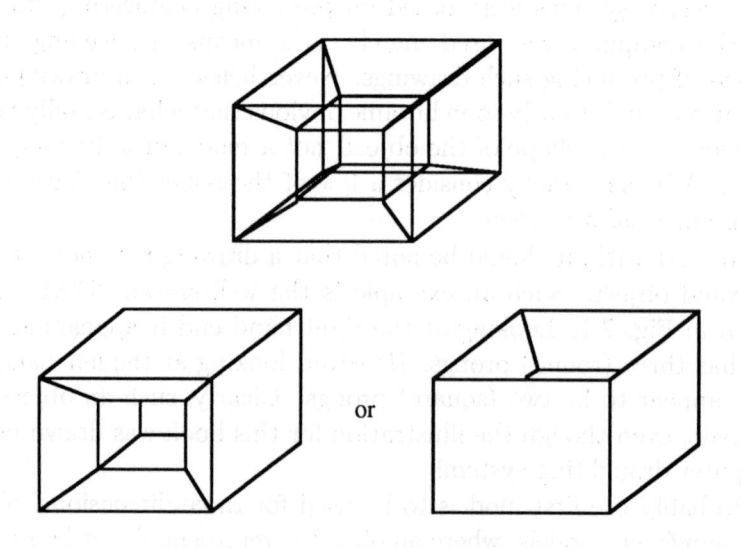

Figure 7.2: An ambiguous wireframe

or vertical through hole respectively. Thus, a wireframe is not sufficient by itself to define where the faces of the object are, or where there is solid and where there is air.

Even though ambiguous wireframes of this type are fortunately not that common, the whole concept of wireframes breaks down once we start to consider objects with curved surfaces. For example, if we take a circular wireframe loop, this can quite obviously be the boundary of a flat disk, or of a hemisphere.

For the reasons given above, it is evident that more rigorous methods of representing solids are required.

7.2 Solid Model Representations

7.2.1 Introduction

The fundamental question that any method of solid representation should be able to answer, given a test point and a solid model, is

- Does the given point lie inside the solid?

Further important properties of the modeller are implied by the following questions:

- Does every model which can be constructed by the modeller represent a unique, valid three-dimensional object?

- Can every solid object of the type we wish to consider be represented by some model?

- Is the model corresponding to a given solid unique, or can the same solid be represented by more than one model?

While we obviously desire a positive answer to the first two questions, a negative answer to the final question is not so bad. It may just make it more awkward to tell if two objects are the same, as when we wish to decide if a stored model of an object exactly matches an object model deduced by a vision system, for example.

Other issues of importance for choosing between types of model include compactness of representation, speed of construction and modification of models, and ease of extracting desired information from models. Below, we review various types of solid model, and then evaluate which is most suited for our purposes in computer vision. For further details, two excellent introductions to solid modelling can be found in the book by Woodwark [243] and the collection of papers edited by Rooney [206]. More advanced details of solid modelling are provided by Hoffmann [118] and Mäntylä [155].

7.2.2 Spatial Enumeration

Let the solid object under consideration be contained within some region of space. Probably the simplest general method of representing the object is to divide the region of space up into elements using a regularly spaced grid. By analogy with the word *pixels* for picture elements, the name *voxels* is used to denote the individual volume elements, which are typically small cubes. Each element is marked as either full or empty, showing which regions of space are occupied by the solid.

The disadvantage of this scheme is readily apparent. Supposing we wish to inspect a typical engineering object with a diameter of about 1m, we may need to describe it to a tolerance of 10^{-5}m, which implies that 10^{15} bits of information are needed. Such a scheme is clearly not practical with current computer technology, both in terms of amounts of memory required, and time needed to access this much memory.

A more sophisticated version of spatial enumeration is based on recursive subdivision, and uses a data structure called the *octree*, the three-dimensional equivalent of the *quadtree* used in computer graphics. Again we start with an initial cubical region of interest containing the solid object. At each stage of the description, if the region of interest is completely full (contains only solid) or completely empty (contains only air), we record this fact. Otherwise we divide the region of interest into halves in each of the coordinate directions, giving us eight subcubes. Each of these is then considered in turn as the region of interest, and so on. This method leads to a tree structure where leaf nodes record full or empty regions of space, while other nodes have eight children (hence the name *octree*), one for each subcube. We stop the process of recursive subdivision when the subcubes are equal to or smaller than the spatial resolution with which it is desired to describe the solid. The first stage of subdivision of the initial cube of an octree into its octants is shown

7.2. SOLID MODEL REPRESENTATIONS

in Fig. 7.3, while a front view of the first four subdivisions required

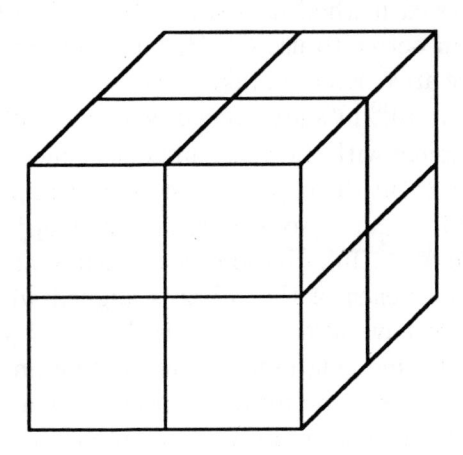

Figure 7.3: First stage of subdivision for of an octree

in finding the octree corresponding to a triangular prism are shown in Fig. 7.4. Note that some of the undivided nodes are completely outside

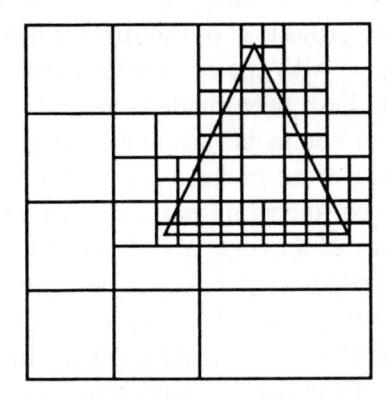

Figure 7.4: Front-on view of an octree

the triangular prism, while others are complete within it.

An octree obviously uses much less storage than a voxel based model. We expect much subdivision to occur in an octree near the boundary of the solid, as this is the region of space where the solid to air transition occurs. (Again, see Fig. 7.4.) Well inside the object or well outside probably corresponds to leaf nodes. Using this as a guideline, if a voxel based method uses storage of $O(N^3)$ bits, we can very crudely expect an octree to use $O(N^2)$ bits, in the ratio of the solid's volume to surface area. Even so, taking the typical figures quoted above implies that about 10^{10} bits are needed, which is still too high.

A further problem with the voxel and octree methods is that higher level information about the object is lost. For example, it is not immediately apparent from an octree description of an object how its surface is divided up into faces. Information about surface normals is to a large extent lost, because each of the cubes is aligned with the coordinate axes. Also, as we have already seen in Section 4.6.3 the process of region splitting and merging used for segmenting an image and which can produce a quadtree or octree decomposition of an image, some real world objects cannot be accurately represented using an octree unless a high resolution octree is employed. This leads us to similar problems to those of the voxel model representation.

Furthermore, although input from a vision system is typically of a low-level nature as are voxels, comparisons may generally not be made directly between pixels in an image and voxels in a model as the two-dimensional data is a projection of the three-dimensional object, not a slice through it aligned with the coordinate axes.

Various attempts have been made to improve octrees, by allowing the nodes of the tree not to be just full or empty, but to also be of other types, such as surface nodes, edge nodes, and vertex nodes. A surface node is allowed to contain part of a face of the object; an edge node can contain part of an edge; a vertex node may contain a vertex (*i.e.* a corner where edges meet). Any cube which contains more than one vertex, or more than one edge and no vertex, or more than one surface and no edge must still be subdivided. In general, storage requirements for this type of scheme (sometimes referred to as a *polytree*) are expected to be far less, at the expense of much greater complexity both of the contents of nodes, and the algorithms for constructing and manipulating the tree.

One advantage of polytrees is that some higher level information is kept. For example, we now know which faces run through a node, and their surface normals. On the negative side, it is possible to construct problematic cases where the storage requirements are still high unless

7.2. SOLID MODEL REPRESENTATIONS

yet further types of nodes are allowed, and the data structure is made even more complex.

Further details of the theory and implementation of polytrees can be found in the papers by Brunet [45, 44], Carlbom [50] and Dürst [69]. In some ways, polytrees can be considered to be a hybrid between octrees and boundary representation, which will be discussed in Section 7.2.4.

7.2.3 Set-Theoretic Modelling

This modelling method is also widely known as *computational solid geometry*, or *CSG* for short, although the name of *set-theoretic modelling* better describes it.

Here, the idea is that various primitive shapes, such as rectangular boxes, spheres, cylinders and cones, are used in conjunction with set operators to build up more complex objects by means of a tree structure, as shown in the simple example in Fig. 7.5. The model of the L-shaped

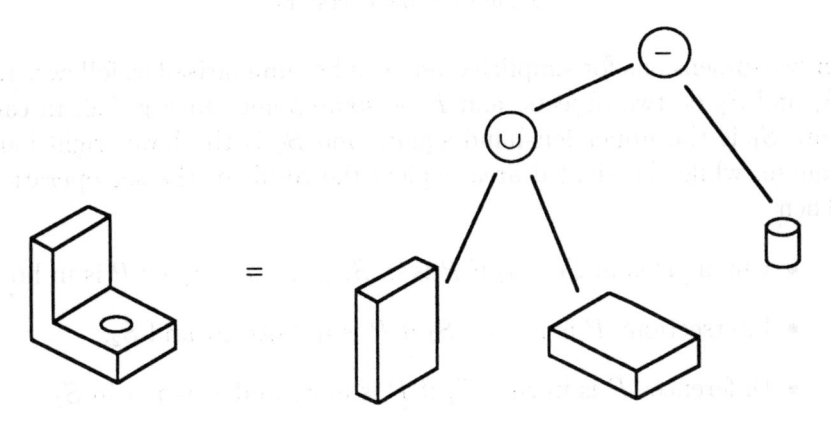

Figure 7.5: A set-theoretic model

bracket with a hole in it on the left is built up by firstly forming the set union of the two rectangular blocks shown on the right to make an L-shape, and then by subtracting the cylinder to make the hole.

Each primitive has associated with it a transformation which gives its overall size, position and orientation with respect to world coordinates, and is applied to each primitive before it is combined with others. The transformation information is stored in the leaf nodes of the tree together with the description of the type of primitive, and its defining parameters — for example, height and semi-angle for a cone.

The set operators used in the tree are similar to those of Boolean logic. The principal ones used are *union*, *intersection*, and *difference*, which in terms of more familiar Boolean logic correspond to *or*, *and*, and *and not* respectively. Their meanings are illustrated in Fig. 7.6, drawn

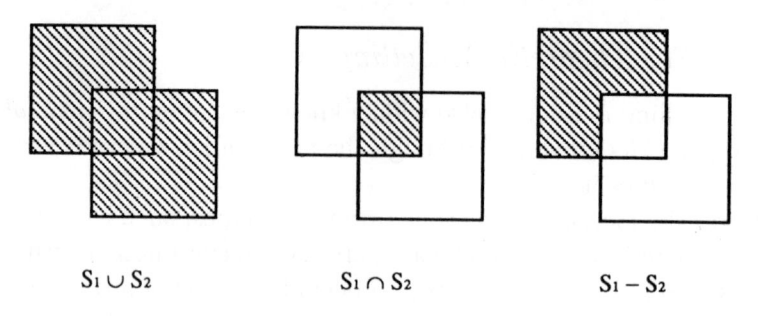

Figure 7.6: Set operators

in two dimensions for simplicity, and can be summarised as follows. Let S_1 and S_2 be two objects, and P be some point. In Fig. 7.6, in each case S_1 is the upper left-hand square and S_2 is the lower right-hand square, while the shaded area depicts the result of the set operation. Then

- Union: P is in $S_1 \cup S_2$ if P is in S_1 or P is in S_2 or P is in both.

- Intersection: P is in $S_1 \cap S_2$ if P is in both S_1 and S_2.

- Difference: P is in $S_1 - S_2$ if P is in S_1 and P is not in S_2.

In fact, the set operators used are not quite as simple as this would at first suggest. Above, it was stated that the fundamental question which could be asked of a solid model is whether a test point is inside the solid. In fact, we need to be more subtle, and rather than just classify a point as inside or outside, we should classify it as *inside*, *outside*, or *on the surface of* the solid. Consider the intersection of the L-shape and the rectangular block shown on the left in Fig. 7.7. If we just consider points on the surface of the object to be inside the object, the result will be as shown on the right, which includes a *dangling face*.

7.2. SOLID MODEL REPRESENTATIONS

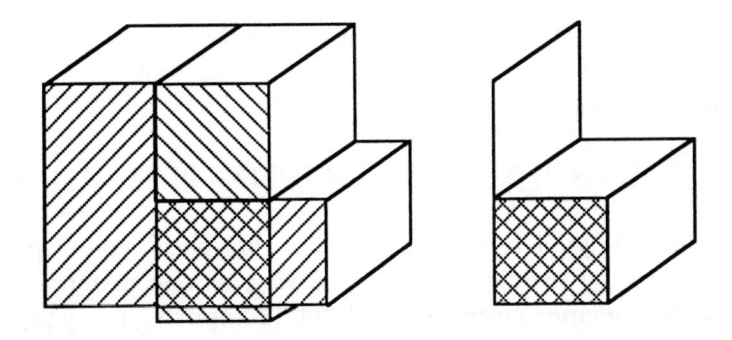

Figure 7.7: A dangling face formed by intersection

The resulting object is not what we want, and contains an element of a lower dimensionality. Dangling edges may also arise in a similar way to dangling faces. When combining objects with set operators, a more careful treatment is required to cope with cases in which some points lie on the surface of *both* objects. The key to resolving this problem is not only to consider the points themselves, but small neighbourhoods of the points, and to keep careful track of which part of the neighbourhood is air, and which is solid. Doing this in such a way as to prevent dangling edges and faces from arising is referred to as *regularising* the set operators.

Although the user starts from the primitive objects, in many set-theoretic modellers these are broken down within the model into *half-spaces* whose boundaries are implicit surfaces. An implicit surface (see Section 4.6.1) of the form

$$f(x, y, z) = 0 \qquad (7.1)$$

divides space into two regions, the region for which $f(x, y, z) < 0$, and the region for which $f(x, y, z) > 0$. One of these regions is used to represent solid, while the other represents air. Each of these is referred to as a half-space; at least one of the half-spaces is infinite in extent.

For example, a unit sphere can be represented in implicit form as

$$x^2 + y^2 + z^2 - 1 = 0; \qquad (7.2)$$

the points contained within it are those for which $x^2 + y^2 + z^2 - 1 < 0$.

On the other hand, a unit cube is bounded by six planar faces, for example, $x = 0$, $x = 1$, ..., $z = 1$, and can be represented in half-space terms by the combination

$$(x > 0) \cap (x < 1) \cap (y > 0) \cap (y < 1) \cap (z > 0) \cap (z < 1). \quad (7.3)$$

The advantage of using half-spaces is that each type of primitive solid can now be handled in the same way, as one or more implicit surfaces. In principle, given an overall solid comprising many primitives combined in a tree, it is now simple to tell if a given point lies within that solid. Firstly, we classify the point with respect to each of the half-spaces, a simple matter of evaluating the surface function for the given values of x, y and z, and looking at the sign of the result. Secondly, these classifications are combined according to the tree describing the object.

A set-theoretic description can be a very compact way of describing quite complicated objects. However, it is quite difficult and time consuming to extract most types of information about the object from such a model. If we wish to draw the object, for example, we shall typically wish to know the positions and extents of the faces and the edges. Such information is not directly available in the model, and must be calculated. For example, to find the edges of the object, one direct method is to find all edges which correspond to the intersection of every pair of half-spaces in the model, and then use the point classification method to break each of these up into edge segments which lie wholly inside, on, or outside the solid.

7.2.4 Boundary Representation

In contrast to set-theoretic models, boundary representation models are a more explicit representation. The object is represented by a complicated data structure giving information about each of the object's faces, edges and vertices and how they are joined together. Some of the faces, edges and vertices of the previously described L-shaped bracket with a hole are indicated in Fig. 7.8.

The description of the object can be conveniently broken down into two complementary parts. The first part is referred to as the *topology*

7.2. SOLID MODEL REPRESENTATIONS

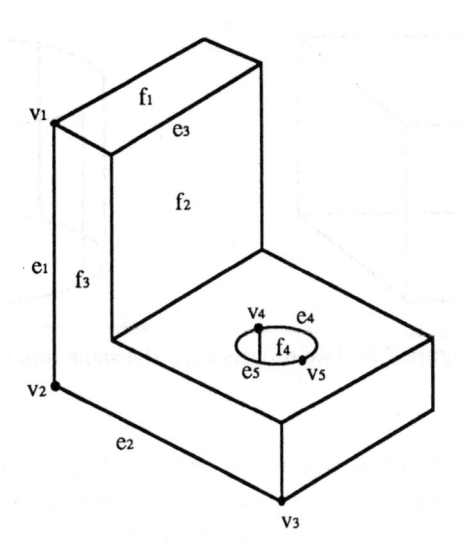

Figure 7.8: Faces, edges and vertices

of the object, although its meaning here is somewhat different to the more conventional mathematical meaning.

The topology of an object records the connectivity of the faces, edges and vertices by means of pointers in the data structure. Thus, each edge has a vertex at either end of it (e_1 is terminated by v_1 and v_2 in Fig. 7.8), and, for manifold objects, each edge has two faces, one on either side of it (edge e_3 lies between faces f_1 and f_2). In practice, more connectivity information is stored than the bare minimum necessary to completely describe the topological relationships, for reasons of efficiency in traversing the data structure, when passing from one topological element to adjacent ones of the same or another type. One widely used data structure is the *winged-edge* data structure, devised by Baumgart [11]. Further possibilities are considered and reviewed by Weiler [237].

It is important to note that two quite different objects can have exactly the same *topological* description, as evidenced by Fig. 7.9. Both the block and the cylinder have six faces, twelve edges, and eight vertices which can be put into one-to-one correspondence, where corresponding elements in each object are connected together in the same

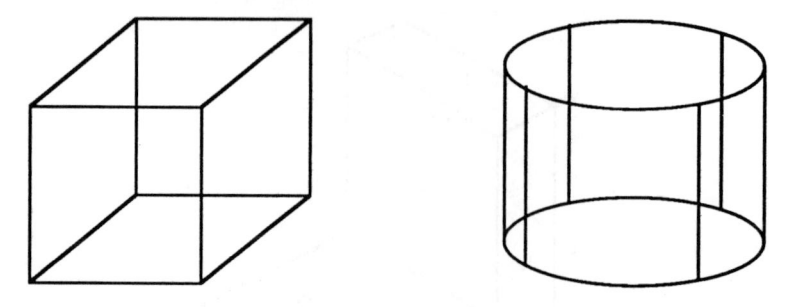

Figure 7.9: Two objects with the same topology

way.

The other part of a boundary representation of an object is its *geometry*. This described the exact shape and position of each of the edges, faces and vertices. The geometry of a vertex is just its position in space as given by its (x, y, z) coordinates. A face is either represented as an implicit surface of the type mentioned when considering set-theoretic models in Section 7.2.3, or as a parametric surface of the form (see Section 4.6.1)

$$\mathbf{r} = (x(u,v), y(u,v), z(u,v)). \tag{7.4}$$

For example, a planar face like f_1 in Fig. 7.8 could be represented by an implicit equation of the form

$$ax + by + cz + d = 0 \tag{7.5}$$

for suitable values of a, b, c and d, while a cylindrical face like f_4 could be represented in parametric form as

$$x = r\cos u \qquad y = r\sin u \qquad z = v. \tag{7.6}$$

Edges may be straight lines (like e_1), circular arcs (like e_4), and so on. Their geometry may be described by a parametric curve of the form

$$\mathbf{r} = (x(u), y(u), z(u)). \tag{7.7}$$

Alternatively, they may not be stored, but may be found by computing the intersection of the two surfaces on either side of them when required.

Returning to Fig. 7.9, in each case, the top face has the same geometry, as do the vertical edges. Nevertheless, the vertical faces are planar

7.2. SOLID MODEL REPRESENTATIONS

for the block, while they are cylindrical for the cylinder. Also, the edges round the top face are straight lines for the block, while they are circular arcs for the cylinder. These differences in *geometry* of the two models, even though the topology is the same, causes the two models to describe different solids.

As already stated, boundary models are a much more explicit representation than set-theoretic models. On the one hand, they have higher storage requirements, but on the other hand, as faces and edges are directly stored in the model, drawing a picture and other similar computations from a boundary model are much easier.

Note that set operators can be used as an input method even for boundary modellers. Here though, each set operation is explicitly carried out and evaluated as it is specified. For this reason, boundary representation is often referred to as an *evaluated* model representation, as opposed to *unevaluated* set-theoretic models, where all the set operators are left in the tree. The set operations in this case are only carried out when some question is asked about the model.

One final point which may be worth making here is that amongst the geometric modelling community, the word *feature* is used in a rather different sense than we have used it in this book. Here, a feature is usually taken to mean a point, edge or surface extracted from an image. In modelling terms, however, a feature is the name used for a collection of faces, points and edges which bound some significant part of an object, such as a slot or pocket, which it may be useful to identify for functional or manufacturing reasons. Thus, in these terms, a feature is something between the lower level entities of points, edges and faces, and the higher level entity of a complete object.

7.2.5 Model Validity

Let us return to the question of validity of solid models. It is quite clear that any voxel or octree model must represent a valid object by its very nature.

For set-theoretic models, it is not difficult to show that provided we start with valid three-dimensional primitive objects, and only use regularised set operators, it is only possible to build up valid three-dimensional objects.

In the case of boundary representation models, the question of validity is more complicated. The *Euler-Poincaré formula* states that for a collection of solid objects

$$V - E + F - H = 2(M - G) \qquad (7.8)$$

where V is the number of vertices, E is the number of edges, F is the number of faces, H is the number of hole loops, M is the number of objects and G is the genus (the total number of through holes). A hole loop is a ring of edges lying completely within a face of an object, while a through hole is typified by the hole between the handle and body of a cup. For example, the object shown in Fig. 7.10 has one through hole,

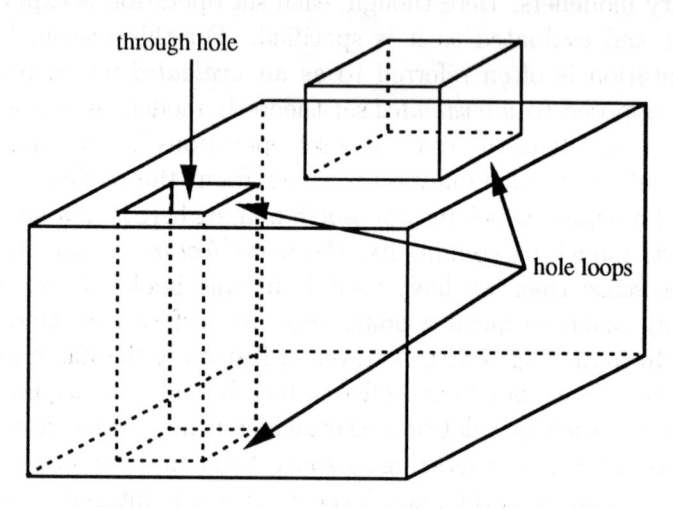

Figure 7.10: A through hole and hole loops

and three hole loops, one at either end of the through hole and one around the base of the protrusion. For this object

$$V = 24, \quad E = 36, \quad F = 15, \quad H = 3, \quad M = 1 \quad \text{and} \quad G = 1, \qquad (7.9)$$

which satisfies the Euler-Poincaré formula as expected.

It is possible to ensure the topological consistency of a boundary representation model by means of so-called *Euler operators*. Euler operators are used to build up objects by taking a consistent model as created so far into another, new, consistent model. For example, an operator which adds a single edge and a single vertex at the same time to a model has this property, because E is increased by one, and so is V, which leaves the left hand side of the Euler-Poincaré formula unchanged. By exclusively using Euler operators when topological changes are made to a model, topological consistency can be guaranteed.

However, from a geometrical point of view, we must still be careful. For example, the two-dimensional objects shown in Fig. 7.11 both have

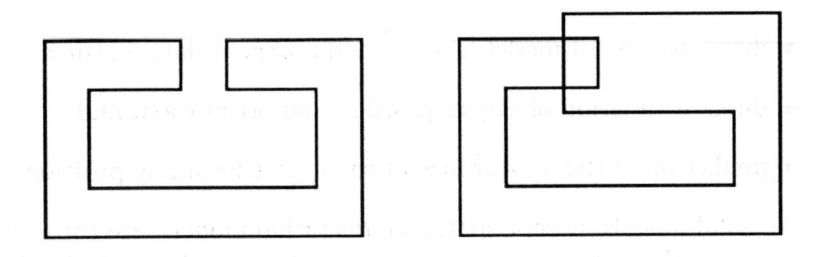

Figure 7.11: Geometrical invalidity

the same topology, and are topologically valid objects. Nevertheless, they have different geometry, and the one on the right intersects itself. The only way of avoiding geometric inconsistencies of this type is to check for unwanted intersections between the geometric elements.

7.3 Solid Models for Computer Vision

For a computer vision system to be able to accurately recognise, locate and inspect an object a three-dimensional geometric model of the object must be available for interrogation within the vision system.

Many different three-dimensional geometric modelling methods have been discussed in this Chapter. Most of these methods were originally developed for computer graphics and computer aided design applications where producing a picture of an object is important, although other applications have included the machining of solid objects. However, for object recognition and inspection tasks, geometric reasoning about certain feature characteristics is more important than producing a picture of the object. Not surprisingly, some of the modelling methods are not wholly suited to such tasks. The applicability of the geometric modelling methods described above to computer recognition tasks is discussed here. Let us consider the properties that are desired from a solid model for such tasks.

7.3.1 Desirable Model Properties for Vision

Typical tasks required when using a solid model for vision purposes include

- direct pairing of model features and observed data features,

- direct estimation of object position and orientation, and

- prediction of the appearance of the object from any position.

All solid models, except spatial enumeration models, are capable of representing three-dimensional structure and feature interrelationships that are needed for efficient matching and recognition.

However, some modelling methods yield this information more readily than others. Thus, in order that a model may be efficiently interrogated to produce the desired data, we may infer that it should satisfy the following requirements. These have been compiled from a variety of sources [16, 55, 83, 84, 85, 89, 156, 196, 199, 216, 227, 230] as well as from our own observations, some of which have already been stated in this Chapter.

- A solid model should represent the three-dimensional object itself rather than its appearance from a *fixed set* of viewpoints. It should be possible to predict the appearance of the object from *any* given viewpoint from its shape description.

- The geometric information about an object required during matching and inspection must be explicitly represented so that it is readily available rather than as a result of heavy computation. In particular, in order to reduce the complexity of the matching problem, the features represented by the model should be compatible with those extracted from the scene. We have far greater freedom of choice in selecting the model representation than the scene representation, since only certain features can be reliably

7.3. SOLID MODELS FOR COMPUTER VISION

extracted from a scene (see Chapter 4). In particular, since depth maps best provide sampled data of an object's surfaces, the model should readily yield surface information about the object.

- The modelling technique should be capable of representing a wide range of real world objects.

- The representation of an object should preferably be unique. This allows for efficient model interrogation strategies to be employed, reducing the time taken to resolve any ambiguities during the matching and inspection processes. If a unique representation is not provided, the number of potential different representations for a solid object should be as small as possible.

- Representation of fine levels of detail for some parts of the object should not have to mean that all parts of the object have to be represented at the same level of detail.

- Representation of fine levels of detail should not make it difficult to extract higher level descriptions of features or groups of features from the model.

- The model must be suitable as a basis for generating synthetic images that are similar to real sensor data, for purposes of calibration, testing of algorithms, etc.

- For purposes of inspection, it must be possible to extend the basic models to include representations of geometric tolerances (see Chapter 11).

7.3.2 Which Representation is Best?

Given the above discussion of model representations, it is clear that only the set-theoretic and boundary representations have the potential to be developed for computer vision use. We shall now consider them a little further, with emphasis on their uses in computer vision.

General set-theoretic modellers have the drawback that objects cannot be uniquely modelled, although some have been developed [2] that, by constraining the representation structure, do not suffer from this problem. However, methods for extracting geometric surface information, necessary for vision tasks, are computationally intensive when using set-theoretic modellers[16]. It is also very difficult to represent geometric tolerance information [200] in set-theoretic representation.

Boundary representations explicitly and compactly represent surface information which greatly simplifies any computations required during the matching stages. It therefore seems logical to adopt a boundary representation scheme to model objects which are to be matched and inspected. This observation is further justified in that intensity images are strongly dependent on object surface geometry [156] while depth images are also directly related to the surface geometry. Therefore for any general matching or recognition method using depth or intensity images, boundary representations appear to be the natural choice. However the use of parametric spline based patch representations of surfaces in the modeller should be avoided [83]. Matching to such sets of patches would require deducing where one patch stops and the next one starts. Because such patches would typically be smoothly joined together, deciding which points in the image belong to which patch would be a difficult, if not impossible, task. Thus, where possible, each face of the object should be represented by a single implicit equation. This point ought, hopefully, to be taken taken into consideration when designing objects which are to be manufactured and then recognised and inspected automatically by vision systems.

In conclusion we deduce that for most computer vision applications boundary representation provides the most promising method of modelling solid objects. We shall return to the topic of solid modelling strategies in Chapter 11 where methods for representing geometric tolerance information will be discussed in connection with inspection of objects.

Chapter 8

Model Based Matching

8.1 Introduction

Probably the most important stage in any intelligent vision system is the recognition stage. Here the system seeks to match inferences that have been made about the observed data to stored facts about objects or classes of objects that it expects to see. In the cases in which we are interested, these facts are stored in the form of three-dimensional geometric models of the type described in Chapter 7, leading to *model based matching*. Usually these models will have been directly designed using computer aided design systems, or perhaps in simple cases will have been constructed by hand from engineering drawings. An alternative approach, however, is to compile models by presenting sample objects to the vision system, using a learning method.

Over the last few years many model based matching strategies have been developed. These have been primarily concerned with recognising instances of objects from a database of stored object models. However, most also produce information about the position and orientation of the observed object in the form of a transformation matrix that maps between the model coordinate system and the scene coordinate system. Here we discuss a range of such methods. Those chosen reflect basic trends in matching strategies employing different primitive segmentations, as discussed in Chapter 4. Our aim is to give a brief overview of the fundamental approaches that have been applied to model based matching. A more complete discussion of current strategies may be found in the review papers by Besl and Jain [16], Chin and Dyer [55] and Brady *et al.* [31].

As already mentioned, object recognition is a problem of matching a scene description to a model description. At some stage in this process there comes a point where we must *pair off* corresponding scene and model features which may match. This introduces one of the major problems of model based matching: there are many possible pairings, which in turn means that a very large number of combinations of pairings must be considered in seeking the correct full match. Thus, a major issue is to find ways of *efficiently* searching for the solution.

The methods that we shall describe have been developed to solve a variety of application problems. Consequently each method will be suitable for only a few tasks that may require a recognition system. Some methods only work for simple object types such as polyhedra, while others work for fairly complex objects. Such factors clearly determine the nature of the segmentation data required and also the type of object model adopted. The nature of the three-dimensional (or two-dimensional in some cases) acquisition methods also affects the approach to the matching method. All matching methods must allow for imperfect sensor data and consequently must allow for possible errors in matching.

Another factor that affects the choice of method is the number of objects expected to be present in the scene, which again is dependent on the intended application. Some approaches need to deal with fairly cluttered scenes containing several different objects out of which one or more may need to be recognised. Others may operate in fairly constrained environments where only one or a few objects are visible. In any cluttered environment the problem of *occlusion* arises where features of one object may obscure features of another object from the view of the cameras. When objects with concave edges are present, features may obscure other features of the same object — this is known as *self-occlusion*.

In the following Sections we shall give brief algorithmic details of some popular matching strategies. In the next Chapter we shall give further practical details of the implementation of some of these strategies. We shall also develop a matching algorithm applicable to geometric inspection and related tasks, the use of which will be described in

Part 3 of this book.

8.2 The Nevatia and Binford Method

This method [178] was one of the first to consider object recognition using three-dimensional data. The method uses generalised cylinders (described in Section 4.6.1) both to describe the object model and the surfaces extracted from the scene.

The object model and scene features are represented in a *relational graph structure*. Relational graphs are a popular way of representing and recognising objects in computer vision and have been applied to many applications. We shall firstly briefly describe the relational graph structure in general terms and then return to the particular application of this structure to the Nevatia and Binford matching strategy.

Graph models can readily represent objects. A *graph* consists of a set of *nodes* connected by *links* (also called *edges* or *arcs*). In this application of graphs, each node represents an object feature (for example, a surface) and can be labelled with several of the feature's properties (such as size, shape, area, compactness, type of surface *etc.*). The links of the graph then represent relationships between features. Examples of these might be distance between centroids of the features, or the *adjacency* of the features, defined as the ratio of the length of the common boundary between the two features to the length of the perimeter of the first-named feature. Note that a boundary representation model as described in Section 7.2.4 is thus one particular kind of graph representing a solid object.

An example of a picture of an object (a mug) and a simple graph representation of it is shown in Fig. 8.1. The bottom of the cup is not visible so it is not included in the graph. We should also note that some relationships are *two-way*, such as distance, in that the relation does not depend on the direction of the link. Other relations, such as adjacency, do depend on the direction.

The recognition of an object is a matter of matching two graphs — the graph of the object model to the graph of the scene containing the object. Graph matching methods must take into account occlusion and overlapping objects. Clearly a graph derived from a solid model of the

(a) Picture of a mug

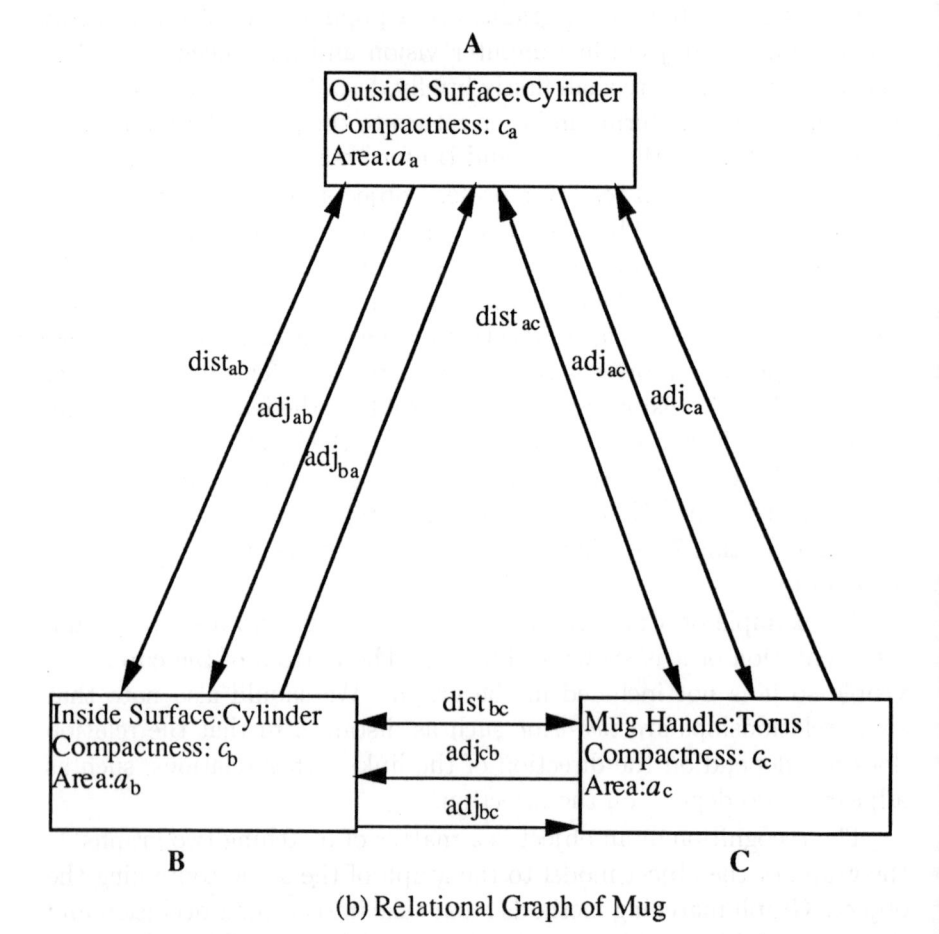

A
Outside Surface:Cylinder
Compactness: c_a
Area: a_a

B
Inside Surface:Cylinder
Compactness: c_b
Area: a_b

C
Mug Handle:Torus
Compactness: c_c
Area: a_c

$dist_{ab}$, adj_{ab}, adj_{ba}

$dist_{ac}$, adj_{ac}, adj_{ca}

$dist_{bc}$, adj_{cb}, adj_{bc}

(b) Relational Graph of Mug

Figure 8.1: Picture of a mug and its simple graph representation

8.2. THE NEVATIA AND BINFORD METHOD

mug would contain the bottom which is missing from the view shown. In such cases the graph produced from the view would be a *subgraph* of the complete graph derived from the solid model. We need to find this subgraph to obtain a match. More strictly, the nodes and links form a subgraph, but because of occlusion and related problems, the labels on the nodes and links may have different values to those for the complete graph. For example, a reduced area will be seen for any face which is partly occluded.

Let us return to consider how Nevatia and Binford perform their matching strategy. They use a surface model with generalised cylinders as features, arranged in a relational graph. The features are collected into sets called *ribbons* which contain groups of three-dimensional edges. Each group is hypothesised to form the boundary of a single generalised cylinder.

A region where several ribbon connect is represented by a *joint*. The representation of each joint contains a list of connected ribbons. This list is ordered, placing the longest or widest ribbons first. The longest (or widest) ribbon is said to be *distinguished*.

For each view of the object a relational graph is constructed with joints as nodes, and with links corresponding to associated ribbons. These links store various geometric properties of the ribbons, including axis length, average cross-section width, elongation and class (cone or cylinder).

The graph representing the view is then matched to the model description in two steps:

1. For each distinguished ribbon three properties are chosen to drive the matching. These are:

    - the *connectivity* of the ribbon,
    - the *type* of the ribbon — long or wide,
    - the *class* of the ribbon — conical or cylindrical.

   Initial matches are sought by looking for distinguished features in the image. If more than one match is found then the above properties are used to decide which matches are most likely and hence should be considered first.

2. The matches are then grown to include other ribbons so long as the consistency of the connectivity relations is retained. A graph formed from a view is, however, allowed to match the model graph if some ribbons in the model graph are not present in the view. This amounts to finding subgraphs and allows for partial occlusion.

8.3 The Oshima and Shirai Method

This method [181, 182, 216] was one of the first methods developed to deal with a variety of classes of objects and to be able to recognise a scene described by depth maps containing multiple objects in any orientation. Objects are considered to have planar surfaces, smoothly curved surfaces or both. The classes of smoothly curved surfaces allowed include ellipsoids, hyperboloids, cones, paraboloids and cylinders.

This method operates in two stages.

8.3.1 The Learning Phase

This method does not use any stored solid model representation to perform matching. Instead models are constructed from images of the objects to be recognised. Thus, a learning phase takes place in which known objects are individually shown to the system in different poses. A description of each object is built up, in terms of properties of surfaces and their relationships, to create a model of the object.

The description is made using the segmentation of curved and planar surfaces from individual images. The segmentation method employed by Oshima and Shirai [181] is similar to the one described in Chapter 4 except that many surface properties and interrelationships are also calculated. These then are represented in a relational graph. Typical surface properties for each region include:

- surface type (plane, cylinder *etc.*),

- number of adjacent surfaces,

- area,

8.3. THE OSHIMA AND SHIRAI METHOD

- perimeter,

- minimum, maximum and mean radii from the surface centroid to the boundary, and

- the standard deviation of the radii from the surface centroid to the boundary.

The relationships between surfaces are characterised by:

- distance between surface centroids,

- angle between best fit planes to this and neighbouring regions (if the surface is curved the best fit plane is still used for this measure),

- type of intersection — classed as convex, concave, mixed (if intersection is not concave or convex) or no intersection.

Only a discrete set of views are used to model an object. Thus, for a complete model description of an object to be compiled, all unique views of the object must be presented to the system. In other words, the object must be viewed from each possible *general position* as defined in Section 5.2.

8.3.2 The Matching Phase

This phase operates by trying to match the set of visible regions observed in a scene to each object view in a database of learned object descriptions. Not all surfaces present in the corresponding model view of the scene may be present in the observed scene description because of occlusion or poor segmentation, so a procedure that restricts incorrect matchings is required. This is achieved by initially driving the search with observed data and then, when a suitable set of scene features have been successfully matched with the model, model data is used to efficiently guide the rest of the search (see Fig. 8.2 which is based on Fig. 14.22 of Shirai's book [216]).

The image data driven part of the search involves constructing one or more *kernels*, each of which contains one or two regions that provide the best potential matches to an object model. The chosen regions

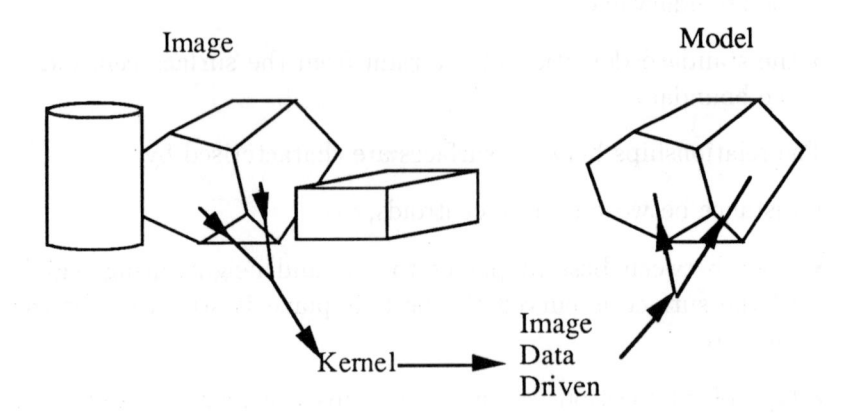

(a) Image driven first phase of match

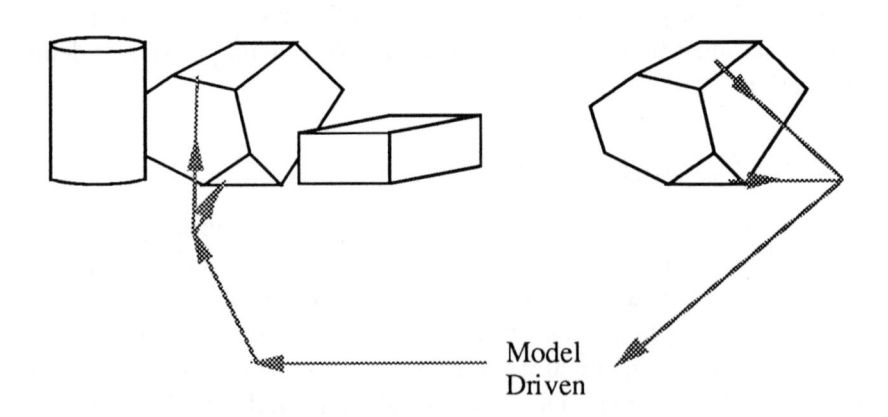

(b) Model driven second phase of match

Figure 8.2: Oshima and Shirai matching method

8.3. THE OSHIMA AND SHIRAI METHOD

are selected according to the criteria that they should, if possible, be planar, have large area, and have many neighbours. The criteria are relaxed until ideally two regions or at worst one region for the kernel is found.

The model database is then searched for matches starting with the best kernel. If no match is found then another kernel is selected. If only one match is found using this kernel then the model driven matching phase is invoked. If more than one match is found then other possible kernels are also considered. The system suspends any further matching for kernels which give more than one match until all other kernels providing a single match have been considered by the model driven phase. Note that at the end of the matching process there could be more than one interpretation for a given kernel.

The model driven matching process of the second phase involves searching the scene around the kernel for regions corresponding to those in the model. The search is controlled by a similar type of depth first tree search to that which we shall describe in detail later in Sections 8.6 and 8.8.

At all stages regions are considered to be matched if a *similarity function* (see Oshima and Shirai's paper [182] for details), based on region properties and relationships, is below a certain threshold.

When a match for a particular kernel has been found, a candidate body is said to have been found in the scene. The matching process continues until all scene regions have been tested for participating in a match. If only one candidate body is found in the scene then it is accepted as the desired match. Otherwise the scene requires further interpretation. If there are multiple candidates, they are examined to filter out inconsistent interpretations: two candidate bodies must be inconsistent if they share a region. For each kernel found a candidate body is rejected if one or more of its features are included in another candidate body which is consistent with respect to the same kernel. At the end of this process, hopefully one or perhaps more sets of mutually consistent interpretations will remain. Each such combination gives a possible interpretation of the scene from which position and orientation information can be calculated.

8.4 The *3DPO* Method

The *3DPO* (Three-dimensional Part Orientation) vision system [26, 27, 121] was developed by Bolles and his colleagues to recognise and locate parts from depth maps for the purposes of directing a robot arm to grasp a wide range of moderately complex industrial parts. To achieve this goal they argued that most current vision strategies did not provide the required generality, being based upon simple generic object types such as polyhedra, cylinders, spheres or surface patches. Instead they adopted an entirely different approach that could recognise objects using only a few features or feature clusters that highly constrain the object (such as a circle with a fixed radius) and consequently the size of the solution search space. These ideas are now known as *local feature focus* methods.

The matching process consists of five stages:

Primitive Feature Detection — Here low-level features are segmented and linked together. Edges are detected in the three-dimensional data using discontinuity and zero-crossing techniques (see Chapter 4). Edges are classified as being convex, concave or step edges.

Feature Cluster Formation — The aim of this stage is to form clusters of (primitive) features that can be used to hypothesise the type, position and orientation of the object. Thus coplanar edge clusters are formed and circular arcs are separated out, for example.

Hypothesis Generation — A search is made of the database of models to find a match. The strategy used here is to test individual clusters first and if these are not sufficient then groups of clusters are tested. It is at this stage that the feature focus approach is used by searching for object specific clusters such as circles of a certain radius.

Hypothesis Verification — Next, any hypotheses formed are tested by looking for additional features in the image that are consistent with each given hypothesis.

Parameter Refinement — In this final stage a given verified hypothesis is interrogated again in order to produce a more accurate location for the part. Averaging or least squares techniques (see Appendix B) are applied during this step.

This matching process uses an extended solid model. Here a form of boundary model (see Section 7.2.4) is augmented with a *feature classification network*. The feature network classifies features according to type and size, and groups features that share common surface normals, common axes (for cylinders) *etc.*

8.5 The *ACRONYM* Method

Brooks [43, 40, 41] has developed a vision system, *ACRONYM*, that recognises three-dimensional objects occurring in two dimensional images. The examples given in the papers involve recognising aeroplanes on the runways of an airport from aerial photographs. He uses a generalised cylinder approach to represent both stored model and objects extracted from the image. The representation is similar to that described in Section 4.6.1, restricted to simple polygonal and circular cross-sections. As in Nevatia and Binford's method in Section 8.2, a relational graph structure is used to store the representation. Nodes are the generalised cylinders and the links represent the relative transformations (rotations and translations) between pairs of cylinders.

The system also uses two other graph structures, which are constructed from the object models, to help in the matching strategy:

Restriction Graph — This graph represents constraints on classes of objects, in order to make generic part descriptions more specific. An example given in [40] considers classes of electric motors. These can be described by a generic motor type which is then divided into more specific classes of motors such as motor with a base, motor with flanges. These can then be further described in terms of functional classes (dependent on use) such as central heating water pump or gas pump. Additional constraints are allowed to be added to the graph during the recognition process.

Prediction Graph — Links in this graph represents relationships between features in the image. The links are labelled *must-be*, *should-be* or *exclusive* according to how likely it is that a given pair of features will occur together in a single object.

Since the image is only two-dimensional account must be taken of the projection of the three-dimensional object features into two dimensions. Brooks uses two descriptions for two-dimensional features:

Ribbons — See Section 8.2. These are used to describe a projection of a body made of generalised cylinders. Note that Nevatia and Binford use a three-dimensional form of ribbon.

Ellipses — These are used to describe the projection of the ends of a generalised cylinder. For ends of a circular cylinder the projections are exactly ellipses. For polygonal cross-sections, ellipses can provide a description of the ends by fitting the best circle or ellipse through the vertices and noting the projection of this shape.

The matching process is performed in two stages:

1. The image is searched for local matches to ribbons derived from the stored model. Such instances of ribbon matches are grouped into clusters.

2. The clusters are checked for global consistency in that each match must satisfy both the constraints of the prediction graph, and the accumulated constraints of the restriction graph.

This work has been extended by Kuan and Drazowich [148] to permit use of three-dimensional images. The basic principles of *ACRONYM* are adhered to but surface properties are incorporated. Also, numerical feature measurements are used during matching (whereas there is a purely symbolic use of features in the original method). The surface properties are included in a multilevel representation of objects which is used in the prediction phase. The levels and the features represented at each level are:

Object Level — stores, for example, spatial relationships between object components, object dimensions, view point characteristics such as minimum and maximum size, and occlusion relationships between features.

Generalised cylinder level — stores cylinder contour, cylinder position and orientation, cylinder dimensions, relationship between edges *etc*. This level is very important since generalised cylinders are the basic representation of objects in the system. Specific cylinder properties are therefore kept separate from more general surface characteristics.

Surface Level — describes surface type, edge boundary information, and spatial surface relationship.,

Edge Level — records edge types (step, convex, concave *etc*.).

8.6 The Grimson and Lozano-Perez Method

This method was originally designed to match sparse oriented surface points [93, 107, 108, 109, 110, 153] but has also been extended to match straight edges by Murray [52, 172, 173, 174, 175, 176] and others [189, 190, 191]. Here we discuss the method in terms of matching straight line edges in a scene to straight line edges of a model, based on the work by Murray, as edges are more suitable than points when dense depth data is available. However, the discussion is equally valid for the case of oriented surface points. The only differences are that the points are matched to planar model faces and the implementation of some geometric constraints is slightly different.

The method requires the following assumptions:

- The image data have been segmented to provide edges that are accurate to within some specified error. This error is specified in terms of the positioning of the end points and an uncertainty cone for the direction of the edge. Oriented surface point errors are specified in terms of a small volume of error.

- The objects to be recognised can be modelled as polyhedra.

This method is divided into two stages:

1. An exhaustive set of *feasible interpretations* of the sensed data is produced. Interpretations consist of pairings of each sensed edge with some stored model edge. Interpretations inconsistent with local edge constraints derived from the model are simply discarded. The edge constraints are said to be local since they are derived from knowledge of neighbouring edges only and not from overall knowledge gained from the whole view or object model.

2. Each feasible interpretation is tested for consistency with the object model. An interpretation is allowable if it is possible to provide a rotation and translation that places each sensed edge in correspondence with a model edge to within some specified error.

The two stages are now described separately.

8.6.1 Generating Feasible Interpretations

If an object model contains m edges and there are s sensed edges in the scene then there are m^s possible sets of complete interpretations which match each sensed edge with some model edge. Only a few interpretations lead to satisfactory solutions. Even for simple three-dimensional objects the number of possible interpretations is very large. For example, a view of a cube can show up to nine out of a possible twelve edges giving $12^9 = 2,985,984$ possible interpretations of which 24 are plausible. It should be noted that this number can be very much larger greater still when sampled oriented points are employed instead of edges. Clearly it is not possible to exhaustively test all these interpretations for a match so the basic idea is to exploit some simple, local geometric constraints on the sensed edges to minimise the number of interpretations that must be tested explicitly.

The possible pairings of sensed and model edges are represented in the form of a search tree [93], called an *interpretation tree*.

At each level of the tree, one of the edges from the scene is matched with each of the m possible edges in the model. Each node has m children representing taking the match proposed so far together with all

8.6. THE GRIMSON AND LOZANO-PEREZ METHOD

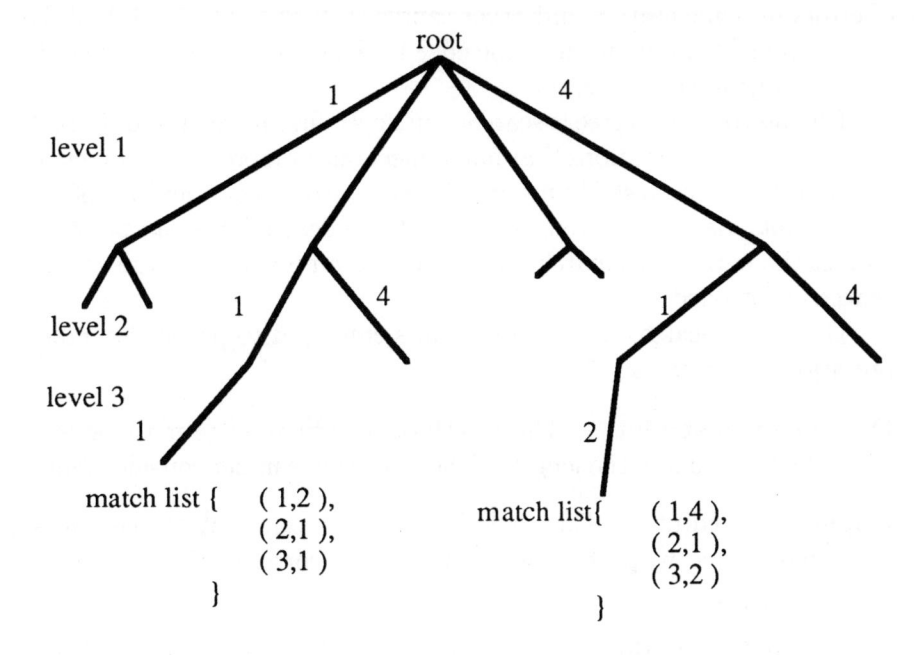

level 1

level 2

level 3

root

match list { (1,2), (2,1), (3,1) }

match list { (1,4), (2,1), (3,2) }

Figure 8.3: Interpretation tree

possible matches for the current scene edge. The result is a tree where complete paths to the deepest level correspond to complete matchings of object and scene edges. In the example shown in Fig. 8.3, we assume (unrealistically) for the purposes of illustration that the model has 4 edges and 3 edges have been found in the scene. Thus the pairings $((1,2),(2,1),(3,1))$ corresponds to edge 1 in the scene matching edge 2 in the model, edge 2 in the scene matching edge 1 in the model and edge 3 in the scene matching edge 1 in the model. Note that because of errors in segmentation and other causes such as partial occlusion, it may be possible for more than one edge in the scene to correctly match a single edge in the model.

The interpretation tree is searched in depth first manner and pruned by rejecting interpretations that fail to meet one or more local geometric constraints. If the test is unsuccessful the current node and all of its descendants may be rejected as none of these descendants can possibly correct the current mismatch. In this way the potentially large search space is controlled.

The three local geometric constraints employed to prune the interpretation tree are:

Distance Constraint — The length of the sensed edge must be less than or equal to the length of the model edge under consideration.

Angle Constraint — The angle between two adjacent sensed edges must agree with that between the two corresponding matched model edges.

Direction Constraint — Let d_{ab} represent the range of vectors from any point on sensed edge a to any point on sensed edge b. In an interpretation which respectively pairs sensed edges a and b with model edges i and j, this range of vectors must be compatible with the range of vectors produced by i and j.

The constraints used if matching oriented surface points to planar model faces are simpler [110, 109]:

Distance Constraint — The distance between each pair of sensed points must be a possible distance between the corresponding faces.

8.6. THE GRIMSON AND LOZANO-PEREZ METHOD

Angle Constraint — The angle between each pair of oriented sensed points must be compatible with that between the known surface normals of the corresponding faces.

Direction Constraint — The vector linking two sensed points must lie in a direction which lies within the range of all possible vectors going between the respective surfaces they belong to.

An advantage of this constraint based approach is that all the model information required by the constraints can be computed off-line and stored in look up tables. Any information that is required for comparisons is then simply referred to as needed.

8.6.2 Model Testing

This second stage involves interrogating the sets of possible interpretations which are left after the first stage in order to firstly, estimate the transformation (rotation, scale and translation) between scene and model coordinates, and secondly, to check that all edges or surface points under the estimated transformation lie within acceptable bounds of the assigned edges or assigned faces.

The transformation is estimated by fitting the scene's description to the matched model edges. Two approaches can be applied to solve the problem. The first approach involves simply averaging the estimates of the transformation obtained for each matching pair. Details of the mathematics involved in this approach can be found in Grimson and Lozano-Perez's original paper [110]. Alternatively, to produce more stable results, an approach similar to the estimation of rotation and translation presented in Section 9.2 may be employed. This method is due to Faugeras and Hebert [77] and has been employed by Murray [174, 175, 176] and in the *TINA* vision system [191] (discussed in Section 8.7).

Related Methods

Apart from the Murray matching algorithm, several other approaches have been developed over recent years based on the Grimson and Lozano-Perez method. In particular Sheffield University's AI Vision

Research Unit (AIVRU) have extended this strategy [189, 190, 191] for their *TINA* vision system. This will be described in the next Section.

Lozano-Perez and Grimson have also developed their strategy to take into account overlapping and mismatching faces [153]. This is achieved by incorporating an extra branch at each node in the interpretation tree to allow for *null* matches. This technique is described in detail later in Section 9.4.5.

Ikeuchi [137] has developed a method for object manipulation tasks. He generates an interpretation tree from object views and then proceeds to orientate and locate objects in the scene using extended Gaussian images (see Section 8.14) and surface-edge relationships.

Recently, Brady and his colleagues have proposed a method of recognising objects described by parametrised models [34]. They aim to allow recognition of objects that do not have or do not require explicit or fixed representations. The example cited in the paper is that of a vision system mounted on an autonomous guided vehicle navigating around a factory environment. This system must recognise wooden pallets used for stacking. A pallet can have many valid forms which humans readily recognise. However, a system using fixed object models would only be able to recognise an observed pallet if it were lucky enough to have a model of that particular pallet. To overcome this a parametrised object model is used where each wooden slat's position s_i relative to some reference point is allowed to vary in the model, as shown in Fig. 8.4. The value of each parameter s_i is determined in the matching process along with the rotation and orientation.

8.7 The *TINA* Method

Sheffield University's AI Vision Research Unit (AIVRU) have developed a matching strategy [189, 190, 191] for their *TINA* vision system. Much of their work has recently been published as a book [165] containing a collection of papers related to the TINA vision system and other important aspects of three-dimensional model-based matching. Their method uses an approach similar to that of Grimson and Lozano-Perez except that the depth-first tree search is replaced by a hypothesise and test strategy. This *grows* feasible interpretations in order to improve

8.7. THE TINA METHOD

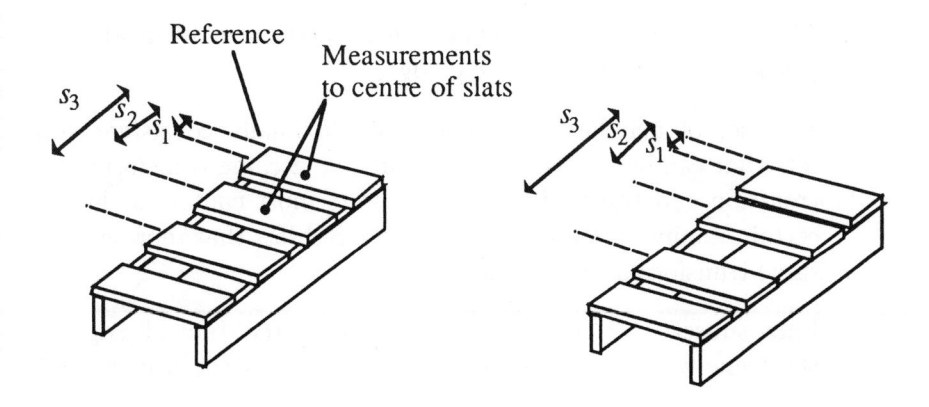

Figure 8.4: Parametrised representation of a pallet

the efficiency of the search. When some information in a scene may be missing (for example, when occlusion of an object occurs) tree searches tend to be combinatorially explosive [93, 107, 109]. The hypothesise and test strategy is driven by selecting features from the image based on feature sizes and neighbourhoods [189]. This hypothesise and test strategy is similar to the feature focus methods used in the 3DPO matching algorithm discussed in Section 8.4.

The system matches three-dimensional edges (obtained by the use of passive stereo) to three-dimensional edge based models. The matching algorithm uses similar pairwise local geometric relationships to those of Grimson and Lozano-Perez and Murray (see Section 8.6) to limit the search space further.

The basic method is as follows:

1. A edge is chosen from the model to *focus* the search on. The edge is determined on the basis of its length and the relative length of other edges within a predefined distance from it.

2. A set of edges close to the focus edge that exceed some predefined length is determined.

3. Potential matches to the focus edge are identified in the image on the basis of their length.

4. This set of potential matches is narrowed down by testing for consistency with neighbouring edges.

5. This set of matches is searched for maximally consistent subsets that can be represented by a single matching transformation (rotation and translation). The subsets must contain at least a certain minimum number of elements. Subsets that represent the same transformation are merged.

6. Each subset is extended by adding new matches for all edges in the scene which are consistent with the current set of matches.

7. The extended subsets are ranked on the basis of number and length of their elements.

8. A final transformation is recovered using the least squares method of Faugeras and Hebert (see Section 9.2).

8.8 The Faugeras and Hebert Method

This method matches primitive surfaces extracted from a scene to ones in the model. In the original approach [77] the method is described as coping with planar and quadric surfaces.

The basic approach is to try to match each primitive surface from the model, say m_i, to one segmented from the scene, s_i, to obtain a match pair (s_i, m_i). This is achieved using a tree search similar to the interpretation tree described earlier. However, an estimation of the transformation (rotation and translation) describing the mapping from model to scene is used as a constraint to control the search as it proceeds, in contrast to the previous methods based on the Grimson and Lozano-Perez method (see Section 8.6) where the transformation is only estimated after matching is completed.

From a small initial set of match pairs, the match list, an initial estimate of the transformation is made. Pairs of primitives from the scene and model are then added to this list if they are compatible with the match so far. Each time a successful match is found a new estimate of the transformation is calculated, and the process is continued until all scene primitives have been matched. If at any stage no possible

8.8. THE FAUGERAS AND HEBERT METHOD

match pair can be found then a backtracking mechanism is used. A match is considered to be successful if it is consistent with the existing transformation within an allowed tolerance.

Thus, the method exploits the *global* constraint of *rigidity* under the transformation to control the potentially combinatorially explosive search space. This is a major difference from the Grimson and Lozano-Perez method. The resulting paradigm of the Faugeras and Hebert method is that of recognising whilst locating, which is controlled mainly by global geometric constraints and not by testing a set of locally generated feasible interpretations. The use of global geometric constraints to control the search is generally more efficient [16, 80].

At each step the estimation of the transformation is found by solving a least squares minimisation problem, which will be further discussed in Chapter 9.

A cheap constraint is also employed at each level of the search in order to eliminate grossly mismatching pairs more quickly. The purpose of these tests is to provide a rough estimate of the geometric consistency of the new pair with the current match list. This test can be either or both of:

1. The angle between a matched primitive in the scene and the current scene primitive is equal to the angle between the corresponding primitives in the model to within a given tolerance. Any previously matched scene primitive can be used in principle to implement the constraint although usually one of the first matched primitives is employed since we can order scene primitives before the matching process to make the matching process more efficient. This can be done by choosing the best segmented features or the features most likely to define the matching transformation completely early on in the search or both. This is discussed in more detail in Section 9.4.1.

2. The angle between the current scene primitive and the current model primitive rotated by the current estimation of the rotation is equal to zero, to within a given tolerance.

This method has been used in many different computer vision applications. It has been employed in matching in the two-dimensional domain [4], matching different depth maps [5] and matching using simple occluding (or jump) edges and surface primitives [114]. In the last case Hebert and Kanade argue that because segmentation of high order surface patches is unreliable, low level primitives should be used first to obtain a small number of possible solutions which are then tested using a detailed surface matcher. They employ simple occluding edges which are easily and reliably calculated to limit the number of possible interpretations.

The basic Faugeras and Hebert method will be examined in greater detail later in Chapter 9, as it forms the basis of our inspection system described in Part 3 of this book.

8.9 Relaxation Labelling Methods

We have already seen some applications of relaxation labelling in Chapter 5 on reasoning with images. The relaxation labelling approach to model based matching is no different. Matching is posed as a labelling problem, where a model primitive m_i is labelled with (*i.e.* matched to) a scene primitive s_i. Region based measurements of both a numerical and topological nature are normally used to assess the goodness of match. Each primitive is given a quality measurement, normally a probability, for the likelihood of it labelling each model primitive. Starting from these initial measurements, the goal of the relaxation technique is to reduce iteratively the ambiguity and disagreement of the initial labels by employing a coherence measure for the set of matches. This may be achieved by the compatibility of each primitive with its neighbourhood, and iteratively increasing the size of the neighbourhood. Relaxation labelling problems may be easily visualised as graph labelling problems. Many methods have been designed using the basic labelling techniques [19, 75, 128, 208]. We shall describe one due to Bhanu [19] further.

This method takes a scene description consisting of segmented planes and curved surfaces, and matches it to a model constructed from multiple depth maps of known objects. The surfaces are matched using a relaxation scheme called *stochastic labelling*. Using surface features derived during the segmentation the initial face labelling probabilities are computed. The surface features used are similar to those employed by the Oshima and Shirai method in Section 8.3. A sorted surface neighbourhood table is used where neighbours are specified in order of their area (largest first).

Two relaxation stages are then applied. The first relaxation stage involves iteratively maximising a compatibility measure based on a single largest area neighbour function. The second relaxation stage uses a two largest area neighbour function to iteratively optimise the labels obtained from the first stage. Both compatibility functions are based on a weighted sum including the distance between neighbouring face centroids, the ratio of the areas of neighbouring faces, and the difference in face orientations. After the second stage is completed translation and rotation information is computed.

8.10 The *SCERPO* Method

The matching techniques in the *SCERPO* (Spatial Correspondence, Evidential Reasoning and Perceptual Organisation) vision system, developed by Lowe [150, 152, 151], are based on psychophysical observations of the human visual system. Lowe argues that *perceptual groupings* of features of interest suggest something *non-accidental* about an image. Lowe also suggests that often too much emphasis is placed on three-dimensional images and that humans can perform recognition when depth information is not present: much three-dimensional information can be inferred from two dimensional scenes. For example, we have seen similar approaches in line labelling in Section 5.2: if we see an L-junction of edges in a scene we would expect the three-dimensional edges to meet in a similar manner. However, if the edges meet in a T-junction, in three dimensions this generally means that one edge occludes the other.

Another important aspect of Lowe's work is that he notes that practical (human or machine) vision necessarily involves a very large database of objects and consequently much effort is required to search for matches. He suggests that perceptual groupings may provide an efficient way to search such a database.

The basic approach of the *SCERPO* matching method [151] is as follows:

Edge Detection — Potential edges are detected in of a two-dimensional, greyscale image using the Laplacian of a Gaussian (LOG) function to detect edges occurring at zero crossings in the image, as discussed in Section 4.5.3. The Sobel edge operator, also discussed in Section 4.5.3, is then applied to the pixel array produced by the LOG operator and the resulting estimate of the gradient value is stored for each edge point. This is because if only low changes in gradient intensity occur across the original image many zero crossings may occur which do not all correspond to actual edges. The Sobel gradient estimate is used detect those zero crossings which are more likely to correspond to edges, being those with high gradient magnitudes.

Line Extraction — Straight lines are extracted using edge linking techniques similar to those described in Section 4.5.4. Only lines that have a sufficiently low error in the fit of points to the line are retained. The lines are also indexed according to their endpoint locations and orientation in the scene.

Perceptual Grouping — The lines are grouped according to collinearity, connectivity (endpoints of lines lying within a certain distance of each other) and parallelism.

Model Matching — Each perceptual grouping is compared to structures in the three-dimensional object model that are likely to produce such groupings in the two-dimensional image. In order to reduce the large number of possible matches the groupings are ordered. Groupings containing more line segments are likely to produce fewer matches. Thus only groupings containing more than three lines are considered first. Groupings that completely define the viewpoint from the initial match are also chosen first as these provide tight constraints that speed up the subsequent verification process. We shall develop ideas similar to these, but applied to planes, in the next Chapter.

Model Verification — Having obtained a set of initial matches, the match set is extended by using the object model to predict lo-

cations of other features in the image and to verify t
of these features. This is performed in a similar way
described earlier in Sections 8.4 and 8.7.

8.11 The *IMAGINE* Method

IMAGINE is an object recognition system developed by Fisher [82, 84, 83, 85, 86, 87]. The system recognises objects from a surface description obtained from segmentation routines, although the *IMAGINE* system itself does not perform any segmentation — it merely assumes that this process has been carried out previously.

In some ways this method can be seen as combining a surface approach similar to that used by Faugeras (Section 8.8) with recognition techniques similar to those used by the of the *ACRONYM* (Section 8.5) system.

For matching, the scene is required to have been segmented into regions of similar principal directions of curvature and labelled boundaries between regions as discussed in Section 4.6.7. The boundaries are labelled according to whether a depth, surface orientation or surface curvature direction discontinuity occurs. Object models are based on quadratic surface patches.

Using the above data, the *IMAGINE* system first tries to group surfaces together to form higher order descriptions and then matches to the model as follows:

Surface Completion — The idea behind this stage is that the initial segmentation may not have yielded complete surface segments so this process attempts to reconstruct parts of the surface that have been obscured or not segmented properly. This is achieved by joining or extrapolating surfaces across occlusion boundaries. Fisher [85] argues that this enables more information to be deduced about the shape of each surface thus giving better evidence during the recognition stage.

Surface Clustering — Here surfaces are grouped together with the aim of using the clusters to aid the subsequent model matching process. This process is performed without any knowledge of any object identity to which surfaces may belong. A surface cluster is formed by finding closed loops of boundary segments.

Extraction of Three-Dimensional Properties — Having formed surface clusters, three dimensional-surface properties (*e.g.* descriptions of surface curvatures and area) and relationships between surfaces in the clusters are determined.

Model Invocation — The database of object models is searched for possible instances of the surface clusters formed above. This is achieved using a relational graph where the nodes hold surface information and the links represent geometric constraints.

Hypothesis Construction and Verification — An attempt is made to recognise the whole object at this stage. The invocation provides an initial hypothesis on model to image correspondence. Object pose (orientation and translation) is estimated by considering matched model–image pairs of surface vectors (surface normals and principal axes of curvature, which are assumed to be approximately constant for the regions produced by segmentation) and other surface features using methods similar to those described in Chapter 9. From this estimate, object features that are likely to be occluded, partially occluded or visible can be deduced. The image data is then searched for evidence that the remaining image features not yet recognised do correspond to features expected to be visible and that features predicted by the model to be invisible are indeed not present.

Fisher has produced two versions of the system, *IMAGINE I* being described above. *IMAGINE II* is a new, redesigned system based on experience gained from the former. For *IMAGINE II* Fisher has developed a solid modelling system [84] solely designed for recognition tasks, in which model invocation, frame reference (relative positioning and orientation of objects) and matching are more important than being able to provide pictures of the object. Fisher's system takes the principles discussed in the Chapter on modelling (Chapter 7) somewhat further by allowing the model to be *suggestive* rather than explicit. In particular, it is only meant to provide an overall clue to the shape of an object and not to provide great geometric detail. In particular, surfaces may be approximate in shape, and only aspects relevant to computer vision tasks are modelled. (These have been discussed in Chapter 7.) Features that are difficult to segment or minor in detail are omitted. Finally, the model does not necessarily have to be completely closed or connected.

Other improvements made for *IMAGINE II* [85, 86, 87] are concerned with improving the representation of and reasoning with descriptions of surface features throughout the various phases of the system. In order to achieve this a new representation VSCP (Volume, Surface, Curve, Point) is incorporated in the modelling system [84, 85]. This is basically an extension of the standard surface, edge, vertex representation used by boundary representation modellers.

8.12 Fan's Method

This method [73, 72] has various similarities with many other methods described in this Chapter. The method matches a model to images that have been segmented to produce surface patches and boundaries or edges. Two types of boundaries are used, *jump* edges where surfaces are discontinuous, and *crease* edges where surface normals are discontinuous.

A relational graph structure (see Section 8.2) of the whole scene is formed in which nodes represent the surface patches and the links give their adjacency. The nodes are labelled with surface properties such as area, orientation of the surface and classification of the surface based on its curvature. The links are labelled with the type of boundary and an estimate of the likelihood that the two surfaces lie on the same object.

Object models are also made up of graphs and multiple graphs are defined for each model, each one specifying a distinct view of the object.

By considering the likelihood that the surfaces lie on the same object the graph from the image is divided into subgraphs each of which is hypothesised to form a single object in the scene. The first stage of the algorithm identifies likely model graphs for the particular view the scene depicts. This is done by ordering the model view graphs according to a quick and simple comparison to the scene graph. Clearly, it is undesirable to waste time considering an extremely large possible number of scene to model matches at this stage. Thus, a heuristic search is used in which each model view graph is compared to the scene graph and differences in the number of nodes of all surface types, the number of planar nodes and the area of the largest visible node are used to order the model graphs.

A tree search is then invoked to try to exactly match the scene graph with each object model subgraph. This is done by a pairwise comparison of model and scene nodes in a similar manner to that adopted in the Grimson and Lozano-Perez (Section 8.6) and Faugeras and Hebert (Section 8.8) approaches. The algorithm works as outlined below:

Generation of Pairs — For every possible model-scene pair of surfaces a check is made to determine whether the pairing is compatible. This achieved by comparing the visible areas and average curvature of the pair of surfaces.

Creation of Matching Sets — This is done using a depth first tree search going down to a depth of four levels. If by this level the compatibility between the pairs for a given matching set is not high enough the set is rejected. Otherwise the set is kept for further consideration. Rejection of sets at this stage is done so that the subsequent search is only focussed on the most promising paths.

Full Depth First Search — The small sets retained in the previous step are now fully searched in depth first fashion to find complete matches. Both here and in the previous stage the search is constrained using compatibility constraints between pairs.

Fine Modification — The estimate of the matching transformation formed in earlier stages is refined by applying the stronger constraint of rigidity, and by applying stronger bounds on the allowable error of fit for each match pair, excluding from the match faces not satisfying these new limits.

Rejection or Acceptance — Based on the measure of goodness of match a decision is made to accept or reject the match or to choose the best match from a set of matches.

Throughout the above process various constraints are applied. These are similar to those discussed in earlier Sections of this Chapter and are based on relationships with neighbouring features such as direction and distance.

A final stage uses the matching information to partition the graph of the scene into distinct objects.

8.13 Hough Transform Methods

The Hough transform has been popular in pattern recognition problems for many years and its basic operation is discussed in Section 4.5.4. Initially it was designed to recognise two-dimensional straight lines but it can be generalised to recognise curves in two-dimensional space. Following this it is natural to extend the Hough transform into three-dimensional space for recognising three-dimensional objects. A few techniques have been designed to achieve this [6, 30, 218].

8.14 Extended Gaussian Image Methods

The theory of extended Gaussian images (EGI) has been applied in attempts to solve the matching problem [42, 122, 124, 136, 138, 137]. We have discussed these matching strategies in Section 5.5.2. Some researchers like Ikeuchi [137] have used EGI in conjunction with other surface properties to attempt object matching (See Section 8.6)

8.15 Summary

We have detailed quite a selection of matching methods in this Chapter. From the above discussion it is clear that they share quite a few common strategies.

The combinatoric explosion of possibilities to be considered when searching for the solution is a major problem. Two data structures have been typically used to represent the possible solution space. These are the relational graph and the interpretation tree. However, blind searching of either of these structures makes the matching problem intractable. Consequently, geometric constraints are applied to *prune* the search space and to make the problem more manageable. This approach can be taken a stage further, and a hypothesis and test search strategy can be used. Here we apply some knowledge (*a priori* or otherwise) to predict what we expect the data to represent after taking into consideration our present beliefs about the scene. If the prediction is correct we are, hopefully, a step nearer the solution.

Another common thread shared by some of the more recent algorithms is the use of edge information, which is relatively cheap to compute compared to surface information, to initially reduce the search space. Only then are more robust surface matching routines applied to obtain the final, accurate match.

No method we have described is perfect for every possible application of model based matching. Some of the limitations are summarised below.

Classes of Object Represented — Some methods can only reason with polyhedral models (the Grimson and Lozano-Perez method and derivatives), generalised cylinders (the Nevatia and Binford and ACRONYM methods) or restricted surface patch types (most of the other methods).

Sensitivity to Occlusion — Some methods cannot deal with occlusion very well or at all. For example, the extended Gaussian image approaches are sensitive to occlusion and are generally only suitable for convex objects.

Sensitivity to Segmentation Output — Some of the relaxation labelling techniques are particularly prone to problems caused by poor segmentation. This is basically because no global information is used in these methods. Global compatibility functions would be difficult to define and use. They are also very sensitive to initial matches because of this. Other methods such as the Grimson and Lozano-Perez method that also rely only on local feature information can suffer from related problems [109].

Nevertheless, each method may also have its advantages. For example the extended Gaussian image approach is fast and fairly straightforward. Many of the above methods, especially *ACRONYM, IMAGINE,*

8.15. SUMMARY

and *3DPO* have contributed ideas towards effective geometric reasoning methods that should prove useful in many vision applications in the present and future.

Chapter 9

Practical Model Based Matching

9.1 Introduction

This Chapter uses many of the concepts discussed in the preceding chapter to develop a model based matching algorithm. The ideas are taken further where needed and practical implementation details are considered. The particular application the algorithm is intended for is our automatic inspection system which will be described further in Part 3 of this book. In the discussion of the merits of our algorithm and other matters we therefore shift our emphasis towards the requirements of inspection. The major need is for obtaining accurate measurements and accurate positional information. For most general mechanical engineering inspection tasks we require features to be measured to within .01inch (0.254mm) and therefore our vision system, including matching algorithms, must be able to achieve this.

Whilst our major emphasis is on accuracy to satisfy the requirements of inspection tasks we stress that our matching algorithm is applicable to a wide range of model based matching needs. In many respects suitability for inspection can be regarded as the ultimate test of the accuracy of such a vision system, although recognition of objects is easier in inspection systems, as an object model is normally available. Thus the recognition phase merely involves testing the compliance of the scene data to the model, matching of scene and model features, and determining the position and orientation of the observed object. Even so, these are still not trivial tasks. The techniques we describe to enhance the accuracy of the matching algorithm add no real extra computational burden. Many other developments we describe generally improve the performance of the matching algorithm.

As we have already discussed in Chapter 4, planar surfaces are likely to be the most accurately segmented features from a depth map. We therefore propose to drive the matching algorithm using planar features. In consequence, we note that an approach based on the Faugeras and Hebert matching technique (Section 8.8) is most likely to give sufficiently accurate results in an efficient way, and we have therefore based our matching algorithm on this approach. However, an analysis of the approach shows that various improvements on the method can be made in order

- to improve the accuracy of the method,

- to improve the efficiency of the method, and

- to extend the method for reliable matching of non-planar surfaces.

The basic approach of the Faugeras and Hebert method was discussed in Section 8.8. This Chapter gives further implementation details, and then considers the modifications described above. Some of these are adaptations of ideas proposed by other authors; some of them are our own suggestions. Finally, practical results including the effects of these improvements are presented. Also, the quality of the output of the segmentation routine we use is considered.

Many of the modifications we will describe are not particular to any matching technique, especially if it adopts an interpretation tree (see Section 8.6) approach. In particular, in the next section, we describe a method of estimating the position and orientation of an object in the scene. This method can be used by any matching technique and is indeed presently employed in many vision systems.

Firstly we will briefly review the original Faugeras and Hebert matching approach. As described earlier, their method may be used to match planar and quadric surfaces. Their method for matching planar primitives is discussed below. The discussion in Chapter 4 pointed

9.1. INTRODUCTION

out that matching quadric surfaces is not desirable for many vision purposes, especially for high tolerance work like inspection. Details of their approach to quadric surface matching may be found in Faugeras and Hebert's original paper [77] and also [80].

The approach is to produce a list of matching pairs of planes, one from the scene, the other from the model, that contains all of the planes extracted from the scene. To start the process, two non-parallel match pairs are initially required to estimate the rotation between scene and model coordinates, and three match pairs which correspond to non-parallel planes are required for a translation estimation. Each successive match pair is then considered for addition to the match list on the basis of its compatibility with the current estimated transformation. If this match pair is accepted then the transformation is updated to include information from the new match pair. If the match pair is rejected then backtracking occurs and another match is sought. The whole search strategy can be easily implemented as a tree searching strategy (depth first) and this was discussed in Section 8.8.

In practice, in the original method, only the rotation component of the transformation is used for compatibility testing during most of the search. The translation part is only employed in the lower levels of the search to find the best solution from amongst several potential matchings. The reason for this is that the rotation provides the more powerful constraint and employing only this constraint saves time. The estimation of both the translation and rotation represents the majority of the computation time in the algorithm and for efficiency reasons these calculations should be kept to a minimum. Finding the translation also requires the estimate of the rotation component of the overall transformation. Note that grossly mismatching planes can be more quickly eliminated by using local consistency constraints than by using the rotation estimate constraint as described in Section 8.8. Most other mismatching planes can be eliminated by only employing the rotation part of the transformation with only certain symmetrically related faces of objects not being eliminated because of *rotational inconsistency*.

The method of estimation of this transformation is now considered. This estimation was originally described in Faugeras and Hebert's paper [77]. It has also been adopted by many other vision systems.

9.2 Estimation of the Transformation

At any stage the estimation is treated as a least squares minimisation problem. The transformation is assumed to map object model coordinates to the image coordinate system.

Let the transformation, T, be expressed as $t.R$, where t is the translation component and R is the rotation. This notation implies that the rotation is applied first and the translation second.

Any plane P_i can be represented by two parameters v_i and d_i, where v_i is the unit vector normal to P_i and d_i is the signed distance of P_i from the origin.

Let us assume the scene has been segmented into N planes and that the model to contains N' planes. Let $\{P_i(v_i, d_i) | i = 1 \ldots N\}$ be the set of scene planes and $\{P'_i(v'_i, d'_i) | i = 1 \ldots N'\}$ be the set of model planes.

Consider the matching M, where

$$M = \{[P_i(v_i, d_i), P'_i(v'_i, d'_i)] \, | i = 1 \ldots N\}$$

contains N pairs in appropriate order, linked by the transformation T which transforms the model coordinate system into the scene coordinate system.

Thus, the ideal transformation, T, if there were no errors, linking model plane $P'_i(v'_i, d'_i)$ to scene plane $P_i(v_i, d_i)$ is given by

$$\begin{aligned} v_i &= Rv'_i \\ \text{and } d_i &= v_i t + d'_i. \end{aligned} \quad (9.1)$$

In practice, as there will be errors, we wish to find the best estimate of T that can be obtained from the match set. Thus, values of R and t are found that minimise the error in the estimate of T (in the least squares sense), expressed as

$$\sum_{i=1}^{N} \|v_i - w'_i\|^2 + W|d_i - d'_i - w'_i t|^2. \quad (9.2)$$

Here, W is a weight factor. The $w'_i = Rv'_i$ represent the normals of the model planes after rotation, while the v_i represent the normals of the corresponding scene planes.

9.2. ESTIMATION OF THE TRANSFORMATION

The first term above represents errors in the estimate of the rotation, while the second term represents errors in the translation estimate. Each part is minimised separately. The weight factor is chosen so that the errors given by both minimisations are of the same order. For all practical examples given later in this book a value was chosen based on preliminary results with simple test objects using our vision system. A value of $W = 0.1$ was chosen which reflects that the errors in the translation part were about 10 times greater than the errors in the rotation part, as shown by the results in Section 9.7.

9.2.1 Estimation of the Rotation

The problem is to find an estimation of the rotation $\mathbf{R}$ which minimises

$$\sum_i \|\mathbf{v}_i - \mathbf{w}'_i\|^2. \tag{9.3}$$

A rotation may be represented in various different ways, for example using orthonormal matrices, Euler angles, an axis $\mathbf{a}$ and an angle θ, and *quaternions*. In the latter case the product $\mathbf{Rv}$ is represented as the product of two quaternions. A survey of representations for rotations can be found in the papers by Funda [92] and Rooney [205].

The first representation of a rotation uses a large number of interdependent variables while the second representation requires solution of non-polynomial equations. They cannot therefore be readily utilised in conjunction with least squares methods to estimate the rotation. Faugeras and his colleagues [77, 79, 80] give a method which uses quaternions to reduce the minimisation to a form that can be directly solved by classical least squares methods. This method has also been employed by Murray [172, 176, 173, 175] and in the *TINA* vision system [189, 191] to estimate the rotation part when using Grimson and Lozano-Perez type methods (see Section 8.6). These authors claim more accurate results than when using the original estimation technique. The method is as follows.

A quaternion may be represented by four real numbers. It is often convenient to divide these up into the pair $(\mathbf{w}, s)$ where $\mathbf{w}$ is a vector and s is a scalar. (For a summary of quaternion arithmetic see Appendix C).

A rotation **R** with axis **a** about an angle θ can be represented by the quaternion $\mathbf{q} = (\sin(\theta/2)\mathbf{a}, \cos(\theta/2))$. Note that if the signs of both **a** and θ are swapped, we have an alternative representation of the same rotation. In order to avoid this ambiguity we constrain θ such that $\cos(\theta/2)$ is positive. A vector **v** which is to be rotated is represented by the quaternion $(\mathbf{v}, 0)$.

The application of **R** to a vector **v** to form a vector **v′** is now found by calculating

$$\mathbf{v}' = \mathbf{R}\mathbf{v} = \mathbf{q} * \mathbf{v} * \bar{\mathbf{q}} \tag{9.4}$$

where $\bar{\mathbf{q}}$ is the conjugate of **q** and $*$ denotes quaternion multiplication (see Appendix C).

Assuming the axis of rotation is described by a unit vector, the rotation quaternion must also have modulus 1. The conjugate of **q** also has modulus 1. Thus the minimisation problem can be translated into quaternion space with the new constraint that $|\mathbf{q}| = 1$.

Therefore the quantity to be minimised in Expression 9.3 can be rewritten as

$$\sum_i \|\mathbf{v}_i - \mathbf{q} * \mathbf{v}'_i * \bar{\mathbf{q}}\|^2 \tag{9.5}$$

with $|\mathbf{q}| = 1$.

Since quaternions are multiplicative (see Appendix C), and we have constrained $|\mathbf{q}| = 1$, this may in turn be rewritten as

$$\sum_i \|\mathbf{v}_i * \mathbf{q} - \mathbf{q} * \mathbf{v}'_i\|^2. \tag{9.6}$$

Now $\mathbf{v}_i * \mathbf{q} - \mathbf{q} * \mathbf{v}'_i$ is a linear function of the four components of **q**, so there exists a 4×4 matrix $\mathbf{A}_i$ such that the components of **q** considered as forming a 1×4 matrix satisfy

$$\mathbf{q}\mathbf{A}_i = \mathbf{v}_i * \mathbf{q} - \mathbf{q} * \mathbf{v}'_i. \tag{9.7}$$

Therefore the quantity to be minimised, Expression 9.6, becomes

$$\sum_i \|\mathbf{q}\mathbf{A}_i\|^2 \equiv \sum_i \mathbf{q}\mathbf{A}_i\mathbf{A}_i^T\mathbf{q}^T \equiv \mathbf{q}\mathbf{B}\mathbf{q}^T \tag{9.8}$$

where $\mathbf{B} = \sum_i \mathbf{A}_i\mathbf{A}_i^T$, is a symmetric matrix. (^{T} denotes matrix transpose).

9.2. ESTIMATION OF THE TRANSFORMATION

Minimisation of Expression 9.8 represents a classical problem [100] where the solution q_{min} is the eigenvector corresponding to the smallest eigenvalue of B. The smallest eigenvalue also gives a measure of the error in the estimation of the rotation.

Let us now consider how to compute the matrix B. Let $q = (w, s)$ be a quaternion and v and v' be two vectors in $\Re^3$. Let a quaternion q' be defined by

$$q' = v * q - q * v'. \tag{9.9}$$

Then

$$q' = ((v' + v) \times w + s(v - v'), w.(v' - v)) \tag{9.10}$$

where $\times$ = vector cross product.

We may associate a matrix U^0 with a vector u in $\Re^3$ by defining it to satisfy the relation

$$u \times u' = u'U^0 \tag{9.11}$$

for every vector u' in $\Re^3$. In this case, if $u = (u_1, u_2, u_3)$ then it can be seen that

$$U^0 = \begin{bmatrix} 0 & u_3 & -u_2 \\ -u_3 & 0 & u_1 \\ u_2 & -u_1 & 0 \end{bmatrix}.$$

Also, for a given axis a and an angle of rotation θ, a rotation R of a vector v about a through θ is given by the Euler formula

$$Rv = v + (1 - \cos\theta)w \times (w \times v) + \sin\theta\, w \times v. \tag{9.12}$$

Now, combining the results of Eqns. 9.7, 9.10, 9.11 and 9.12, the matrix A is sought such that

$$q' = qA$$

which gives A, a 4×4 matrix, as

$$A = \left[\begin{array}{c|c} V^0 & (v' - v)^T \\ \hline v - v' & 0 \end{array} \right] \tag{9.13}$$

where V^0 is the 3×3 matrix associated with $(v' + v)$, as defined by Eqn. 9.11.

The rotation matrix **R** is easily recovered once the smallest eigenvector $\mathbf{q}_{min} = (\mathbf{w}, s)$ of **B** has been found by computing

$$\mathbf{R} = (\mathbf{I}_{3\times 3} + (1 - \cos\gamma)(\mathbf{W}^0)^2 + \sin\gamma \mathbf{W}^0)^T \quad (9.14)$$

where $\mathbf{I}_{3\times 3}$ = the 3×3 identity matrix,
γ = $2\cos^{-1} s$ (see Appendix C),
$\mathbf{W}^0$ = the matrix associated with the vector $\mathbf{w}/\sin(\gamma/2)$ (Appendix C)

Note however, that when using this rotation, for example to compute the translation estimate, it is usually simpler to continue using its quaternion form.

9.2.2 Estimation of the Translation

The problem is to find the best estimate of the translation **t** linking scene and model coordinates, according to the criterion obtained from Expression 9.2, which is to minimise

$$\sum_i |d_i - d'_i - \mathbf{w}'_i \mathbf{t}|^2 \quad (9.15)$$

In the original method of Faugeras and Hebert this was done using the pseudo-inverse method [77]. Matrices **A** and **Z** are formed, such that

A is the $N \times 3$ matrix : $[\mathbf{w}'_1, \ldots, \mathbf{w}'_N]^T$, where $\mathbf{w}'_i = \mathbf{R}\mathbf{v}'_i$
and **Z** is the $N \times 1$ matrix : $[d_1 - d'_1, \ldots, d_N - d'_N]^T$.

The error in **t** is minimised by choosing

$$\mathbf{t} = \mathbf{t}_{min} = (\mathbf{A}^T\mathbf{A})^{-1}\mathbf{A}^T\mathbf{Z}; \quad (9.16)$$

the corresponding error as given in Expression 9.15 is

$$e_{min} = \mathbf{Z}^T(\mathbf{Z} - \mathbf{A}\mathbf{t}_{min}). \quad (9.17)$$

In practice, instead of computing the inverse of $\mathbf{A}^T\mathbf{A}$, the linear system of equations formed by $(\mathbf{A}^T\mathbf{A})\mathbf{t}_{min} = \mathbf{A}^T\mathbf{Z}$ is solved.

Although the Faugeras and Hebert method basically provides satisfactory results, various modifications can be made to improve the method so that it is more accurate and efficient. The following sections detail our suggestions in this respect. Many of these are also applicable to other matching strategies.

9.3 Improving the Method's Accuracy

Two improvements can be made to give improved estimations of the translation; both can be applied together. These are described separately.

9.3.1 Replacing the Pseudo-Inverse

The pseudo-inverse method used for minimising the translation part of the transformation given in Expression 9.15 is in practice not very accurate, and is unstable in the presence of noise. The method is also unnecessarily prone to rounding errors.

This is a manifestation of a well-known problem in numerical analysis and arises since the *conditioning* of the matrix $\mathbf{A}^T\mathbf{A}$ is worse than that of $\mathbf{A}$ [180]. This frequently leads to difficulties which cannot be remedied once $\mathbf{A}^T\mathbf{A}$ has been formed. Detailed descriptions and error analysis of the pseudo-inverse method are well documented [10, 101, 102, 103, 104, 129, 180]. The nature of this problem is shown by the following example.

Consider, for the purposes of illustration, the case when

$$\mathbf{A} = \begin{bmatrix} 1 & 1 & 1 \\ \varepsilon & 0 & 0 \\ 0 & \varepsilon & 0 \\ 0 & 0 & \varepsilon \end{bmatrix} \qquad (9.18)$$

where ε is small. Then

$$\mathbf{A}^T\mathbf{A} = \begin{bmatrix} 1+\varepsilon^2 & 1 & 1 \\ 1 & 1+\varepsilon^2 & 1 \\ 1 & 1 & 1+\varepsilon^2 \end{bmatrix} \qquad (9.19)$$

Now, if ε is such that $\varepsilon^2 < \beta$, where β is the machine precision of the computer being used, then $\mathbf{A}^T\mathbf{A}$ will be represented as

$$\mathbf{A}^T\mathbf{A} = \begin{bmatrix} 1 & 1 & 1 \\ 1 & 1 & 1 \\ 1 & 1 & 1 \end{bmatrix} \qquad (9.20)$$

In such a case $\mathbf{A}^T\mathbf{A}$ is singular and, therefore, no inverse of $\mathbf{A}^T\mathbf{A}$ can be computed. Thus no solution to the minimisation problem can be found.

It is important to note that problems of the form shown in the example above are common in practice, since errors arising from the estimation of the rotation are introduced to $\mathbf{A}$, because $\mathbf{A} = [\mathbf{w}'_1, \ldots, \mathbf{w}'_N]^T$, where $\mathbf{w}'_i = \mathbf{R}\mathbf{v}'_i, i = 1 \ldots N$ and the elements of $\mathbf{R}$ may contain errors. Thus, an error may be introduced to each element of $\mathbf{w}'_i$ and, hence, each element of $\mathbf{A}$ [101, 102, 180].

Various methods for solving this type of minimisation problem exist which reduce these errors [10, 101, 102, 103, 129]. The preferred method for minimising Expression 9.15 is described below. In order to obtain a solution N must be at least 3. This is assumed in the subsequent calculations.

This problem may be solved as follows:

1. Factorise $\mathbf{A}$ using *singular value decomposition* (SVD) (see [103]) so that

$$\mathbf{A} = \mathbf{U}\mathbf{\Sigma}\mathbf{V}^T \quad (9.21)$$

where $\mathbf{U}^T\mathbf{U} = \mathbf{V}^T\mathbf{V} = \mathbf{V}\mathbf{V}^T = \mathbf{I}_{3\times 3}$, the 3×3 identity matrix and $\mathbf{\Sigma}$ is a diagonal matrix whose elements are $(\sigma_1, \sigma_2, \sigma_3)$.

In fact, the $N \times 3$ matrix, $\mathbf{U}$ consists of the three orthonormalised eigenvectors (see Appendix A) corresponding to the three largest eigenvalues of $\mathbf{A}\mathbf{A}^T$. The 3×3 matrix $\mathbf{V}$ consists of the orthonormalised eigenvectors of $\mathbf{A}^T\mathbf{A}$. The diagonal elements of $\mathbf{\Sigma}$ are the non-negative square roots of the eigenvalues of $\mathbf{A}^T\mathbf{A}$ (the singular values of $\mathbf{A}$).

Obviously, any straightforward evaluation of the matrices $\mathbf{U}, \mathbf{V}$ and $\mathbf{\Sigma}$ would introduce similar errors to those discussed previously. Thus $\mathbf{A}$ is first reduced to bi-diagonal form using the *Householder transformation*, and secondly the *QR algorithm* is used to find the singular values of $\mathbf{A}$ (again, see [103]).

2. Let $\mathbf{g}$ and $\mathbf{y}$, both N vectors, be defined by:

$$\mathbf{g} = \mathbf{U}^T\mathbf{Z}, \quad (9.22)$$
$$\mathbf{y} = \mathbf{V}^T\mathbf{t}. \quad (9.23)$$

9.3. IMPROVING THE METHOD'S ACCURACY

Then the quantity to be minimised with respect to **t** in Expression 9.15 is

$$\|At - Z\| \equiv \|U\Sigma V^T t - Ug\|$$
$$\equiv \|\Sigma V^T t - g\|$$
$$\equiv \|\Sigma y - g\|. \qquad (9.24)$$

3. Let y_u be the value of **y** which minimises Expression 9.24. Then the value of **t** which minimises Expression 9.15 is

$$t_{min} = V y_u. \qquad (9.25)$$

This method has two advantages over the alternative methods [10, 101, 102, 103]. Firstly, the only iterative part of this method is in the calculation of the SVD — other methods apply iteration over the whole solution. In particular, Golub [101, 102, 103] uses iteration to converge to the solution and at each stage in the process also calculates the SVD by iteration. The adopted method therefore finds the solution more quickly and with no loss in accuracy. The other advantage of this approach is that it works for cases when **A** is *rank deficient* [129]. This case will arise if parallel planes are extracted from a scene and used in the matching process, for instance.

9.3.2 Using the Centres of Gravity of Regions

A second technique for improving the translation estimate will now be considered. If the rotation estimate inaccurately estimates the orientation of the plane then the estimate of the translation will be in error as well, as illustrated by Fig. 9.1. Here, the rotation estimate has resulted in the orientation of the model plane normal, v'_i, relative to its pairing with the scene plane normal, v_i, being in error. The effects of this rotation error on the translation, with all measurements taken relative to the camera coordinates (shown in the figure), are now considered.

Let θ be the error in the estimate of the rotation between scene and model plane normals. If the distance of the scene plane from the origin is d, then the error in the placement of the model plane within its own plane is $d \tan \theta$ and the error in the placement of the model

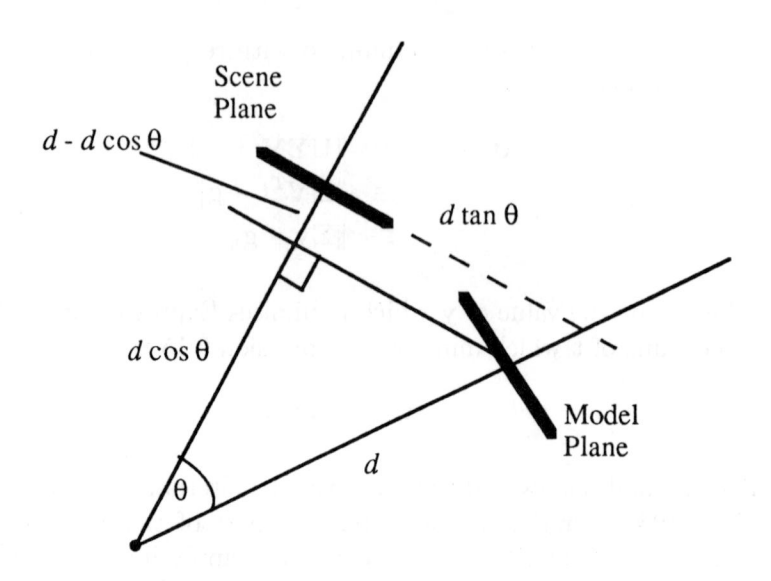

Figure 9.1: Errors in translation estimate

plane along its plane normal is $d(1 - \cos\theta)$. Given that d can be of the order of 500–700mm large values of θ are not required for the error in the placement (in particular within the plane) to become significant.

The least squares approximation of a plane to a set of points always passes through the centre of gravity of set of points. Thus if the model plane is transformed such that it passes through the centre of gravity of the scene plane to which it is matched then the errors in the translation estimate will be kept to a minimum. Previously the translation was approximated by minimising (Expression 9.15) the distance along the (misaligned) direction of the normal of the rotated model plane as shown in Fig. 9.1.

After transforming coordinates to the centre of gravity, the translation is estimated by minimising the perpendicular distance between the centre of gravity of the scene plane and the model plane along the direction of the model plane normal, as measured by

$$\sum_i |\mathbf{w}'_i \mathbf{c}_i - \mathbf{d}'_i - \mathbf{w}'_i \mathbf{t}|^2 \qquad (9.26)$$

where $\mathbf{c}_i$ is the translation transforming vectors to the centre of gravity of the scene plane. This is then minimised using singular value decomposition as described in Section 9.3.1.

The centre of gravity of a plane may be easily calculated during the segmentation.

9.4 Improving the Search Efficiency

As we have already discussed, the number of possible matchings between image and object features for even the simplest of objects is very large. Consequently any method that performs model to scene matching must take this into account. We have already introduced some strategies for finding solutions to the matching problem in an efficient manner in Chapter 8. Here we discuss some of these strategies in greater detail as well as introducing some new ideas. Once again the methods are discussed in the context of the Faugeras and Hebert method, although most of the ideas can be applied to many of the methods we have discussed.

9.4.1 Order of Matching Pairs

The order in which feasible matching pairs are considered has an effect on the speed of solution and on the eventual outcome of the result as will be discussed shortly. Many of the matching strategies we have met have applied this observation in some form (see descriptions of Murray's method, the *TINA* vision system, Nevatia and Binford's method and the *SCERPO vision system* in Chapter 8). Faugeras and Hebert [80] have also investigated this approach. The method we describe below is based on a hybrid of these aforementioned approaches and provides details of our own implementation.

Consider the case where primitives from the scene, $s_1, s_2, s_3, \ldots$, are matched to primitives from the model, $m_1, m_2, m_3, \ldots$ to give a set of match pairs $\{(s_1, m_1), (s_2, m_2), (s_3, m_3), \ldots\}$. Assume the features considered early on in the search have large errors in their description. For example, poor segmentations due to noisy data would give poor estimates in a planes and other surfaces in the scene.

When the first few match pairs $\{(s_1, m_1), (s_2, m_2), \ldots\}$ are considered then the estimate of the rotation they provide (the only component of the transformation evaluated at this stage) will also be in error. It is possible that the error in this estimate could be above the allowable threshold, which would terminate the search early. Consequently, subsequent more reliable match pairs would not be given a chance to improve the estimate of the rotation provided by the match (in a least squares sense). Thus this could lead to us rejecting a solution we may wish to keep.

However, if match pairs with scene primitives of low error are chosen first then the errors in the estimation would start small and only grow as further match pairs are considered. Therefore we can use tighter thresholds for rejecting poor matches early on which will therefore mean that more time is spent considering matches likely to lead to the solution, and less time is spent considering potential matches which will be ultimately rejected.

What is needed is a quick, reliable way of finding the best possible scene primitives for matching. By taking measures for each surface of its surface area, the quality of fit of the data points to the surface, the type of surface and the number of neighbouring surfaces, a value can be assigned to each surface, where larger values indicate a better possibility of a good match. Thus, at each stage of the search the scene primitive with the largest value is chosen as the next candidate for matching.

Each of the measures has a weighting factor associated with it. Our experimental results have shown that the weighting factors should be chosen to satisfy the following criteria.

- The face of largest area should be taken first so long as it is not subject to worse segmentation errors than a face of smaller area.

- Planar, cylindrical and spherical faces should be taken in that order as long as a face of simpler type does not have a worse fit than a more complex face.

- A face with more neighbours should be chosen first between like faces.

These results are intuitively reasonable and reflect some of the conclusions drawn from the analysis of features suitable for matching and

9.4. IMPROVING THE SEARCH EFFICIENCY

inspection given in Chapter 4 and the discussion of current matching strategies in Chapter 8.

The value of a surface is calculated as follows :

$$V(a, e, t, c) = w_1 a + w_2/e + w_3/t + w_4 c \qquad (9.27)$$

where w_1, w_2, w_3 and w_4 are weighting factors,
    a = area of surface,
    e = error of fit to surface,
    t = surface type = 1   if the surface is planar
                           = 2   if the surface is cylindrical
                           = 5   if the surface is spherical
                           = 10 if the surface is of any other type,
    c = number of neighbours.

The error of fit of the surface, e, is the root-mean-square distance of the points fitted to the surface from the surface, as provided by the least-squares fitting procedure. The values of the weighting factors we use were chosen from experiments on sample objects (such as the widget shown in Fig. 2.5) in order to approximately satisfy the criteria given above. In practice w_3 and w_4 are set to one, w_1 is set to the reciprocal of the area of the largest segmented surface and w_2 is set to the maximum value of e for any segmented face.

These measures can be calculated during segmentation at little extra cost.

It was also found that best results are obtained when the rotation is completely defined early on in the search. In this case the search space can be reduced since mismatching features can be more effectively eliminated. For this complete rotation to be defined at least two of the initial surfaces should be linearly independent. In the case when planes are chosen the surface normals can be readily used to find the rotation and planes are more likely to provide reliable data, although as we have described in Section 4.6.1, non-planar surfaces can be represented by linear quantities such as axes of symmetry or principal axes which can be used to determine the rotation. Thus, the preferred case is that the initial surfaces used in matching are planes which are as near orthogonal as possible. This is achieved by searching the sorted list for almost orthogonal planes and only allowing sufficiently reliable almost orthogonal candidates to be used for initial matches. Clearly, for complex objects there may well be no orthogonal or even nearly so primitives and therefore the search for suitable initial surfaces will have to be less strict.

9.4.2 An Additional Local Constraint

It was noted that even for relatively simple objects such as the widget (shown in Fig. 2.5) that certain incorrect partial matches are followed for some length of the search path before being rejected. This is due to the fact that since only local consistency and the estimate of the rotation is used to constrain most of the search, certain symmetries cannot be eliminated (see Section 9.2). The basic problem arises when two parallel planes occur in the model. This is a common occurrence, particularly when symmetries occur in an object, causing ambiguities of 180° to arise. Each plane could be potentially matched to a scene plane, but, obviously, only one of the two planes is visible and hence a candidate for matching. A match with the wrong parallel plane causes the subsequent rotation estimate to be 180° out. Thus, some time may be wasted trying to match other planes under this incorrect assumption. Similar mismatching problems also arise when more than two parallel planes exist, as illustrated in Fig. 9.2.

The solution adopted to help solve this problem and hence to increase the speed of the search is to use the area of the respective faces as an aid in discarding a wrong match. This test is performed at the same time as the other local consistency constraint test which compares the angle between the current scene primitive and a previously matched scene primitive to the angle of correspondingly paired model primitives to eliminate grossly mismatching faces (Section 8.8). The areas of the faces in the scene are already known and the areas of the model faces can easily be precomputed. It should be noted that the scene faces will usually be seen obliquely; the calculation of the areas and matching need to take this into account. If a scene face is greater in area than that of a model face then there is no possibility of them matching. The exact constraint of the areas being equal cannot be used since occlusion of scene faces or over-segmentation (one model face is

9.4. IMPROVING THE SEARCH EFFICIENCY

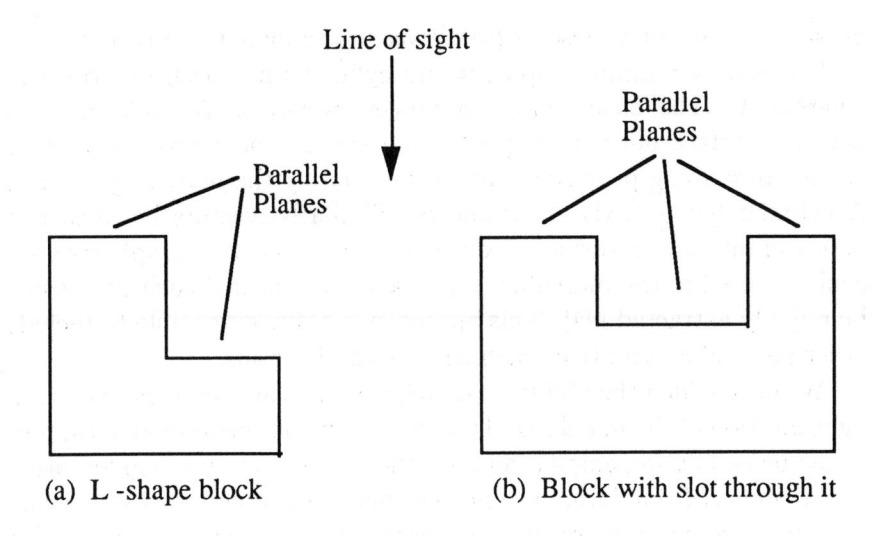

(a) L-shape block

(b) Block with slot through it

Figure 9.2: Examples of multiple parallel planes

segmented as more than one scene face) may occur.

In summary, the constraint is very cheap to implement and does not greatly increase the computation required before performing matching. Many grossly mismatching faces can be quickly and effectively eliminated.

9.4.3 Extension to Primitives Other Than Planes

By restricting attention to just planar surfaces the potential for inspecting or even just identifying objects is severely limited.

Segmenting scenes into general higher order primitives that are reliable is often very difficult and time consuming, as discussed in Chapter 4. Even using quadrics, in some sense the simplest type of general surfaces after planes, segmentation is unreliable so we can assume that any approach to recognition that relies on surfaces of this type, or any other higher order surfaces, will be unreliable.

As discussed in Chapter 4, a vast majority of parts can be modelled solely using planes, cylinders and spheres. Our allotted task is inspection. Since many of the objects that it may be required to inspect fall into the former class of parts [35], it is natural to explore the possibilities of using these as primitives in the matching process.

The representation of spheres and cylinders has been described in Chapter 4. The basic matching process works on the relationships between surface normals of planes and there is no reason why other vectors describing primitives cannot be used by the matching process. A cylinder has an axis of symmetry. Similarly, vectors between two points of interest on surfaces (centres of gravity, centres of spheres etc.) could be used in the matching but it is not yet clear if such points can be reliably extracted [94]. This approach is easily extensible to include other particular primitives such as cones and toroids.

We have added the ability to match cylinders and spheres to our implementation of the matching algorithm. Tests on theoretical data, produced by simulating angular and distance errors from the segmentation of real and artificial data, have proved that the strategy works in principle, giving comparable results to those to be discussed in Section 9.7.1 (using planes segmented from real and artificial data) for objects including cylindrical and spherical primitives, such as the widget with a cylindrical hole shown in Fig. 2.5. However the segmentation errors assumed for the cylinders and spheres were based on planar segmentation errors since reliable segmentations of the non-planar primitives are not yet available to our system. It is expected, from the analysis of these features in Chapter 4, that the errors in segmenting cylinders and spheres will in practice be worse and in consequence so will the results obtained from matching. It is difficult to predict the exact size and effects of such errors. Work is being carried out to achieve the accurate and reliable segmentation of cylinders and spheres [94] and when such segmentations are available further investigations into their use in matching will be possible.

9.4.4 Further Search Control

For reasons given below it was found that the location and orientation of the object is well defined after most of the segmented primitives of good fit have been considered for matching. Using primitives with a worse fit which come later in the search in the estimation of the transformation does not help, and indeed contributes unnecessary errors. Thus a method that bases the fitting on only the good primitives

9.4. IMPROVING THE SEARCH EFFICIENCY

will not only produce better transformation estimates but will also be quicker since no updating of the transformation to include information from the poorly fitting primitives need be performed.

As mentioned in Section 9.4.3 and Chapter 4 only reliable segmentations of planar surfaces are currently available. Thus, we restrict the range of parts we consider suitable for recognition by our system to those for which at least three good planar primitives can be extracted from the scene. This at least allows us to form an acceptable initial estimate of the transformation. This restriction appears to be no real drawback for the intended inspection purposes in that a survey [35] has shown that many of the mechanical parts considered do meet this restriction. The term good is applied in the same sense as described in Section 9.4.1.

We feel that due to the reliability problems of non-planar surfaces, it is best to match and locate the object using planes only if possible. After this has been done, the location of other unmatched primitives is then posed as an *hypothesis and test* problem (see Chapter 8). The position of these primitives can be hypothesised from the object model; they can then be looked for in the scene. Moreover, by seeking a match in this way, we can be highly selective about the nature of the features we are looking for (for example, spheres of a given radius or centre). Thus, we can obtain more accurate surface descriptions from the scene than we could by attempting to find general classes of surface when performing the segmentation. In particular, this technique is especially useful in inspection tasks of the types we will be discussing at greater length in Chapter 12, where we can obtain a *match* of optimum accuracy using planar features and then *inspect* the other features using a hypothesis and test strategy.

Only minor modifications to the existing matching method are required in order to implement the strategy of restricting matching to reliable planes.

- A *cut-off* is introduced when the order of matching pairs is evaluated, as described in Section 9.4.1, such that if there exist more than three planes which provide a satisfactory estimation of the transformation then no non-planar primitives need be considered for estimating the transform. This cut-off can be extended further to exclude poorly segmented planes from consideration when estimating the transformation. It should be noted, however, that as many good primitives as possible should be retained. This will provide more data for the system of linear equations used to estimate the transformations.

In practice once more than three scene planes have been matched, subsequent planes are only used for updating the transformation if the error in their fit (see Appendix B) is below a certain threshold. It has been found, using test objects and artificial data as described in Chapter 2, that satisfactory results are obtained when the value of the threshold is 0.1mm for normal inspection accuracy.

- A small modification to the segmentation procedure is required. The segmentation process may, if not carefully controlled, make mistakes and decide that a curved surface is many small planar facets. Since inclusion of such facets in the matching process could cause failure or otherwise be a source of error, it is important that they should be rejected early on. It is therefore required that a plane must be larger than a certain specified area for it to be considered by the matching step. If for some reason a larger planar facet of this type avoids detection it is anticipated that any fitting errors will be so large that it will fail the cut-off constraint described earlier.

9.4.5 Rogue Faces

This idea was first introduced by Grimson and Lozano-Perez [109, 153] in an extension to their matching method (see Section 8.6) to allow for overlapping parts.

If a *rogue* face is obtained from the scene, when a match is sought with the model, no suitable match may be found. Possible sources of rogue faces include planes belonging to the background or other equipment in the scene such as a robot or the turntables. Under poorly controlled conditions, the segmentation could also be a source of such errors. A different type of possibility is that the object being viewed has been incorrectly manufactured. In such cases it is desirable that the matching process is continued and does not just stop. For example,

9.4. IMPROVING THE SEARCH EFFICIENCY

other features correctly identified can still be inspected and the whole object need not be immediately rejected as not matching. The incorrect faces can be clearly flagged during the inspection phase.

Accommodating such mismatching faces in the matching strategy is straightforward. As well as having n branches in the interpretation tree coming from each node, where n is the number of model faces, an additional *null* branch is added to each node. The tree search proceeds as described before except that if no model primitive matches the scene primitive then the scene primitive is assigned to the null branch and the search continues.

Clearly, it is possible that a serious mismatch of model and object could occur resulting in a large sequence of successive null branches throughout the tree. This would be seriously inefficient in terms of both computational time and space. Thus we do not permit the set of null branches in the search tree to grow to more than half the total possible depth of the search (*i.e.* half the number of scene primitives). This is not unreasonable since if more than half the scene primitives have not been matched then a satisfactory match is unlikely.

9.4.6 Position Recovery From Insufficient Data

As mentioned previously our particular matching strategy requires that the positions of three linearly independent planes (or in general three linearly independent vectors) are known in order to estimate the translation and thus completely define the transformation (two linear independent planes are sufficient to define the rotation) from model to world coordinate frames. However, because of

- the relatively narrow field of view of the cameras, or
- a poor segmentation of the scene, or
- the class of surfaces of the viewed object,

it is possible for the scene of the viewed object not to contain a set of planes that are linearly independent in three dimensions. This is illustrated in Fig 9.3 where each individual camera can see three linearly independent planes but the resulting depth map (where only points common to both views have depth estimates) does not.

View from master camera   View from other camera

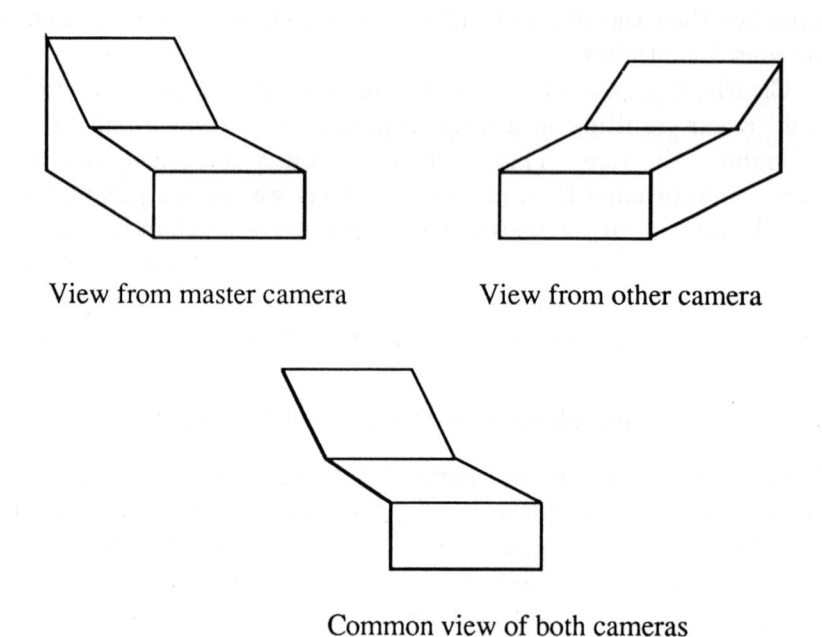

Common view of both cameras

Figure 9.3: Insufficient linearly independent planes

9.4. IMPROVING THE SEARCH EFFICIENCY

However, so long as two linearly independent primitives are visible then a cruder estimate of the translation is possible and a method to compute this will shortly be discussed. Whilst this estimate may not be sufficient for performing accurate tasks such as inspection, it may still be useful for various purposes such as:

- Attempting to recognise the object.

- Obtaining a rough idea of the position of the object for the hypothesis and test phase.

- Providing positional information required to a relatively low accuracy, for such purposes as pick and place.

- Repositioning the object to a new view likely to provide more accurate positioning information with some degree of certainty rather than some arbitrary or random position that does not guarantee better results.

We will now describe our method for estimating the translation from only two linearly independent vectors. It uses the singular value decomposition (SVD) methods described earlier in Section 9.3.1.

In the estimation of the translation, the SVD of the matrix $\mathbf{A}$ gives an orthogonal basis for the set of vectors represented in $\mathbf{A}$, as described in Section 9.2.2 — the eigenvectors of $\mathbf{A}^T\mathbf{A}$ (matrix $\mathbf{V}$ in the SVD, see Section 9.3.1) and the corresponding eigenvalues (matrix $\mathbf{\Sigma}$ in Section 9.3.1). For each zero eigenvalue in $\mathbf{\Sigma}$ the number of linearly independent vectors of $\mathbf{A}$ (its *rank*) is reduced by one. If there are three non-zero eigenvalues there exits a linearly independent triad of vectors and the translation may be determined as previously explained. If there are two non-zero eigenvalues then the space of vectors represented by $\mathbf{A}$ only spans a two-dimensional space. In this case the translation is estimated as follows:

1. Find a third vector perpendicular to the two linearly independent eigenvectors by forming the vector cross product.

2. Work out the components of the translation vector in the two linearly independent eigenvector directions by using the SVD of

A projected into the two dimensional space defined by these two eigenvectors. This amounts to simply dropping the third eigenvector from the calculations in Section 9.3.1 since the translation vector may be expressed as:

$$\mathbf{t} = \lambda_1 \mathbf{e}_1 + \lambda_2 \mathbf{e}_2 + \lambda_3 \mathbf{e}_3 \qquad (9.28)$$

where $\mathbf{e}_1, \mathbf{e}_2, \mathbf{e}_3$ are the three eigenvectors and $\lambda_1, \lambda_2, \lambda_3$ are to be estimated. We now use the vector formed above in place of $\mathbf{e}_3$; λ_1 and λ_2 are estimated from the SVD and λ_3 is estimated separately.

3. We replace λ_3 by an estimate formed by minimising the distance between the respective centres of gravity of the matched scene and model primitives along the direction of the vector replacing $\mathbf{e}_3$.

9.5 A Practical Assessment of our Vision System

We will now study the effect of these modifications on the matching algorithm. However, the results of any matching algorithm depend on the data gathered by the vision system and the segmentation of this data into higher order features. We have described our vision system and the problems in acquiring accurate data in Chapter 2. Our segmentation approach was described in Chapter 4 but we have yet to address its performance on three-dimensional data.

In this Section we firstly recap the results of our assessment of our vision system from Chapter 2. We then present a fairly complete study of the performance of the segmentation algorithm on both simulated and actual depth data. Finally we study the performance of our matching algorithm based on these segmentation results.

9.5.1 Our Vision Acquisition System Reviewed

Our vision acquisition system capable of producing depth maps was described in Chapter 2. The typical size of the depth maps is 256 × 256 pixels for an area of interest 300–500mm square. A method of producing simulated depth maps was also described for analysing theoretical working conditions in Section 2.6.

By analysing the performance of the calibrated system (see Section 2.5.2) it has been shown that depth measurements are not accurate enough to achieve inspection to the desired tolerances. However, it has been shown that measurements can be achieved that are accurate to within 0.5mm which is not much greater than accuracies required in practice.

The reasons for the inaccuracies of the system are currently being investigated and may be due to several sources — errors in the calibration procedure, an insufficiently detailed camera model, or poor camera performance. Whilst it is difficult to deduce the exact sources of the inaccuracies it is not unrealistic to expect this problem to be resolved in the near future. Work to this end includes developing the calibration procedures further and investigating the use and development of potentially more accurate equipment.

9.6 Segmentation Algorithms in Practice

This Section describes results obtained from segmentations of both artificial and real depth maps using the segmentation procedures described in Section 4.6.6. Initially the overall segmentation approach is assessed, then results from artificial and real data are discussed separately.

9.6.1 Our Segmentation Approach

The segmentation approach described in Section 4.6.6 consists of four stages:

1. A stage in which a mask is passed over the whole image and small planar approximations are made.

2. A *coalescing* stage which rapidly reduces the initially large number of regions by growing regions together.

3. A *forcing* stage which performs a rigorous region growing procedure.

4. A *featuring* stage where various feature parameters are derived for each segmented region.

The results obtained from the various stages of the segmentation are very encouraging and compare very favourably with those obtained from greyscale and colour images [184] to which similar segmentation techniques have been applied.

The coalescing algorithm has proved itself to be be efficient in reducing an initially very large number of regions (up to 65,000) to about 1000 regions. The forcing method has also proved that it can deal with a large number regions (up to 10,000) and accurately reduce this number down to the correct number of large planes present in test scenes.

There are two minor drawbacks present in the current segmentation strategy. Firstly, as described in Section 4.6.8, poor approximations are obtained for small planar regions near edges. This effect can be seen in the region maps shown in Fig. 4.23. However, the areas of these regions are very small, so they do not contribute serious errors to the much larger final planes. However, note that this also means that accurate edge tracing from the region map is not possible. Secondly, the forcing procedure can be slow (taking about 4 minutes) to converge to the final solution when very noisy data is present. This time could easily be reduced if it were implemented on a parallel computer, which is currently being investigated.

9.6.2 Results From Artificial Data

The reasons for studying the effect of the segmentation procedures using artificial depth maps are twofold. Firstly, it is easier to study the behaviour of the segmentation methods using controlled noise. Secondly, subsequent matching and inspection strategies also need to be studied in detail. Since any errors introduced during the segmentation stage will be passed on to the later stages of the overall inspection process, such errors need analysis in order to assess the suitability of the whole approach.

9.6. SEGMENTATION ALGORITHMS IN PRACTICE

Results in this Section are presented using the synthetic depth images of the three objects shown in Chapter 2 (Fig. 2.5). The smallest distance from the master camera to the nearest face of the cube was 1212.4mm and to each of the *widgets* was 707.1mm. The cube was placed further from the camera than the widgets in order to evaluate the effect of varying the distance between cameras and object. Gaussian noise was added to the images as described in Section 2.6.

The following graphs (Figs 9.4 and 9.5) summarise the results obtained from segmentations of six different views of each object. The views were taken at 30° steps of a 180° degree rotation. The average errors in the estimation of the measured surface normals and the displacement from the origin are given. These errors are easily calculated since the precise values of the normals and displacements are known when the artificial depth map is produced.

As can be seen from these results, the segmentation algorithms perform well beyond the performance required for inspection when presented with simulated errors of a size anticipated with the actual imaging equipment (see Chapters 2 and 10). For typical inspection tasks depth measurements have to be accurate to within 0.254mm and so the segmentation provides sufficiently accurate data, even allowing for further system errors downstream from segmentation. Note that the segmentation process does not give errors in position exceeding typical tolerance limits until the added noise has depth errors of 0.9mm or more which implies that a cheap, less accurate, version of the system could be used for measuring positions of objects for assembly tasks. Alternatively, an error of 0.9mm corresponds to moving the cameras approximately 2.5 times further away from the object since the error is proportional the square of the depth.

The slightly better accuracy of fit to the faces of the widget compared to the widget with a cylindrical hole arises because of the method of fitting planes to 3×3 groups of pixels in the first stage of the segmentation process. Clearly, all pixels within a 3×3 region of the hole will have some pixels that are contained in the hole, leading to errors being introduced to larger regions when the merging process occurs.

Similar errors also occur at edges, where pixels within a 3×3 region of an edge contribute to approximation errors as described above. This is why some small fitting errors exist even at zero noise levels.

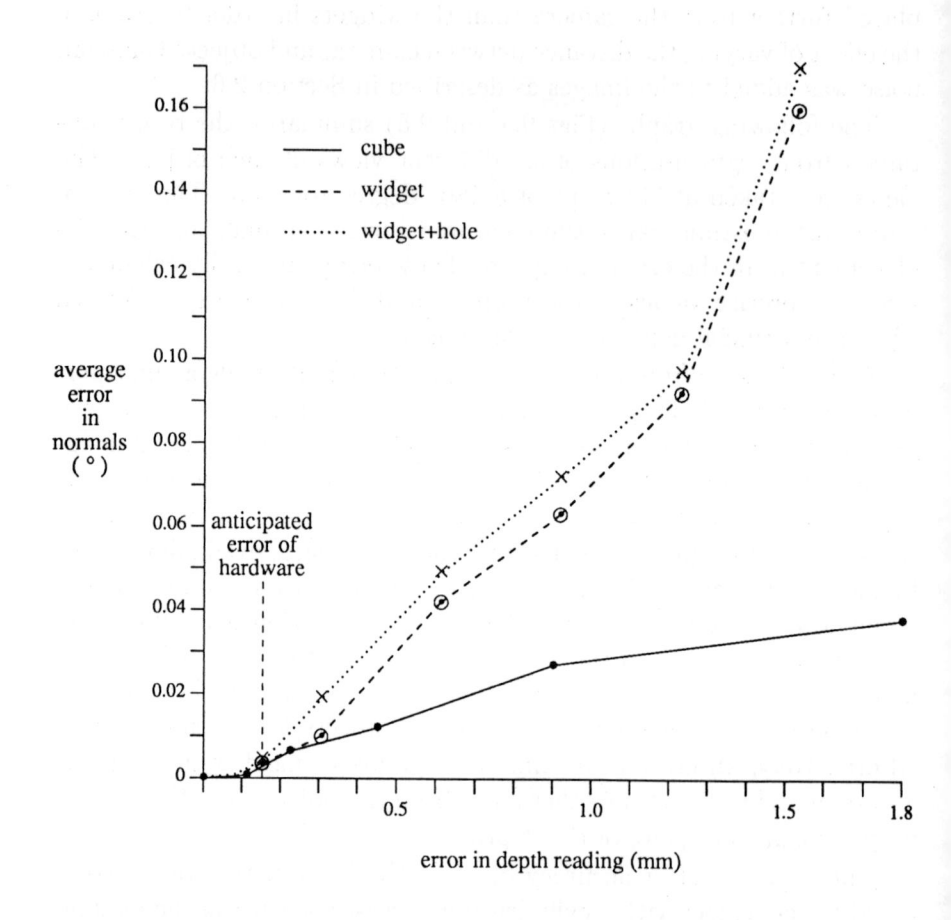

Figure 9.4: Errors in estimation of surface normals

9.6. SEGMENTATION ALGORITHMS IN PRACTICE   267

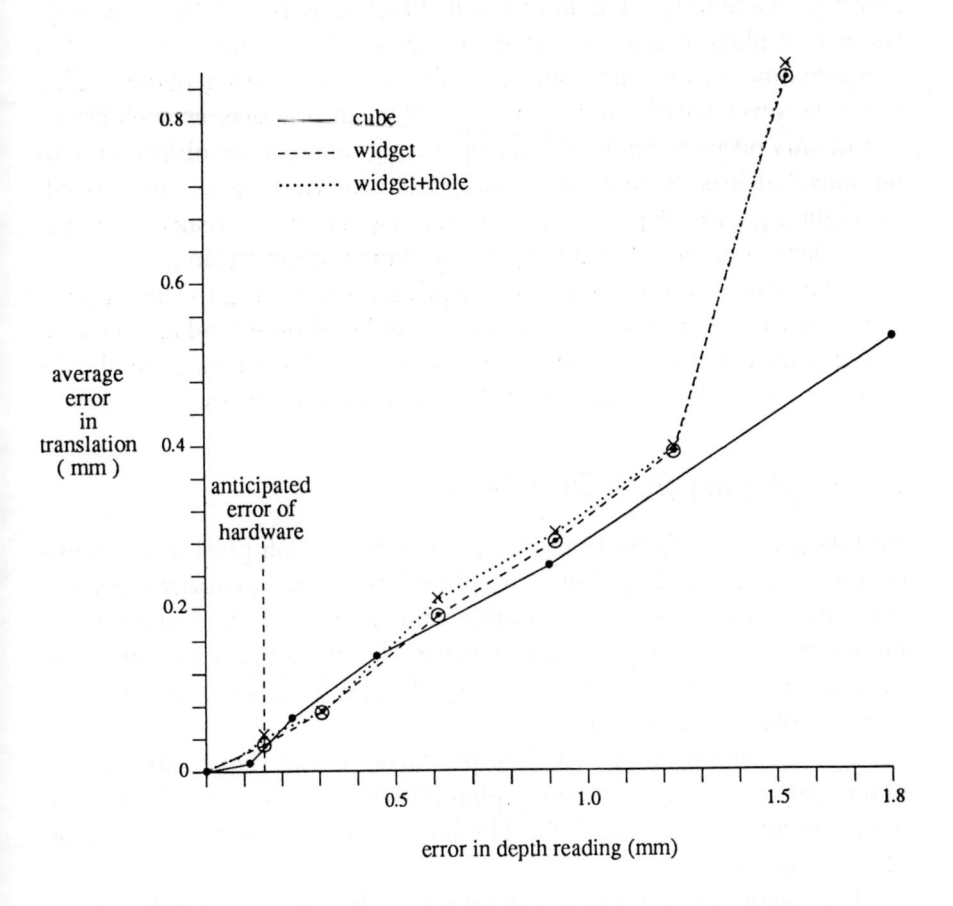

Figure 9.5: Errors in estimation of translation

It can also be shown that after the error in position reaches about 3.0mm, oversegmentation occurs. Thus, a given actual plane may be described as several smaller planes by the segmentation, if large noise values in the approximation of the pieces of plane prevent complete merging. Generally what happens in practice is that a large piece of the actual plane is approximated by a good fit and then the rest of it is approximated by a number of smaller less well fitting planes. This is not however anticipated to be a problem in practice, as such errors would only occur at an unrealistic operating range, if the object were to be placed at least 3 times as far away from the cameras as anticipated. At slightly greater depth errors still, the segmentation breaks down due to the large number of differing initial plane approximations.

In summary, as long as the real depth data presented to the segmentation process correspond to the values modelled here (and expected in practice from the equipment) then the results of segmentation should be more than adequate for general inspection procedures.

9.6.3 Results from Real Data

Results given in this Section are based on 256 × 256 pixel depth maps of views of the *widget* (Fig. 2.6) placed between 0.5–0.7m from the cameras. Since the exact orientation and location of the widget is not known comparisons of the type made in the Section 9.6.2 on artificial data are not possible. However, the following analysis of the segmentation procedure is possible.

The segmentation of real data produces results that contain small planar regions at edges and large planar regions elsewhere as illustrated in the region maps in Fig 4.23. The large regions correspond to actual planar regions.

The average root-mean-square error in the fitting of depth data to the planes is 0.28mm. However, the average error in the angles between the segmented planes on the widget is 1.2°. This error in angles is larger than the error in relative angles of about 0.32° produced using artificial depth data at the largest tolerable noise levels (noise added to depth readings > 1.0mm). There is clearly a discrepancy between these two results as the root-mean-square position error of the planes corresponds to simulated added noise of 0.3mm but the angle errors correspond to

9.6. SEGMENTATION ALGORITHMS IN PRACTICE

added noise more than three times as large.

Since the segmentation results have been shown to be stable for large artificial errors and the errors in the real depth data are small it is reasonable to assume that the errors of the angles between segmented planes are not due to a failure of the segmentation process. It has already been shown in Chapter 2 that calibration errors in the camera set up can lead to errors in locating known points. The camera set up has a root-mean-square error of 0.5mm in world coordinates and a radius of ambiguity of pixel placement of 0.4mm. It is not unreasonable to assume that calibration errors account for the errors in the relative angles of planes observed above. This fact is made plausible by the following two-dimensional example using realistic values for our system.

Consider Fig. 9.6. Assume that all measurements are taken relative to the coordinate frame of the master camera with origin O. Let $O_0 = (x_0, z_0)$ be the origin of the other camera. Let α be the angle between the z axis of the master camera and the base line between the origins of the two cameras. Assume that the measured position of a point M is (x, z). Let β be the angle between the base line between the two cameras and the measured point and let δ be the angle between the z axis of the master camera and the measured point.

Let dx and dz be the errors in the x and z displacements of the measured point due to calibration errors and let $d\phi$ be the angular error in the positioning of the point. Thus the point appears to be at the position $M' = (x - dx, z + dz)$. By simple calculation, angle θ is given by

$$\theta = 180 - (\alpha + \beta) \tag{9.29}$$

and dz is

$$dz = \frac{dx}{\tan \theta}. \tag{9.30}$$

Now $dx = z \tan d\phi$. Therefore

$$dz = \frac{z \tan d\phi}{tan \theta}. \tag{9.31}$$

It can also be shown that

$$\sin \beta = \frac{z \sin(\alpha + \delta)}{D \cos \delta} \tag{9.32}$$

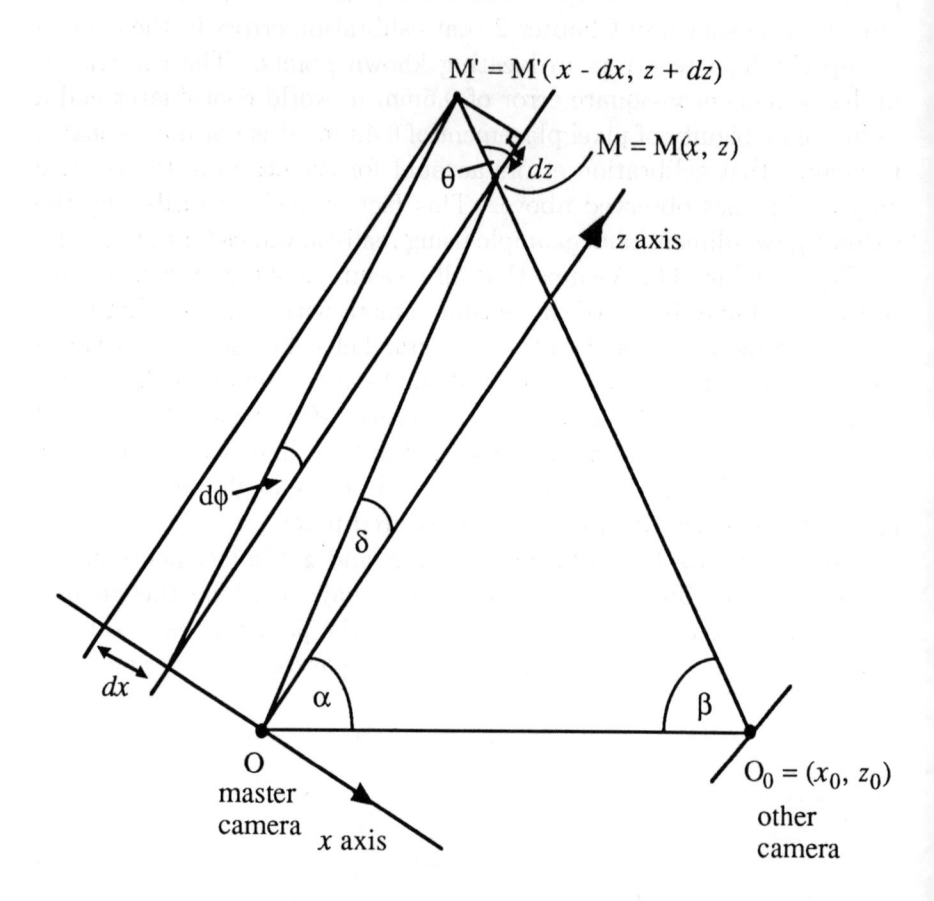

Figure 9.6: Errors in depth measurement due to calibration errors

9.6. SEGMENTATION ALGORITHMS IN PRACTICE

where $D = \sqrt{((x-x_0)^2 + (z-z_0)^2)}$.

A typical value for α is $45°$ and a typical position for the origin of the other camera in our actual system is $O_0 = (289mm, 78mm)$. In Chapter 2 it was shown that calibration errors result in a radius of ambiguity of pixel position of 0.4mm when the object is at a range of about 500mm. Therefore this gives $d\phi = 0.046°$.

Taking three points with typical (x, z) values

$$P_1 = (0, 500), \qquad P_2 = (100, 600), \qquad P_3 = (-100, 600),$$

let two lines pass through the pairs of points $P_1 P_2$ and $P_1 P_3$ respectively.

The errors in the (x, z) position of point P_1 are:
$dx = 0.4$mm, $dz = 0.434$mm.
The errors in the (x, z) position of point P_2 are:
$dx = 0.48$mm, $dz = 0.5$mm.
The errors in the (x, z) position of point P_3 are:
$dx = 0.4$mm, $dz = 0.7$mm.

If two points $P_1 = (x_1, z_1)$ and $P_2 = (x_2, z_2)$ are in error then the gradient between the two points, Δ_{12}, is given by:

$$\Delta_{12} = \frac{x_1 - x_2 - |dx_1| - |dx_2|}{z_1 - z_2 + |dz_1| + |dz_2|} \qquad (9.33)$$

where dx_1 and dx_2 are the errors in the x positions of P_1 and P_2 given by the radius of ambiguity and dz_1 and dz_2 are the errors in the z positions of P_1 and P_2.

Comparing the true gradients of the lines $P_1 P_2$ and $P_1 P_3$ with the gradients based on the incorrect positions of the points gives angular errors ω_{12} and ω_{13} in the orientation of these lines:

$$\omega_{12} = 0.57° \quad \text{and} \quad \omega_{13} = 0.8°. \qquad (9.34)$$

Therefore the error in the angle between these two lines is $\omega_{12} + \omega_{13} = 1.37°$. This is of the same order as the actual observed errors in angles between planes output by the segmentation process.

9.6.4 Conclusions

It has been shown that the segmentation of planes from a depth map is likely to yield the most accurate primitives that can be used for matching and inspection purposes. Since accuracy is of great importance in inspection processes it is sensible that planes should be adopted as the primary type of feature to be extracted. A method for extracting planes has been outlined and examined for reliability.

It has been shown using simulated depth map data that the proposed plane segmentation process can extract planes accurately and reliably enough for inspection purposes. It has also been shown that the segmentation process can reliably extract planes from real depth map data to within realistic inspection tolerances (0.28mm). However due to camera calibration problems, errors in the angles between segmented faces occur that are too large for accurate inspection purposes (see Chapters 10, 12, 13). Should these calibration errors be overcome then it is reasonable to assume (by comparing errors in plane fitting from simulated and real depth data) that these angular errors will be reduced and that segmentation of real depth data to sufficient accuracy for inspection will be achieved.

However, in order that complete inspection of a general class of manufactured parts can be achieved, reliable segmentation of non-planar surfaces must also be obtained. Some ideas of how this may be achieved have been presented; this is an area of continuing research. We believe the most promising method to be the use of model information to drive the segmentation process.

It is also hoped to overcome the slight drawbacks in the segmentation method that have been highlighted by this study [94]. The main problem is that errors arise when the initial planar approximation mask falls over an edge, leading to not very accurate predictions of boundary positions. A variable size mask with some method of detecting edges or adapting itself if it falls over an edge could be employed to remedy this problem. This should be relatively simple to implement as the segmentation is already capable of reasoning with edge data during its region growing phase (see Section 4.6.6 and [95]). The segmentation process is also being adapted to segment non-planar surfaces.

9.7 Matching Algorithm Assessment

We now present results on the matching algorithm described in this Chapter. In particular, we compare results of this algorithm with those of the original Faugeras and Hebert algorithm (see Section 8.8) upon which the new algorithm is based.

9.7.1 Matching Segmented Artificial Data

The results presented in this Section are based upon taking the particular examples of segmentation output for artificial data described in Section 9.6, and matching the results to geometric models stored in the computer. The output of the matching routines is a list of matched planes and a transformation (a rotation matrix and a translation vector) relating the object model to the observed object. Since the pose of the objects was known exactly in the production of the artificial depth maps, the exact transformation is known. The errors in the estimated transformation are easily calculated and are shown in the graphs in Figs. 9.7, 9.8, 9.9, and 9.10.

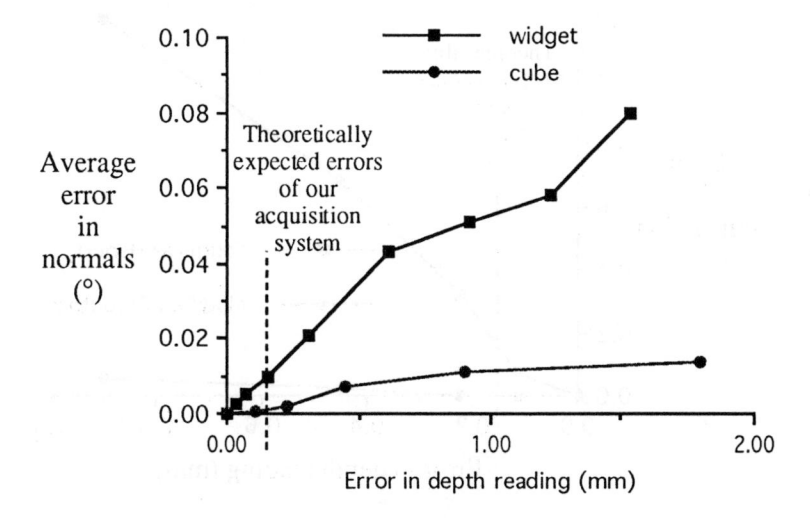

Figure 9.7: Errors in rotation estimate at large depth errors

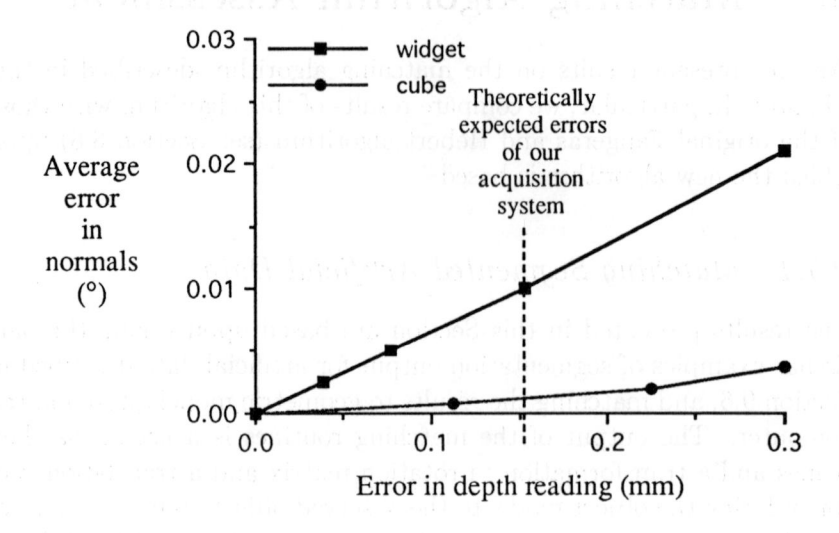

Figure 9.8: Errors in rotation estimate at small depth errors

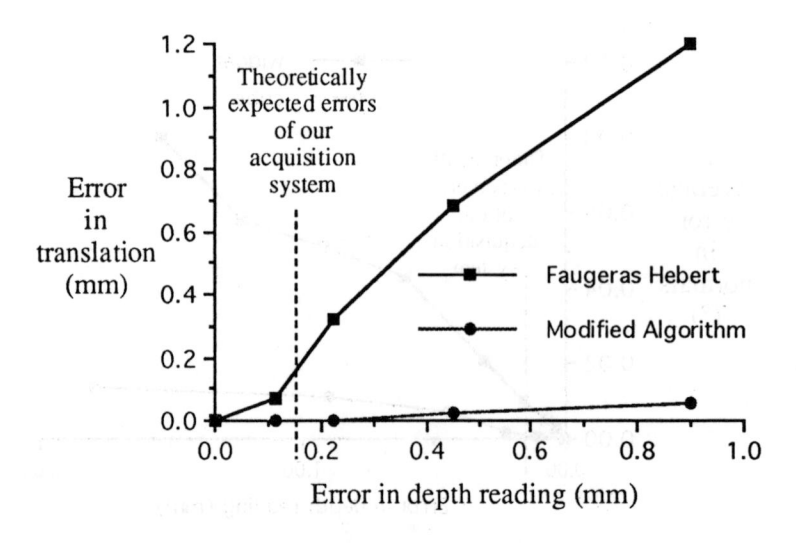

Figure 9.9: Errors in translation when matching a cube

9.7. MATCHING ALGORITHM ASSESSMENT

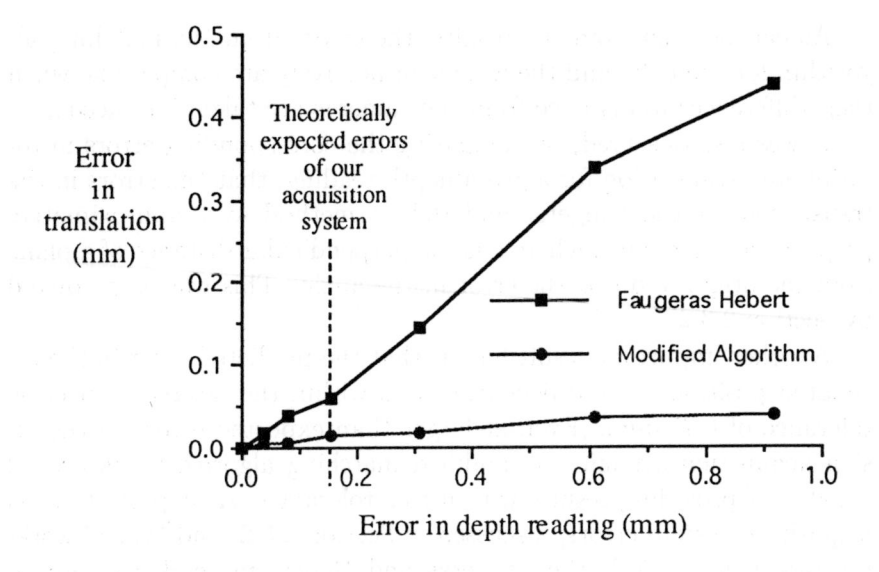

Figure 9.10: Errors in translation when matching the widget

Our modified matching algorithm using all of the constraints (rotational and translational inconsistency, angle between scene primitives and model primitives and area constraint) was applied to the segmentation results as was the original Faugeras and Hebert algorithm. Only planar surfaces were considered by the matching process.

Since both methods adopt the same strategy for estimation of the rotation no comparison has been made between them. The average errors in estimating the rotation are summarised in Figs 9.7 and 9.8 for different ranges of depth error.

Since the estimates of the translation produced by both algorithms are dependent on good estimates of the rotation it is important to observe that the match between model surface normals and scene surface normals has small errors until the noise added to the depth map values is much larger than the expected hardware errors.

The errors in the translation estimate are now considered in turn. The errors in the translation in the matching of model to scene planes are summarised in Figs 9.9 (for the cube) and 9.10 (for the widget). The modelled distance of the cube from the cameras was 1212mm, for the widget it was 707mm.

As can be seen from the results, the errors from the matching algorithm for the cube and the widget respectively are comparable when their differences in distance from the cameras are taken into account.

It was also observed, by comparing the corresponding errors in rotation and translation for a given depth reading, that the errors in the translation for the Faugeras and Hebert method were approximately proportional to $d \tan \theta$, where d is the perpendicular distance of a plane from the origin and θ is the error in the angle. This was as predicted by Section 9.3.2.

Another important result here is that the predicted error in the estimated position of an object feature is within the desired inspection tolerance of 0.254mm. Even at larger than expected errors in the vision acquisition system the modified matching algorithm has proved capable of providing results within this tolerance. At expected vision acquisition system errors, discussed in Section 2.5.2, and typical working range (0.5–0.7m), the Faugeras and Hebert method also proved satisfactory although not as accurate as our modified version of the matching algorithm. However, after the depth errors become greater than about 0.5mm, the errors in the translation estimates produced by the Faugeras and Hebert method exceed the values which will allow inspection to be carried out. Alternatively, this implies that the modified matching algorithm will be more reliable should more strict inspection tolerance limits be required.

The time savings achieved by employing all of the proposed modifications to constrain the search, compared to the Faugeras and Hebert method, were about 50%.

9.7.2 Matching Segmented Real Data

Results in this Section are presented on the same real world views of the *widget* that were used in Section 9.6 for the segmentation analysis on real data.

Once again it is difficult to infer exactly how well the matching methods have performed on real data since the exact positions of objects are not known. However, two analyses are possible.

Firstly, the quality of fit of match of the object model (by applying the complete transformation estimated from the match) to the seg-

9.7. MATCHING ALGORITHM ASSESSMENT

mented data with reference to the camera frame can be compared. The average error in the estimation of the rotation, found by comparing errors in the angles of surface normals between paired scene and model planes, was 0.9°. The average error in the translation estimate, found by comparing the distances from the origin of the paired model and scene planes, was 0.2mm for the modified matching algorithm whereas it was about 4mm for the original Faugeras and Hebert method (and once more was found to be approximately proportional to $d\tan\theta$). As shown in Section 9.6, there is a discrepancy in the angles between segmented faces output by the segmentation method. This accounts for the large error in the estimates of rotation and translation. It is anticipated that, as illustrated by artificial data, should the rotation error be reduced, which will occur with more accurate real data, then the translation error will be substantially reduced thus enabling more accurate measurements. Errors of the type discussed above are illustrated in Fig. 9.11 which shows the visible surfaces of the model of the *widget* positioned in camera coordinates using the matching transformation and superimposed on a depth map of the widget. With the picture suitably enlarged slight mismatches may be observed especially at the edges. (The obvious hump in the depth data in the middle of one face was caused by a coin which was placed on the widget for testing inspection strategies, as will be discussed later).

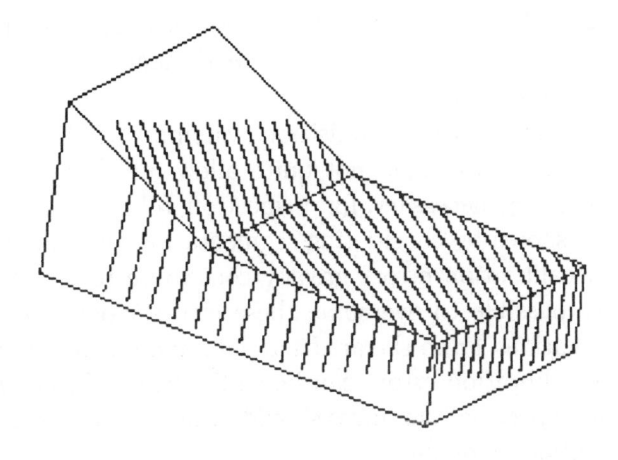

Figure 9.11: Widget with transformed model superimposed on to depth map

A second analysis of the matching results may be performed in world coordinates. The *widget* was placed on the ground plane ($z = 0$) and so the z coordinates of vertices belonging to the base face of the real object were known. The model is transformed using the estimated transformation into scene coordinates, and the z coordinates of these vertices found. The average error in the z positioning of these vertices was found to be 0.5mm. This is consistent with the errors calculated in Chapter 2 for the calibration analysis where the position of the ground plane was found to be in error by 0.5mm. This once more shows that the most likely explanation for the mismatch between the observed data and the model is due to calibration errors, and not matching or segmentation errors. Errors of this type can be seen in Fig. 9.12, which shows the model of the widget positioned in camera coordinates under the matching transformation and superimposed on the observed view of the widget from the master camera.

9.8 Conclusions

In this Chapter we have presented details of a proposed matching strategy and its implementation.

We have adopted a strategy of matching using reliably segmented planar surfaces, and have suggested improvements to previous methods of doing this. By using artificial data we have shown the expected errors from this approach will be sufficiently small to enable the goal of geometric inspection.

With our current vision acquisition system, it has been shown that the depth maps provided do not quite allow matching to within the degree of accuracy necessary for inspection tolerances. However, it has also been shown by considering segmentation and matching results that camera calibration errors greatly contribute to this problem. It should prove possible to overcome these errors; doing so is currently under investigation. From the artificial data analysis it can be assumed that if these calibration errors can be sufficiently reduced then results adequate for inspection tolerances should be achievable by the matching and segmentation algorithms.

We have also demonstrated that the approach of driving the match-

9.8. CONCLUSIONS

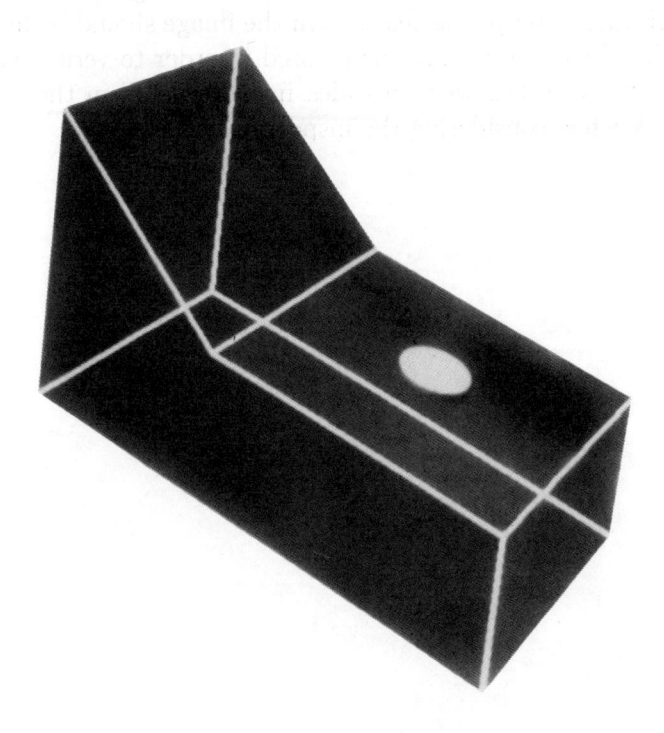

Figure 9.12: Transformed model superimposed onto an actual view of the widget

ing process with planar surfaces is a feasible strategy particularly when it is necessary to minimise errors in estimates of position and orientation of the observed object. It may be the case that planar features by themselves are insufficient for establishing a match between the object model and observed data. In that case, it is suggested that the positions of other, non-planar features in the image should be hypothesised from the object model and then tested in order to verify the proposed match. We will deal with this idea in more detail in the next Part of the book when considering the inspection process.

Part III

Inspection

Part III
Inspection

Chapter 10

Introduction to Inspection

10.1 Introduction

In this last Part of the book we shall now bring together the techniques required for our ultimate goal of an automated vision-based three-dimensional geometric inspection system. In Part 1 of the book we covered much of the ground work in computer vision needed for the rest of the book. In Part 2 we built upon these ideas to develop an object recognition strategy. The particular strategy developed was created not only to match stored objects to the scene, but also to determine very accurately the position and orientation of the observed object, a necessary prerequisite for its inspection. In Part 3 of the book we shall show how the matching stage can be used as part of an overall system for performing geometric inspection of the observed object. We shall also consider other types of inspection task such as surface finish inspection and crack detection since a complete inspection system will need to perform such tasks as well as geometric inspection. We have not implemented any such other inspection methods in our working inspection system, however.

The inspection of a manufactured object can take on many forms. Our particular interest is in inspecting compliance of objects to three-dimensional geometric tolerances. The inspection of an object to geometric tolerances basically involves the following tasks:

- verifying the presence of each expected feature;

- verifying the dimensions of these features (*e.g.* radius and length of a cylinder);

- verifying feature interrelationships (*e.g.* distances between centres of gravity and angles between plane normals).

Inspection in its broadest definition involves testing whether an object meets certain criteria. This involves comparing the object with some object model which describes the relevant features of the object. For many types of data, there are bounds or *tolerances* within which measurements taken from the object must lie in order for the object to be acceptable. In our particular inspection task we use a geometric model of the object, as discussed in Chapter 7, together with a suitable description of the tolerance information. Such descriptions will also be discussed in detail in this Part of the book.

In the rest of this Chapter we shall firstly consider the current methods used in industry for geometric inspection of objects using both visual and non-visual means. We shall then go on to discuss how we can achieve our goal of three-dimensional visual inspection of geometric tolerances. The discussion will include the choice of sensing equipment, segmentation and matching algorithms, using the material from Parts 1 and 2 of this book.

10.2 The Geometric Inspection Problem

10.2.1 Introduction

Industrial practice requires that all or a sample of the products manufactured should be inspected in order to ensure that some measure of quality and reliability is attained. For a complex, manufactured object the inspection process will have many stages, from testing the suitability and compatibility of each component to testing the performance and acceptability of the finished product.

The usefulness of a totally automated inspection system is obvious, as production costs both in time and money can be reduced [9]. However, to date no totally automated three-dimensional geometric inspection systems exist. Let us consider current industrial inspection techniques and their drawbacks.

10.2. THE GEOMETRIC INSPECTION PROBLEM

10.2.2 Current Industrial Inspection

Current industrial three-dimensional geometric inspection methods involve tactile sensing, using *coordinate measuring machines* (CMMs) [22, 98, 194]. These CMMs are slow in operation because a single touch probe is used take many individual readings from the surface of the object under test. Such a machine requires detailed programming for each new object to be inspected. Programs are very time consuming to write and are prone to errors. Thus, especially when small quantities of a particular object need inspecting, this approach is not very cost effective. The alternative of manual inspection — getting a human inspector to perform all measurements using gauges, callipers *etc.* — is also very time consuming and inefficient.

The object is initially fixed in a given position and orientation within the CMM. After this, the location of certain reference points or faces can be measured. By then specifying where to find other features relative to these reference features or *datums*, and measuring their positions accurately, the geometric correctness of the object can be verified. It should be noted that all the geometric information normally comes from engineering drawings, and a human programmer must use this information to generate an explicit program for the CMM. At no stage does any geometric reasoning occur within the CMM. It should also be noted that, normally, until the locations of reference points or faces have been established the CMM is unable to proceed. The reference points are usually indicated to the CMM by the operator carefully guiding the tactile probe to these points. An alternative would be for the probe to sample the entire space occupied by the object. Then, techniques similar to those applied to visual depth data could be used determine the location of the reference points. However, this approach is not generally satisfactory for two reasons. Sampling an adequate number of points with a probe to accurately locate the object would be very slow. Secondly, using the sensitive tactile probe to blindly search the object space may well cause damage to either or both of the object or probe, even though the probe may be fitted with a pressure sensitive safety cut-out.

Thus, even though tactile inspection provides accurate measurement of an object, the CMM approach is not suited to a totally automated method of inspection, both for reasons of speed and flexibility, and because tactile sensing is not always desirable or possible because of the nature of the object being inspected.

10.3 Current Industrial Vision Systems

10.3.1 Introduction

Most general purpose vision based inspection systems used in industry to date have been built to carry out two-dimensional inspection tasks, examining such objects as printed circuit boards and sheet metal components [9, 35], but very few have been built for the inspection of three-dimensional objects. As we have seen in Chapter 2, many three-dimensional vision systems have been assembled recently but these are mainly research systems only and more importantly are not specifically intended for inspection purposes. However, a few production systems *are* now being used in industry for special-purpose applications. For example, a three-dimensional laser system has recently been reported for inspecting positions of door and window housings in car bodies [99].

We will consider two-dimensional inspection tasks later in the book in Chapter 14. However, clearly only a limited class of objects or properties are suitable for inspection in this way.

10.3.2 Three-Dimensional Automated Visual Inspection

Given that coordinate measuring machines are not suited to a totally automated approach to three-dimensional geometric inspection and that two-dimensional vision can only solve a limited class of inspection problems, a logical alternative is to use accurate three-dimensional vision to perform this task [59].

In some respects three-dimensional vision for inspection purposes has certain simplifying aspects compared to some other computer vision problems. For example, the object is known in advance and inspection can be carried out in an carefully controlled environment. However, in terms of operational accuracy, inspection imposes the most demanding tasks on a vision system. Until recently, three-dimensional vision ac-

10.3. CURRENT INDUSTRIAL VISION SYSTEMS

curate enough for typical geometric inspection tasks was not possible due to hardware limitations [59].

Whilst three-dimensional vision may seem well suited to a totally automated approach, any vision system does have drawbacks which should be noted. However, the potential advantages of a vision system outweigh these drawbacks. The advantages and disadvantages of using three-dimensional vision for an inspection system compared with other methods (CMMs in particular) are summarised below. Several of these have already been mentioned.

Advantages

Flexibility — A potentially greater class of objects can be inspected by a non-contact means. Touching a highly polished surface with a probe may not be acceptable, for example.

Speed — Many more readings can be made in a given time using non-contact means.

Reliability — Any inferences made about surfaces in particular should be more reliable because of the much greater number of readings taken. Inspection by coordinate measuring machines typically uses four or five readings to test a plane or a cylindrical hole respectively [35, 194]. As three points define a plane, and four a cylinder, in each case there is only one reading serving to check the goodness of the fit. Using visual data, a feature might be typically sampled at about 1mm intervals over most of its visible area, giving many hundreds of readings for a typical face of the object. Least-squares techniques (see Appendix B) can then be applied to obtain the best fit to the surface, and a measure of the goodness of fit

Automatic Registration — A vision based system can determine the position and orientation of the object before inspection takes place, eliminating the need to place the object in a known position or to register it manually. Whilst this *could* be achieved using a CMM, the time required to gather sufficient data to accurately recognise an arbitrary pose would be very much greater. Also, it may not be desirable for a tactile probe to blindly gather data, as mentioned above.

Increased Productivity — A visual inspection system enables small batch jobs to be inspected efficiently, something that cannot be achieved cost effectively using present technology. Also due to speed increases (over CMMs in particular) many more objects, and hence a larger sample of the articles manufactured, can be tested, thus increasing reliability. The inspection system could also be incorporated *online* into the manufacturing process. Assuming the whole manufacturing and inspection plant is computerised, information about any defects detected could be passed back to the manufacturing stage so that remedies could be instigated.

Disadvantages

Lack of Access — Clearly a visual inspection system can only check visible features. Consequently, any features that can not be seen from any position of the cameras, such as bends in bores through an object, can not be checked by this means. Also it may be difficult to reliably inspect features that can not be seen in their totality from a single viewpoint.

Resolution — Any vision system has a resolution to which it may reliably gather data (*e.g.* a CCD camera has a finite pixel resolution). Thus any features smaller than the resolution of the system such as small holes and screw threads cannot be inspected. A CMM probe is capable of sampling at a much smaller resolution than the size of a pixel.

10.4 Three-Dimensional Data

10.4.1 Introduction

Various different techniques exist to obtain three-dimensional depth measurements from a scene, many of which were discussed in Chapter 2. The suitability of these methods will now be assessed for use in

10.4. THREE-DIMENSIONAL DATA

inspection tasks in particular. Initially, we must consider the requirements we have of a vision acquisition system used for inspection.

10.4.2 Requirements of a Vision Acquisition System

A vision acquisition system that is used to provide input for industrial three-dimensional geometric inspection tasks must satisfy requirements in the following areas.

Flexibility — The system must be capable of providing data from a wide variety of objects.

Depth Resolution — The depth resolution of the system must be capable of providing measurements that are accurate enough to establish whether or not the object under test lies within tolerance limits required by typical inspection procedures (taken here as 0.25mm).

Spatial Resolution — The number of discrete x and y measurements taken across the field of view must be at least large enough to accurately locate the features of the object. Inspection requires not only the carrying out of accurate local measuring tasks, such as finding the diameter of holes, but also the ability to decide whether features of the inspected object are positioned and oriented accurately relative to some global reference such as a datum face.

Ease of Use — The conversion of the sensed information into suitable three-dimensional data must be as simple a task as possible. For any realistic inspection system this task must be performed reliably and quickly since any problems arising at this stage are likely to be unmanageable in the subsequent segmentation, matching and inspection stages.

10.4.3 Possible Data Acquisition Methods

Possible methods for acquiring three-dimensional data fall into one of a few categories which we discussed in Chapter 2.

From our previous study of these techniques it is obvious that most of these methods do not satisfy the requirements of a visual inspection system as outlined above. Laser rangefinding systems do not have sufficient depth resolution. Methods based on structured light and shape from shading are not suited for inspection purposes for the following reasons:

- The methods are too sensitive to small changes in illumination and surface reflectance.

- The methods only work well on surfaces of uniform texture density.

- It is difficult to infer absolute depths of points, and only surface orientation or curvature is easily inferred.

- The depth resolution is generally not good enough or the field of view and close range of operation is not suited to practical applications.

Moire fringe methods are potentially very accurate in terms of depth resolution. However, certain drawbacks such as minor ambiguities in determining surface shape and the small field of view needed for accurate sensing, as discussed in Section 2.4, make the Moire fringe techniques unsuitable for general inspection tasks. However, for detailed inspection of parts of the object to high accuracies the use of Moire fringe techniques has much potential as will be discussed in forthcoming chapters.

Passive stereoscopic techniques involve determining the correspondence of two-dimensional features in each of the two images. This requires reliable extraction of these features from the separate two-dimensional images and the matching of these features between images. Both of these tasks are non-trivial and can be computationally complex. Furthermore, they generally involve only using a few points of interest such as corners, and centres of gravity, or at best, edges. Therefore, depth maps may take a long time to produce with these methods, and even then, may not possess sufficiently accurate information.

The problems of correspondence of passive stereoscopic techniques may be overcome by illuminating the scene with a strong source of

10.4. THREE-DIMENSIONAL DATA

light, normally in the form of a point or line of laser light, which can be observed by both cameras to provide corresponding points in each image. Depth maps can then be produced by sweeping the light source across the whole scene, as described in Section 2.5. The use of such active stereoscopic methods seems to provide the best overall method satisfying the requirements of a vision acquisition system to be used for geometric three-dimensional inspection.

Clearly our active vision system, described in Section 2.5, that provides dense depth maps (see Chapter 2), falls into the last of the above categories. Note that the depth maps obtained are capable of providing a complete picture of a scene with both local and global knowledge of features. Information about features is more likely to be accurate when a large number of measured depth readings per unit area is taken, and when least squares fitting or similar techniques are employed when segmenting these features from the data (Chapter 4).

We now summarise the capabilities of that system with relevant comments on its inspection capabilities.

10.4.4 Our Vision Acquisition System

A prototype vision system capable of producing depth maps has been described in Chapter 2. The typical size of the depth maps is 256×256 pixels for an area of interest $0.5m^2$.

By analysing the performance of the calibrated system it has been shown that depth measurements are not quite accurate enough to achieve geometric inspection to the desired tolerances. However, it *has* been shown that depth measurements can be achieved that are accurate to within 0.5mm which is not much greater an error than that required in practice.

The reasons for the inaccuracies in the vision acquisition system are currently being investigated. Whilst they could be due to a variety of sources the most likely cause would seem to be camera calibration problems, caused by any or all of errors in the calibration procedure, an insufficiently detailed camera model or poor camera performance. Whilst it is difficult to deduce the exact source of these errors it is not unrealistic to expect this problem to be resolved in the near future. Research is currently being conducted into these matters and they will be discussed further in Chapter 15.

10.5 Matching for Inspection

10.5.1 Introduction

Over the last few years many model-based matching strategies have been developed. As we have already discussed in Chapter 8, these have been primarily conceived for object recognition purposes, although most produce information about the position and orientation of an object as a by-product. In this Section we shall discuss the suitability of the various matching methods for use in inspection systems. In fact, many of the details of implementation of a matching algorithm in Chapter 9 were concerned with inspection tasks in mind. Inspection requires highly accurate algorithms, explaining the adherence to techniques that produce accurate rotation and translation estimates in the matching algorithm. The methods described in Chapter 8 will now be assessed for their ability to provide suitable data with sufficiently accurate results to achieve inspection. In Chapter 4 we showed that planar surfaces, rather than other surface types or edges, are likely to provide the most accurate information from a segmentation of an image, and a method for segmenting planes was described. This clearly has a bearing on the choice of matching algorithm. However, even though planes are expected to be more accurately segmented than other primitives, it is still desirable to study matching procedures that can employ other primitives, since a complete matching and inspection strategy may require a selection of techniques capable of handling a range of different surface types.

10.5.2 Suitability of Matching Methods

Many of the matching methods described in Chapter 8 are not suited to inspection tasks, as they suffer from various serious drawbacks. We have already described in Chapter 9 a matching method that is very efficient at finding matches to models, providing very accurate rotation and translation estimates. We will show in this Section that this method is more suited to our intended inspection task than the other

10.5. MATCHING FOR INSPECTION

matching methods we have considered. Firstly we shall highlight some initial drawbacks of other matching methods

Some of the matching methods, such as the Nevatia and Binford method and the *ACRONYM* method, only approximate the actual surfaces in the scene by sets of generalised surfaces, which is obviously not accurate enough for inspection. Others such as the *SCERPO* method infer three-dimensional objects from two-dimensional images, which is both computationally intensive and of limited accuracy.

The Oshima and Shirai method, discussed in Section 8.3, compiles object models offline from selected views of an object and uses this model to match objects viewed from an arbitrary position. This approach has two main drawbacks:

1. For every new object a learning phase is required. In order to reliably and completely describe the object from a series of three-dimensional scenes the number of views required by the learning phase will be large. Also, to a certain extent the need for a learning phase defeats the goal of a totally automated inspection system.

2. The geometry is only implicitly encoded since the method is data driven and therefore requires a stable segmentation method for reliable results (*i.e.* the same feature descriptions must be obtained regardless of variations in the viewpoint or starting point in the image used for the segmentation). Unfortunately, there is no obvious way to guarantee a stable segmentation. In particular, if any view of an object has been incorrectly segmented (or even omitted) during the learning process, this will cause difficulties in matching for *every* object being inspected.

One advantage of the Oshima and Shirai method, however, is that it employs nearly all the information that may be extracted from a scene (numerous surface properties and relationships between surfaces) to drive the matching.

Relaxation methods pose the matching problem as a probabilistic labelling problem, based on feature properties such as area, connectivity *etc.*, and relaxation processes are used to find the most probable match. Such methods have two major drawbacks that make them unsuited for inspection purposes [80]:

1. They are very sensitive to both the initial values chosen for the probabilities and the annealing schedule adopted since local rather than global interaction of features is considered. Thus convergence towards a best match is not always guaranteed and the method may stop at a local rather than global optimum. This means that sometimes only a partial match may be found which may not form a part of the best possible match that could be found.

2. Global information cannot be used to attempt to solve the above problem otherwise the compatibility measures (usually based on local nearest neighbour functions, as discussed in Chapters 5 and 8) become intractable [80, 128].

Hough transform methods are inadequate for solving the matching problem for the following reason. The Hough space for three-dimensional matching is six-dimensional (there are six degrees of freedom in the transformation — three rotational and three translational), and the accuracy of the method can only be guaranteed if the Hough space is divided up into small partitions (requiring a large six-dimensional array). This is far too computationally expensive (both in time and memory) for the level of accuracy required for inspection. Furthermore, the Hough transform does not have an efficient and simple method of employing constraints to reject grossly mismatching primitives which could improve the method.

Extended Gaussian images (EGIs) map surface normals of faces of an object onto a unit (Gaussian) sphere. Objects and models can then be matched by comparing their EGIs. A major drawback with the extended Gaussian image approach is that a unique EGI can only be guaranteed for convex objects. As we have seen in Section 5.5.2 and Fig. 5.12, there are an infinite number of non-convex objects that possess the same EGI. Some approaches [122, 124] have attempted to resolve this problem by employing surface normal histograms, as described in Section 5.5.2, for every view belonging to a discrete set of views. Matching is then done between the EGIs and histograms, but this increases the complexity of the matching problem.

Grimson has shown [109] that interpretation tree searching methods, as discussed in Chapter 8, are very efficient at finding a match when only a single object is present in a scene and it is being compared to a single model. Thus, techniques based on such tree searching techniques seem most appropriate to inspection problems, as this criterion is easily met. However, some such tree searches like the original Grimson and Lozano-Perez, Murray and *TINA* methods use only point or edge based data. We have shown in Chapter 4 that using points or edges is not a good idea due to errors in the sensing of laser stripes at an object's physical edges due to specular reflection of laser light. Other matching methods, such as Fan's method and the *IMAGINE* method, use edge information as an initial stage for hypothesising possible groups of features that can be used as initial data to help reduce the complexity of the matching process. Again, problems with extracting edge information mean that these methods are not well suited to inspection tasks, although some aspects of both these methods are similar to those used in our system.

Finally, then, our suggested matching method for geometric inspection purposes is to use a tree searching approach to obtain a match of planar features in the scene. This provides a very accurate description of the position and orientation of the object. We then suggest the use of a hypothesis and test strategy to accurately determine if non-planar features are present and correct in the test object. In more detail, it is easy to adapt the Faugeras and Hebert approach for use with non-planar primitives [77, 80]. This is not the case with the Grimson and Lozano-Perez method and other related methods which were designed only for planar objects. For these reasons, the Faugeras and Hebert approach is suggested as the best strategy to adopt as a basis for inspection. Details of the implementation of this method with suitable modifications for the inspection tasks and a practical assessment of these methods were given in Chapter 9.

10.6 Outline of Part 3

Most of Part 3 of this book will deal with the design and implementation of a visual inspection system capable of inspecting manufactured parts to typical geometric tolerances. We shall also address the issues of other inspection tasks such as two-dimensional geometric inspection, surface finish inspection and crack detection in this Part.

In order to perform geometric inspection, we need to extend current geometric models of objects from Chapter 7 to include the representation of tolerance information. However, the notion of conventional geometric tolerances as used in engineering drawings and defined in appropriate international or British standards [38] are inherently ambiguous. Chapter 11 deals with these issues and also the problems of actually measuring and testing features for conformance to tolerances.

In the following Chapters 12 and 13 we develop a strategy that will enable the complete inspection of an object once it has been placed within the inspection system.

Chapter 14 considers various other inspection topics which our system does not address, such as inspection of surface finish and two-dimensional inspection. We shall briefly describe some methods for carrying out such types of inspection visually, and address some of the problems involved in such tasks. We shall also consider various application areas of these techniques.

The final Chapter 15 provides an assessment of our work to date, and also points to future directions of work in the field of computer vision, both in general, and with specific reference to automated visual inspection.

Chapter 11

Geometric Tolerances

11.1 Introduction

Tolerances are important since they define acceptable errors in parts that arise due to imperfect manufacturing processes. Thus, we need to incorporate such tolerances into computer stored solid models of objects in order to perform a variety of design, manufacturing and automation tasks. Clearly any such process must have the facility to reason with these inaccuracies as well as to represent them.

Tolerances are applicable to many part attributes from surface finish and texture to part dimensions and operational properties (*e.g.* operating temperature). We concern ourselves only with *geometric* tolerances. In particular, we are interested in variations in the dimensions of features of the part (both in terms of shape and size) and variations in inter-feature relationships (such as the angle between two planar faces).

In this Chapter we discuss a method of representing tolerances in solid models of the type described in Chapter 7. We then develop a method for measuring the conformance of objects to desired tolerances using three-dimensional depth data obtained from a vision acquisition system. Practical results of using such a system will be presented in the next Chapter after we have further considered inspection strategies.

11.2 Tolerance Representation

The representation of tolerance information in a geometric model is essential not only for inspection tasks but also for many other manufacturing processes, including part manufacture planning, tight part assembly and part design. It is perhaps surprising, given the current large scale popularity of computer aided design systems for creating and storing solid models, that the problem of representing tolerances has not been given greater attention. However, most modelling systems are still not able to adequately support tolerancing information [89, 198, 199, 227, 231]. This is due in part to the fact that current industrial tolerancing techniques, employing a mix of conventional tolerancing (plus-or-minus tolerances, see Fig. 11.1) and limited geometric tolerancing, lead to inherent ambiguities. These ambiguities are normally resolved by humans, usually with reference to implied data that can not be explicitly stated (*e.g.* the mechanical function of the part).

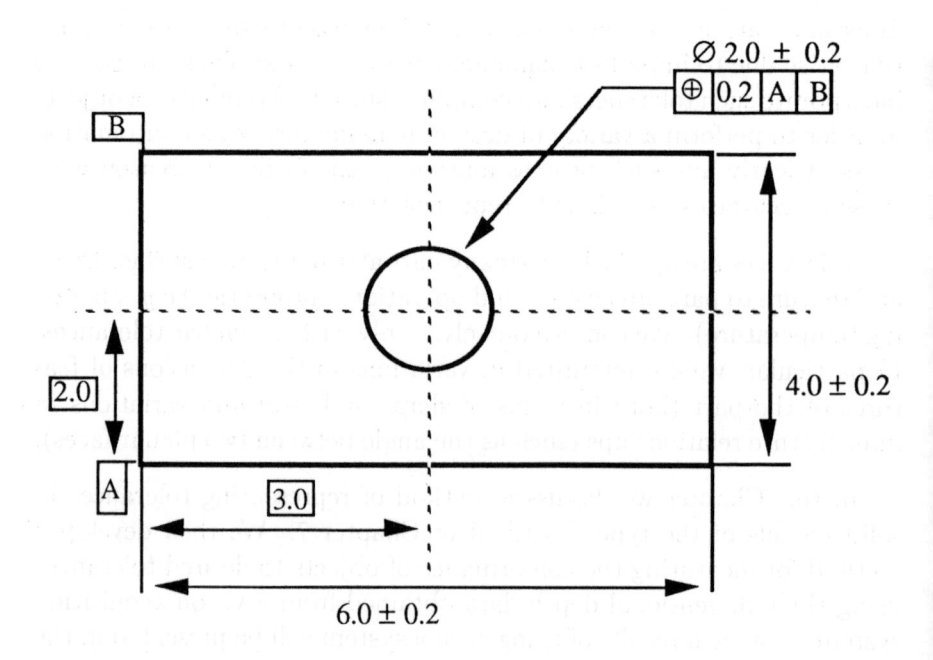

Figure 11.1: Conventional tolerancing

11.2. TOLERANCE REPRESENTATION

Traditional plus-or-minus dimensional tolerances can be interpreted in two (and perhaps more) ways as illustrated by Figs. 11.2 and 11.3.

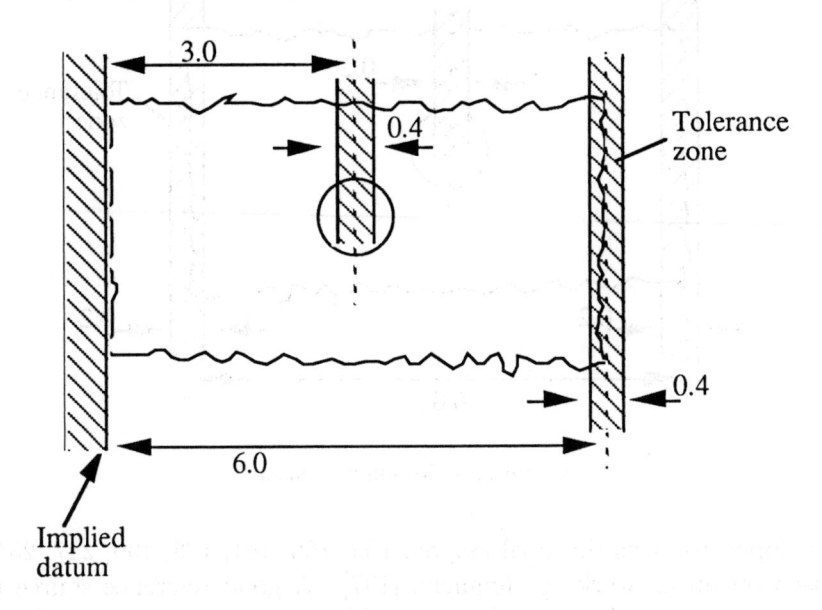

Figure 11.2: Implied datum

In Fig. 11.2 a plane tangent to the left face of an object has been constructed and is treated as a *datum*. A datum is a reference feature from which measurements can be made, for example of distance or angle, to other features. We shall discuss datum systems in more detail shortly. All features must lie within tolerance with respect to this datum. In the particular case shown the centre of the hole is required to lie within 3mm ±0.2mm from the datum and the right-hand face must lie within the area (the *tolerance zone*) bounded by two planes as shown.

In Fig. 11.3 where no datums are assumed the left-hand and right-hand faces must lie within two tolerance zones as must the hole.

Although initial attempts to describe tolerances in solid modelling systems were based on conventional tolerances [117], it became obvious that a more mathematically robust definition of geometric tolerance representation and measurement was required for automated processes involving reasoning with tolerances. Recently, approaches have been

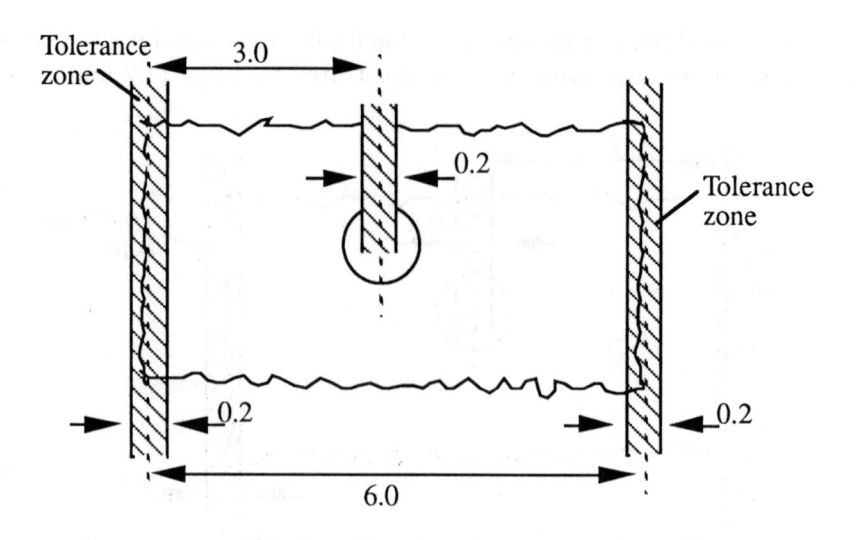

Figure 11.3: No implied datum

developed towards this goal [88, 89, 159, 160, 161, 198, 199, 227, 231], based on initial work by Requicha [197]. A good reference source to most aspects of tolerancing is provided by the textbook by Zeid [245]. The basic theory of geometric tolerancing methods will now be discussed, and developed further for the particular task of inspection.

The meaning of geometric tolerances is defined at length in the appropriate British Standards [38]. Basically, geometric tolerances are concerned with the notions of feature description and size based on cylindricity, roundness, position, perpendicularity, and so on. However, these definitions lack significant robustness in at least two areas [197]:

1. There is no formal definition of the concept of size. For example, if an intended cylindrical hole has been manufactured to be non-cylindrical the way in which its radius is then to be measured is not made clear.

2. There are no precise definitions of the geometric features concerned. For example, there is no method of describing the incorrectness of shape of a cylindrical hole when it has been manufactured to be non-cylindrical, even if it has the correct radius as measured above.

11.2. TOLERANCE REPRESENTATION

In order to overcome such problems methods based on *tolerance zones* are used. Tolerance zones are derived from the model of the perfect object, and they define the boundaries within which each of the surfaces of the imperfect object must lie in order satisfy a particular tolerance constraint. Requicha [197] proposed that each feature should have associated with it three tolerance attributes and thus three tolerance zones. These are discussed in turn using a simple two-dimensional example of a hole:

Size tolerance T_s —A circular hole of radius r satisfies this tolerance criterion if its boundary lies entirely within an annular tolerance zone defined by two concentric circles of radii $r+T_s/2$ and $r-T_s/2$ as shown in Fig. 11.4. The position of this zone is arbitrary so that it can be adjusted so that the surface lies within the zone if possible. For other types of zone, the orientation of the zone may be altered as well as its position when deciding if the feature is in tolerance. Obviously this is not applicable to the case considered, because the tolerance zone has circular symmetry.

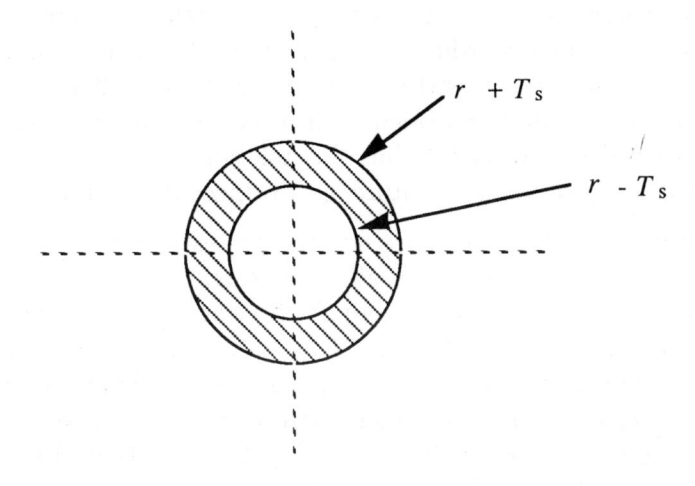

Figure 11.4: Size tolerance

Form tolerance T_f —The tolerance zone is an arbitrarily positioned annulus defined by two concentric circles of arbitrary radii r_1 and r_2 that must satisfy $r_1 - r_2 = T_f$ as shown in Fig. 11.5. This tolerance criterion is used to ensure that the circularity of the hole is within defined limits.

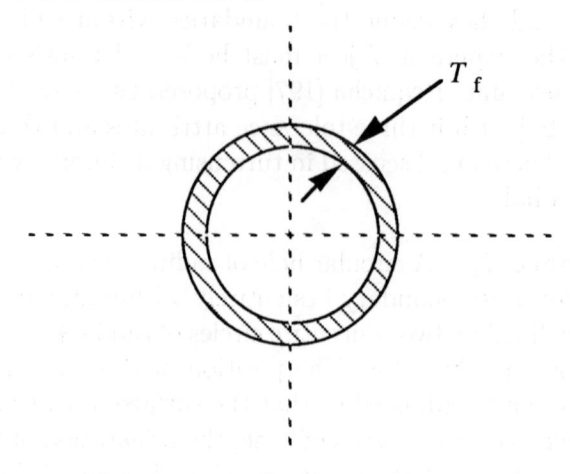

Figure 11.5: Form tolerance

Position tolerance T_p —All points of the circle, measured with respect to some coordinate system, must lie within the annulus defined by two concentric circles of radii $r + T_p/2$ and $r - T_p/2$, correctly located and oriented in that coordinate system, as illustrated in Fig. 11.6. Thus this tolerance criterion ensures the absolute position of a feature with respect to a datum system.

Each of these three tolerance criteria are tested independently and an acceptable feature must pass all three tests.

It should be noted that an object meeting a size tolerance must necessarily satisfy a form tolerance if $T_f \geq T_s$. Also meeting a position tolerance implies the satisfaction of form or size tolerances whenever $T_f \geq T_p$ or $T_s \geq T_p$. However unfortunately, the converses are not true, and in practice it is usual that $T_f \leq T_s \leq T_p$, with typical values, for industrial inspection, being of the order of 0.2mm [89, 159, 160, 161, 197, 227, 231].

The above tolerances have simple meanings in two-dimensions but have to be generalised to three dimensions to be of use with solid objects. In place of two-dimensional tolerance zones, Requicha [197] proposes forming two volumes that define the maximum and minimum

11.2. TOLERANCE REPRESENTATION

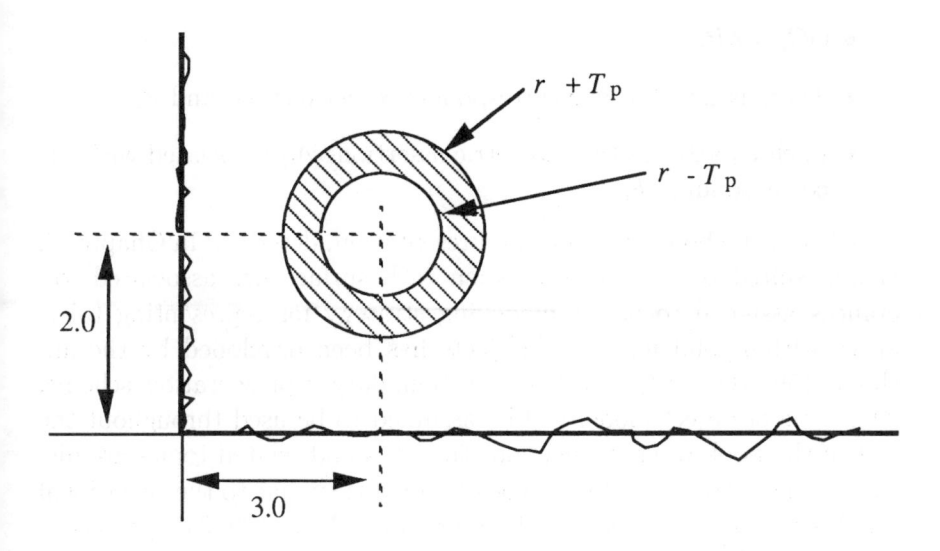

Figure 11.6: Position tolerance

material conditions (MMC and LMC) that a manufactured part can possess to meet the three types of tolerance constraint. These will be described further later in this Chapter.

Requicha [197] proposes that a *tolerance specification* for a solid S must consist of:

- An unambiguous representation of S, as defined in the discussions on general geometric modelling of objects in Chapter 7.

- A representation for a decomposition of the boundary of S, δS, into subsets F_i (where $\cup F_i = \delta S$). These subsets are called *nominal surface features* and should be distinct. Normally these features amount to faces on the object or perhaps collections of faces.

- A collection of geometric assertions A_{ij} about the nominal features of S. These are represented by *attributes* of each feature F_i.

The inspection task for a real part may now be defined as follows. A representation of a physical part P satisfies a corresponding tolerance specification of the part if and only if there exists a decomposition of δP into subsets G_i, called *actual surface features*, such that

- $\cup G_i = \delta P$.

- There is a one-to-one correspondence between G_i and F_i.

- Each G_i satisfies the geometric assertions A_{ij} associated with the corresponding F_i.

Clearly the boundary model representation, described in Chapter 7, is well suited to this purpose when each surface has associated tolerances assigned to it. A modelling strategy for representing tolerances within solid models of objects has been developed by the authors [159, 160, 161] based on the boundary representation scheme. This type of object model will be assumed to be used throughout the rest of the book where inspection strategies and related topics are discussed. The structure of this type of model is similar to the variational graph structure (*Vgraph*) that Requicha and Chan [199, 245] proposed, although their structure is built on top of a set-theoretic modeller rather than a boundary representation modeller.

11.3 Implementing Datum Systems

When verifying the positions of object features, a datum or one or more datum systems must be specified in the model in order to construct coordinate systems in which appropriate measurements can be taken. For our purposes, a datum is taken to be a plane, a straight line or a specific point embedded in or on the surface of an object which can be readily found by the vision system. However consideration of segmentation (Chapter 4) and matching (Chapter 8) of acquired images leads us to generally use only planes as datum primitives for the inspection problems addressed here, unless explicitly required otherwise by the specifications of the object. This is because only planes can be extracted from depth data to the required accuracy.

In this Section we shall discuss some practical details in the use of datum systems. These observations reflect our actual experience of using datum systems as well as addressing some of the theoretical considerations. The choice of appropriate datum systems is important for driving automatic inspection and other processes which will be described in subsequent Chapters.

11.3. IMPLEMENTING DATUM SYSTEMS

At present all datum systems are specified manually in our inspection system. In particular one specific datum is required as a basis so that matching and inspection processes can be driven automatically (see Chapter 13). We assume that this datum can be found by the vision system in order for the inspection process to proceed. This datum is a plane which can be uniquely identified in the object (this might be the plane of largest area in the case of a convex object). The position and orientation of this plane are then used to uniquely define the object's initial position and orientation.

It is important that the segmentation of any datum planes is reliable, otherwise the whole matching and inspection process will be compromised. Any plane used as a datum should possess the following properties.

- It should be easily visible and not occluded to a great extent for a wide range of camera viewpoints.

- It should have a large surface area.

- It should be able to render reliable depth readings (*e.g.* the surface texture should be uniform).

If a single datum plane which uniquely specifies the view and satisfies the above three conditions cannot be found then other datum planes that do satisfy these conditions and are visible in the initial view should be chosen to form a system of datum planes.

The datum or datum system that specifies the initial view of the object is called the primary datum or datum system. All subsequent measurements are taken with respect to the primary datum (or system) unless specified otherwise by the model. Any subsequent datum systems are said to be secondary. Secondary datum systems may be required or helpful if we need to represent specific feature inter-relationships required for the inspection. For example, it may be necessary to determine if two planes which cannot be seen in the same view are parallel.

11.4 Measuring Conformance to Desired Tolerances

The implementation of the theory of tolerancing is described here. Results are presented in the next Chapter where we develop inspection strategies further. The implementation reflects our experience gained with the vision system we have described throughout the book.

As discussed above each primitive has to satisfy three tolerance tests in order to pass the inspection.

It is assumed that the primary datum feature from the model has been matched in the view, so that a unique view of the object is guaranteed. All measurements are then taken with respect to the primary coordinate system, with additional measurements to other, secondary, datum systems being made only when specified by the model.

Instead of forming overall material tolerance zones for a whole object model and testing each view of the model against this, each model primitive is tested for tolerance separately. Any required primitive inter-relationship tolerances such as angles between planes are usually measured with the aid of secondary datum systems. This is often done by choosing one of the primitives under test as the datum.

The testing of the three tolerance requirements is carried out as follows using the depth map points hypothesised as belonging to the relevant feature.

Position tolerance — The position tolerance of a primitive is satisfied if all depth points which are hypothesised as belonging to this model primitive are within the required tolerance zone.

Size tolerance — The size tolerance of a primitive is satisfied if all depth points are within a specified distance of an arbitrarily positioned primitive of exactly the same dimensions as the model primitive. in the case of planes, the best fitting plane is available from the segmentation; the distance of each of the points belonging to the segmented plane can then be calculated to find the worst point. For cylindrical and spherical primitives the same approach is taken except that at the segmentation stage, the radius of the primitive is taken to be the same as the corresponding model primitive and is not approximated by the fitting process (see Exercises 4.10 and 4.11 and Appendix B).

Form tolerance — The form tolerance of a primitive is satisfied if all depth points belonging to the primitive are within a specified distance bound of an arbitrarily positioned primitive of arbitrary dimensions. This tolerance is tested for in the same manner as the size tolerance except that the radius of the cylinder and sphere are now considered to be variables which must also be approximated (see Chapter 4).

Since a position tolerance can imply both a size and form tolerance and a size tolerance can imply a form tolerance use of this fact can be made to make the testing of the tolerances more efficient:

- If the distance of all depth points measured for the position tolerance satisfies either of the size or form tolerances then these need not be tested separately.

- Similarly, if the form tolerance is already satisfied by the size tolerance measurements then the form tolerance does not need testing independently.

11.5 Points Lying Outside Tolerance

Since the errors in acquiring depth readings can be approximated by a Gaussian distribution (see Chapters 2 and 4, [55]) it is to be expected that occasionally a depth reading will lie outside the tolerance specifications. The question arises as to whether an object should fail a tolerance test because of a few data points which are in error.

Throughout the discussion of inspection we shall generally assume that it is better to fail a good object than to pass a defective object. In many manufacturing applications (*e.g.* aeroplane manufacture) the passing of a defective component could have drastic consequences. The vision system can easily classify objects which do not pass inspection as *certain failures* (objects with features missing or grossly out of tolerance) or *possible failures* (objects measured as being marginally out of tolerance). Consequently, it is assumed that if an object is placed into the latter category by the vision system it could be inspected more closely (perhaps with human assistance) later if desired to avoid unnecessary wastage.

The solution we adopt to occasional points out of tolerance is as follows. Since the distance across the surface being inspected between successive depth readings is at best 1mm, if two or more depth readings lying next to each other are out of tolerance then it is highly likely that this indicates a defect with a size of at least 1mm and it is possible for surface defects such as dents or even small holes to be of such a size. We do not wish to allow objects with such problems to pass the inspection process.

However, if one data point lies outside tolerance and its nearest neighbours lie within tolerance then, rather than immediately failing the object, the bad point is considered further to see if it fits in with the overall distribution of points, using the statistical theory of *outliers* [8] to solve the problem.

The theory of outliers can been applied to many problems to decide if certain points fit in with the overall distribution of points or if they lie outside the distribution. Many methods have been developed [8] to solve such problems for many different statistical distributions and for varying numbers of simultaneously outlying points. Since only one outlying point at a time is being considered in the current application to inspection, and the assumed (Gaussian) distribution is symmetric, a simple outlier test may be applied [8].

The basic method for deciding if a point is an outlier and hence can be ignored is:

1. A sample of points in the neighbourhood of the suspected outlier is chosen. We use a region of 5×5 depth readings centred on the suspected outlier. If the region overlaps an edge then the region is translated so that the overlap does not occur but the suspected outlier is still contained somewhere in the region.

2. The sample mean, μ, and sample variance, s^2 are calculated. In this case

$$\mu = \frac{1}{n} \sum_i \sum_j z_{ij}$$

11.5. POINTS LYING OUTSIDE TOLERANCE

$$s^2 = \frac{1}{n-1} \sum_i \sum_j (z_{ij} - \mu)^2 \qquad (11.1)$$

where the summations are over all $n = 25$ depth readings, z_{ij}, in the neighbourhood.

3. A distribution statistic is calculated and compared to a standard value. If the distribution statistic is less than this value then the point lies within the distribution otherwise it is an outlier. An appropriate test statistic ξ for the point z_{ij} being a single outlier is (the *internally Studentized distribution from the mean* [8])

$$\xi = |\frac{z_{ij} - \mu}{s}| \qquad (11.2)$$

The relevant standard values for different significance levels are given in Appendix D. Thus, the test point is classified as an outlier if ξ is greater than the standard value given in the table. The probability of this classification being correct is given by the significance level.

Chapter 12

Geometric Inspection and Single Scenes

12.1 Introduction

We shall divide our consideration of the strategy for performing geometric inspection into two parts: inspection of the features visible in a single scene, which we will discuss in this Chapter, and complete object inspection using multiple views, which we shall describe in the next Chapter.

The discussion quite naturally splits in this way, because a method of inspecting a single view accurately is the underlying basis for any more complex inspection techniques employing multiple scenes. On the other hand, quite different issues must be considered to develop an inspection strategy that takes a suitable set of views of an object necessary for complete inspection of the object.

In this Chapter an overview is given of geometric inspection strategies. Next, a method is given for inspecting a single view of an object to decide if the features it contains are within geometric tolerance. The effects that errors arising from the various components of the system have on the outcome of inspection are also discussed. Finally, results are presented using the adopted inspection strategy in our experimental system.

12.2 Geometric Inspection Strategies

Clearly not all object features can be present in the same single view of an object. Therefore it is impossible to completely inspect an object from a single view. We must build up a set of views to give as complete a description of the object as necessary. There are two ways in which this can be done.

1. A set of views of the object is taken and compiled into a three-dimensional description. This can be done by moving either the cameras or object (*e.g.* by using a turntable) by known amounts relative to some coordinate system. Knowing the transformation between scenes, the relative positions of segmented features can be determined as the complete three-dimensional description is assembled. This description is then compared to the stored object model to perform inspection of the object.

2. A single view is taken at a time. Each view is inspected by comparing its features with the stored object model. The model is interrogated offline to compile the set of views to be used for inspection. The vision system is instructed to obtain these views and inspect them one-by-one.

Various methods exist for compiling complete three-dimensional descriptions from multiple views and matching these descriptions to three-dimensional models [19, 47, 76, 182, 192]. Some of these methods have been described in Chapter 8. These methods typically take views at regular intervals around the object (at $15° - - - 30°$ steps) and whilst such a set of views is generally sufficient for a single axis description of the object (we of course cannot obtain any surface information for faces lying downwards), such a set is not necessarily suitable for inspection purposes for the following reasons.

- In inspecting an object it will be necessary to measure relative positions or orientations of various pairs (or sets) of features. It may be possible that a single view exists which may contain both of features of a given pair, but that none of the set of regularly spaced views does. Alternatively, a regularly spaced set of views

12.2. GEOMETRIC INSPECTION STRATEGIES

may contain poor depth readings for some features, or some features may be partially or completely occluded. It is clearly necessary, for inspection purposes, that carefully chosen views should be used which allow the accurate measurement of single features and accurate relative measurements of pairs of features. There is no guarantee that arbitrary views can achieve this.

- Several of the views taken at fixed intervals may be redundant in that no new information is present in a given view that is not present in various other views. All of the stages involved in inspecting a scene are time consuming and clearly inspecting redundant scenes is extremely inefficient. For example, to measure all adjacent pairs of vertical faces on a cuboid object, four views 90° apart are sufficient compared to perhaps 24 views that might be used in a standard fixed rotation method.

- It may be difficult to register two or more views together sufficiently accurately for inspection purposes if the hardware has inadequately fine control over the position of the object or cameras. This may particularly be the case when robot arms are used for repositioning the object. When trying to build up a single model from multiple scenes it is not easy to recover from such problems. If a pair of features whose relative positions need to be measured can be guaranteed to both be present in a single scene, we have a remedy. If it is suspected that positioning problems have arisen, we can simply reregister the current view (by reapplying the matching processes).

- The nature of a particular vision system may impose conditions on how suitable views are to be chosen, for accuracy or other reasons. For example in our current inspection system (see Section 2.5) which uses laser light and two cameras to obtain depth maps of a scene, it is necessary that features to be measured are as nearly perpendicular as possible to the line of sight of both cameras. This is because the stripes of laser light have to be correlated in the two views. If a feature subtends a glancing angle relative to one of the cameras then large errors in depth readings will result, as described in Chapter 2. Clearly, taking arbitrary views of objects may well lead to many problems of this nature.

In summary, then, what is required is a method that reasons from the stored model of the object and knowledge of the camera positions to decide upon a best set of views to be used. This set should allow accurate determination of relative positions of features, whilst also using a minimal set of views to achieve the desired results. The choice and acquisition of these views is described in the next Chapter.

In the rest of this Chapter we develop and justify some initial inspection strategies (suitable for single scene inspection) before developing strategies that can allow complete object inspection in the next Chapter.

12.3 Single Scene Inspection

As already explained in Section 9.7, we may assume that in general the results of matching will have given the location and orientation of the object to within a small error. Thus, after matching has been performed, visible features extracted from this view of the object can be compared to the stored model using this transformation between the coordinate systems, and their positions verified, as required by inspection. Moreover, this information can be used to drive the inspection of more than one scene. Once the position of the object is known, if accurate manipulation hardware is available, the object can be repositioned to give different views of the object in which the position of the object is already known. Therefore, after the initial match, features that should be visible in subsequent views can be hypothesised from the object model. Matching of these views is simplified to just testing (inspecting) the segmented view for correct occurrences and placements of the hypothesised features.

Using the above scheme it is clear that we can obtain information for each view as we execute the process. We shall now concern ourselves with inspecting a single view of an object. Clearly, when the inspection of a single view is considered only limited inspection can take place in that only features that are visible can be checked.

In detail, the first view is inspected as follows:

1. From the matching of the model to the segmented image a list of matched features and the transformation between scene and model coordinates are found. This transformation is then applied to the model and a list of model features predicted to be visible is compiled. We shall also assume here that the position and orientation of the primary datum or datum system is available, and is either obtained from this view or was obtained from a previous view. This will be discussed more fully in the next Chapter.

2. Knowing the hypothesised position of a visible model feature (other than a datum) and the dimensions of the depth map, the corresponding position and orientation of each feature to be inspected in this view can be postulated. Inspection takes place by testing if the points in the depth map contained within each feature lie within their respective tolerance zones, as discussed in Chapter 11.

12.4 Errors in our Inspection System

Before analysing the performance of our inspection strategy outlined above the errors introduced at each stage of the inspection system are summarised since they have a direct influence on the results obtained. The sources of error are illustrated in the Fig. 12.1.

Each error that occurs will contribute to the error produced by subsequent stages. Thus the whole inspection process is affected by a combination of camera, segmentation and matching errors, and as a whole will be accurate only if each of these individual components is capable of producing results of sufficient accuracy. The source of these errors are now briefly recapped.

As discussed in Chapter 9, the errors due to the matching process can be kept small. Knowing the vision acquisition system camera setup also allows size of the errors in segmentation to be estimated.

The results in Parts 1 and 2 of the book have shown that, using a typical (ideal) camera set up, the predicted errors in the system due to segmentation and matching (determined by using simulated data)

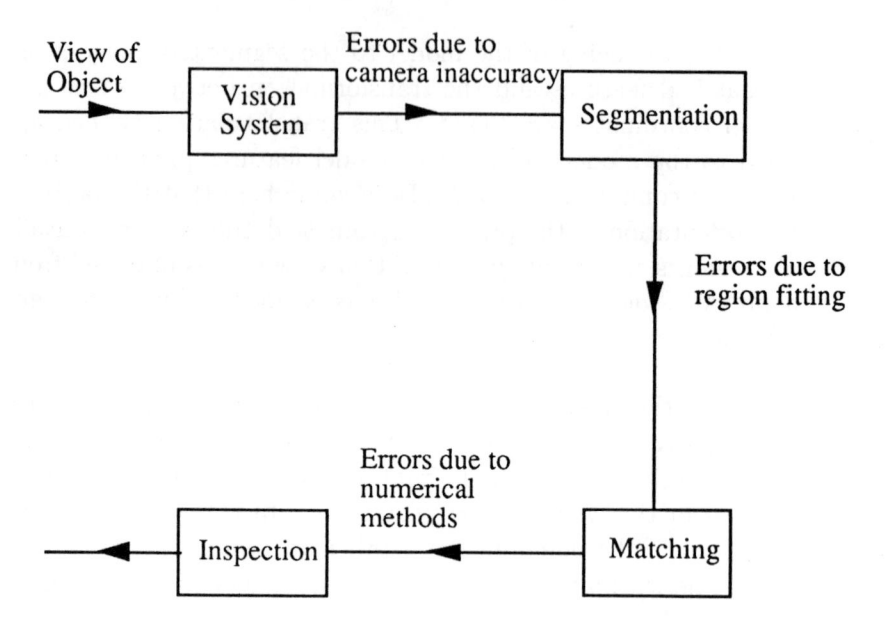

Figure 12.1: Errors in a vision system

are sufficiently small to allow inspection to be carried out. In practice, camera calibration errors make inspection to typical engineering tolerances not quite possible.

12.5 Single Scene Inspection Results

Results on the use of the above outlined inspection strategy, on real and artificial data, are presented in this Section.

12.5.1 Results on Artificial Data

The results presented in this Section correspond to starting with artificial depth maps with added noise, and using each of the processes described in turn to produce an estimate of location and orientation of a model. Finally, this estimate is then used to align the geometric model with the sensed data, and the visible features are inspected using the original depth map. This allows the effects of noise inherent

12.5. SINGLE SCENE INSPECTION RESULTS

in the system to be studied, and the effects of altering camera range. Tests have been performed on the same data as in previous Sections, in particular Sections 2.5, 9.6 and 9.7 which considered camera calibration, segmentation and matching respectively. Three different tolerance values, namely 0.25mm (0.01inch), 0.125mm and 0.025mm, have been used to reflect the fact that typical inspection tolerances are 0.25mm, while detailed inspection tolerances are usually no less than 0.025mm. All three attributes of position, size and form tolerance are set to the same value. Conformance to position tolerance is tested first, and if this is satisfied, the form and size tolerances must also necessarily be satisfied. If it is not, then size tolerance conformance is tested next, and in a similar way, form tolerance conformance is tested last if not implied by the size tolerance test. Note however, that failure of any of these tests implies that the object is faulty. The further tests just serve to indicate how the object is at fault.

The results are summarised in the graphs shown in Figs 12.2, 12.3 and 12.4. Each point is tested for conformance to the tolerance limit. If it is outside tolerance, an outlier test is then used. The bars in these graphs show percentages of points which either pass the first test or are not classed as outliers. They show the inspection of three simulated test objects with different errors in depth readings added as described in Section 2.6. The results of the matching strategy developed throughout this book are used as input to the inspection routine. Results on the accuracy of the matching strategy when presented with the depth data used in this Section are given in Section 9.7.1.

As can be seen from the simulation results using the widget, all points pass the test inspection is possible to within the normal inspection tolerance of 0.254mm when the vision acquisition system produces depth errors of the expected size (0.15mm). In the case of the cube, which was placed at a simulated 1.212m from the vision system, which is further than the working range normally used, the depth errors would have to be slightly smaller than this for successful inspection.

The policy of not using non-planar features for matching but instead only inspecting them seems to work well, as in the case of a widget with a cylindrical hole (see Fig. 2.5), where little degradation of the results from the widget without the hole (see Fig. 12.3) are obtained. Remember that we use the stored object model to predict the location

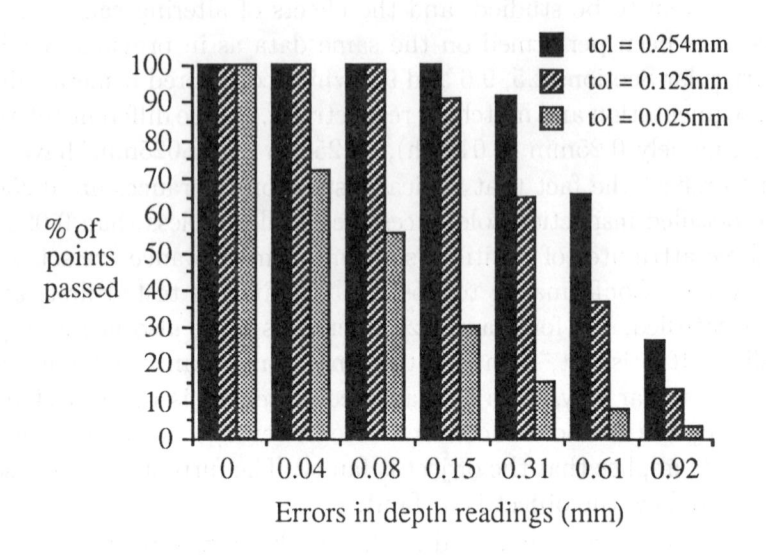

Figure 12.2: Analysis of inspection of a widget

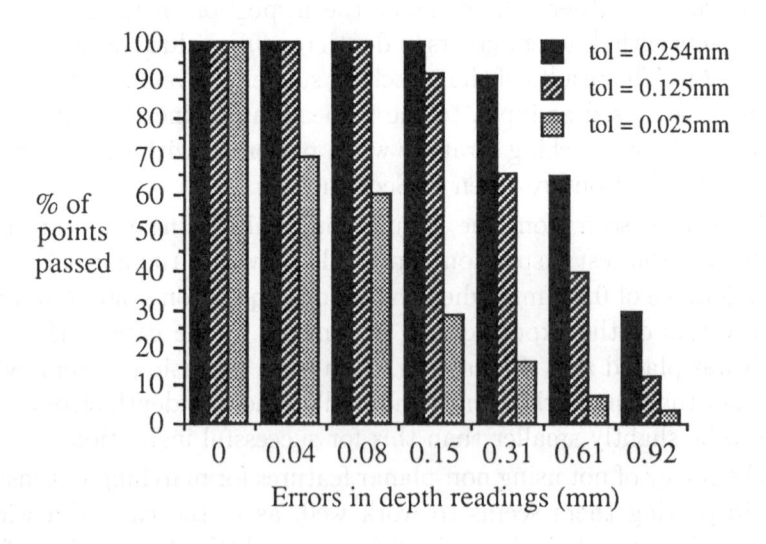

Figure 12.3: Analysis of inspection of a widget with a cylindrical hole

12.5. SINGLE SCENE INSPECTION RESULTS

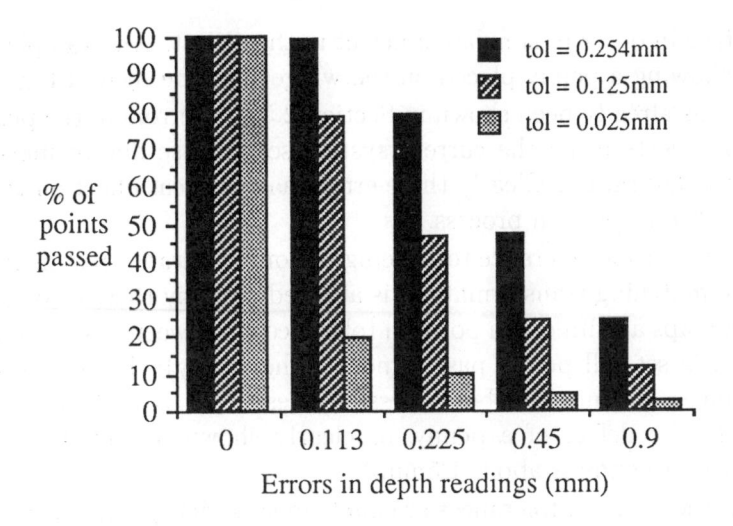

Figure 12.4: Analysis of inspection of a cube

and size of non-planar features. These are *implicitly* segmented, which allows us to decide which depth points are considered to belong to this feature for the purposes of testing conformance to tolerances.

Assuming expected depth errors of 0.15mm and inspecting to normal inspection tolerances (0.254mm), 0.2% of depth points were considered by the outlier test (and passed it). Some outliers are to be expected since the artificial data is simulated by adding errors from a Gaussian distribution. However as either the simulated depth errors were increased (up to 0.6mm), or the inspection tolerances were made stricter (0.0254mm with depth errors of 0.08mm) the number of points tested for being outliers increased to about 1% of the total number of depth points, and some of them failed the test. If the depth errors were increased still further then points outside tolerance became too dense for the particular *single* outlier test to be of use, as many neighbourhoods now contained more than one outlier (see Section 11.5).

12.5.2 Results on Real Data

Results in this Section are presented on real data gathered from several single views of the widget. In some cases a one penny coin was placed on a face in order to simulate a defect in the object. An example depth map showing a penny placed on the widget may be seen in Fig. 2.6.

It has already been shown in Section 2.5 that errors in the positioning of objects using the current system setup arise due to inaccurate camera calibration. Clearly these errors have a great effect on the success of the inspection process.

The position tolerance test, being the only tolerance test dependent on the matching transformation, is affected most by these errors. In all depth maps acquired, if a position tolerance of between 1mm and about 0.8mm is set, all points pass (ignoring the penny). Because position tolerances of 1mm can be successfully tested, it is easy to correctly identify the defect of a penny, as clearly shown in Fig. 12.5. (The height of a penny is about 1.5mm.)

As the position tolerances are made more strict, points begin to fail the test. At a position tolerance of 0.75mm, 2% points are considered by the outlier test and 1% of them fail. At a position tolerance of 0.5mm, 10% are considered by the outlier test and 9.8% of them fail. As above, many outlier neighbourhoods then contain more than one bad point. The basic cause of these bad points is camera calibration errors, leading to errors in the matching transformation as we have discussed in detail in Sections 9.6 and 9.7. When small numbers of points fail the initial tolerance test, they tend to be at the edges of segmented features. As the numbers increase, more are found towards the centres of the features, as illustrated in Fig. 12.5. This effect can be explained by the fact that the normals of segmented planes are in error by around 1° (see Section 9.6 and Chapter 4). The centres of gravity of the segmented planes are fixed at those of the model planes. Thus, if the plane is oriented at an incorrect angle, the further from the centre of gravity a point lies, the larger the error in the estimate of the distance of the point to the plane will be. This explains why points at extremes of the planar faces fail initial tolerance tests first as shown in Fig. 12.5. Errors of this type at present mean that we cannot measure position tolerances to the required accuracy of 0.254mm.

The size and form tolerances are dependent on the quality of the fitting of the data to features rather than on absolute positions and orientations. In the case of planes, the fitting of the planes is performed by the segmentation routines. The average root-mean-square error of fit of

12.5. SINGLE SCENE INSPECTION RESULTS

| Points In Tolerance | Points Outside Tolerance | | Points In Tolerance | Points Outside Tolerance |

(a) Position Tolerance = 1.0 mm
Size/Form Tolerance = 0.5 mm

(d) Position Tolerance = 1.0 mm
Size/Form Tolerance = 0.4 mm

(b) Position Tolerance = 0.75 mm
Size/Form Tolerance = 0.5 mm

(e) Position Tolerance = 1.0 mm
Size/Form Tolerance = 0.3 mm

(c) Position Tolerance = 0.5 mm
Size/Form Tolerance = 0.5 mm

(f) Position Tolerance = 1.0 mm
Size/Form Tolerance = 0.254 mm

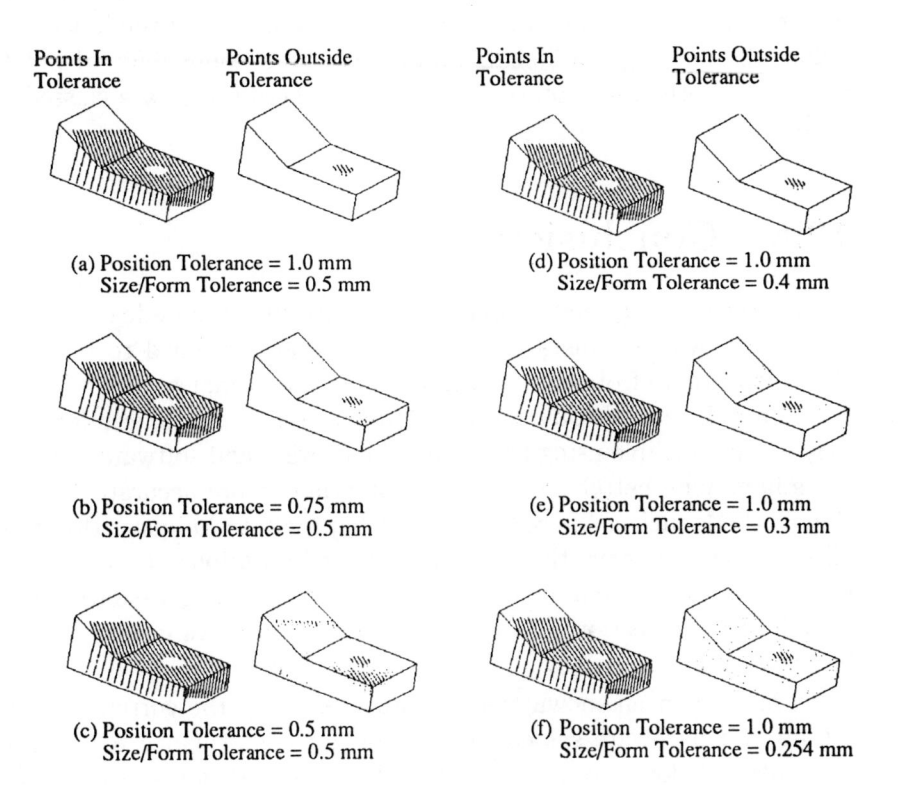

Figure 12.5: Inspection of a single view

planes to depth data has been shown to be 0.28mm in Chapter 9. Thus, it cannot be expected that measurement of objects for conformance to typical inspection tolerances of 0.254mm can be achieved for any type of tolerances. Although all points in the examples can be checked for conforming to size or form tolerances of 0.4mm, attempts to set any smaller tolerance limits results in some points failing the outlier test. At a tolerance limit of 0.3mm about 3% of points failed the initial tolerance test in our experiments and at 0.254mm about 6.5% of points did so. In both cases approximately 1% of these points are classed as outliers.

12.6 Conclusions

It has been shown that under normal expected theoretical working conditions, where camera calibration errors are assumed to be minimal, inspection of objects for conformance to normal tolerance limits (0.254mm) is possible but that detailed inspection to smaller tolerances may not be possible using the outlined hardware and software. In such cases it may be better to use other potentially more accurate acquisition methods such as Moire fringe (see Chapter 2) techniques. However, all such methods have their own particular limitations. It would perhaps be better to use a hybrid of these approaches and use data fusion techniques (see Section 5.3.2) in an attempt to achieve inspection to a higher degree of accuracy.

It has also been shown that, in practice, using the current camera set up and hardware system, inspection for conformance to position tolerances of down to 0.8mm is possible, and also for size and form tolerances of down to 0.4mm.

Since the acquisition of real data is never going to be error free, the theory of outliers has been used to try to decide if data points that lie outside specified tolerances can be ignored. This is done by considering the overall distribution of points within a neighbourhood of the test point. By considering real data it has been shown that under normal working conditions, when tolerance limits of 0.254mm are set, the fraction of outliers ranges between 0.2% and 1% of the total number of data points. The camera calibration inaccuracies have been shown to

12.6. CONCLUSIONS

be the source of errors which limit the accuracy to which inspection can be performed. Should further developments in the calibration accuracy be achieved then the artificial results show that inspection to more realistically useful tolerances can be expected. In detail, we have shown that most of the current outliers occur at the extremities of segmented features which supports our assumption that these will disappear if the calibration errors can be reduced.

Chapter 13

Complete Object Geometric Inspection

13.1 Introduction

The ability of our system in practice to perform three-dimensional geometric inspection to certain limits, on a single view of an object, has been demonstrated in the previous Chapter. However, in order to perform overall inspection of an object an inspection system must be able to acquire and reason with several different views of an object which completely describe it. The ideas discussed in this chapter aim to provide a sound basis for and demonstrate the feasibility of inspection using multiple scenes. However, the approach presented is fairly basic using a very low level of reasoning with geometric data. As a result many problems still need to be resolved before a complete and robust inspection system can be assembled. Two particular problems are:

- Only planes can be used in the reasoning process at the moment. We use linear entities (*i.e.* surface normals) to drive the reasoning processes governing the choice of views. It would be in principle possible to incorporate limited types of higher order surfaces such as cylindrical and even general quadric surfaces using similar linear entities extracted from such surfaces. This approach has not yet been incorporated.

- Camera calibration errors are yet again a source of problems. These cause errors in the determination of the location and orientation of the object in the initial scene. Ideally, having registered the position of the object initially we should be able to manipulate the object within the inspection system (assuming zero error in the manipulation hardware, which for practical purposes is the case for the turntable used in our system) simply with reference to the geometric model. However, we cannot do this since the errors in the initial registration of the object are carried through to subsequent views and worse still these errors are further compounded by the segmentation routines applied to extract the relevant information from each view. To avoid this problem we have to apply the full matching process to re-register the object in each view obtained from the system.

13.2 An Ideal Visual Inspection System

An ideal practical visual inspection system might consist of the following components (Figs 1.1, 2.10 and 2.11 illustrate some of these components).

A Robot Arm — The purpose of the arm is to initially and subsequently position the object on the turntable and to generally move the object around in the field of view of the vision system.

A Turntable — The problem with solely using a robot arm is that the precision with which the object can be positioned is not adequate for inspection purposes. Turntables exist that can be accurately rotated (turntables used in our system are accurate to 1 arc second [3]). The location of the axis of rotation of the turntable in world coordinates can be easily calculated by rotating the turntable through a known angle and observing each position of a reference point on the turntable. Therefore, once the initial position of an object has been determined on the turntable, it can in principle be rotated under registration to give a precisely oriented view of the object. (However, camera calibration errors prevent us from taking this idealised approach in our system.)

13.2. AN IDEAL VISUAL INSPECTION SYSTEM

An Image Acquisition System — This is the means of acquiring the data (see Chapter 2). In order to achieve accurate and reliable results more than one means of image acquisition device may be required. This is discussed later in the Chapter.

A Control and Computation System — The above components need controlling, and various segmentation, matching and inspection algorithms need invoking to perform automated visual inspection. Typically a minicomputer or a workstation would be used for these purposes.

The basic operation of such a system is fairly straightforward. The object is initially placed on the turntable in front of the vision system in an arbitrary position by the robot arm. The position of the object may, or may not (for example, if the object has come from a conveyor belt), be approximately known from the pick and place operation of the robot. However, robots cannot in general accurately place objects with sufficient accuracy for inspection purposes. We thus assume that we have no knowledge of the object's initial position.

The initial position of the object is registered by matching this view to the stored object model. What can be seen in any subsequent views obtained from accurate rotations of the turntable can be safely hypothesised from the model and simply verified, thus eliminating the need for carrying out the matching process on every view. A set of views of the object is selected that will allow all possible features of the object to be inspected simply rotating the turntable. The features in these views are inspected by a process of prediction and verification. The robot arm is then used to reposition the object in a different orientation to allow inspection of features not previously visible. Due to the inaccuracies of the robot arm, any hypothesis about the new position cannot be assumed with accuracy. Therefore each view obtained after robot arm positioning requires re-registering by matching to the object model. Overall, then, the process of manipulating the object by means of rotations and robot arm positioning is continued until all necessary views of the object have been obtained and inspected.

Normally, most of the object can be inspected using the series of views provided by the initial placement of the object on the turntable. Only a few more views of the object are expected to be required for complete inspection, basically for features initially lying face downwards on the turntable. Only this initial *single axis inspection* phase is discussed further in this Chapter. Similar ideas can be applied for the other views.

However, methods for picking up, moving and repositioning an object using a robot arm are non-trivial, and will be discussed in Section 13.5. The associated problems have not yet been addressed by our present practical system.

13.2.1 Choice of Vision System for Inspection

Visual data from one source, in our case an active stereo laser striping system, may not provide enough data in a sufficiently reliable and accurate form for all inspection tasks. Previously, Chapters 9 and 12 have shown that this particular input system may be suitable for general inspection requirements. It is probably the best single type of source for sensing the environment in order to give overall location and position of objects and for performing general geometric tolerance inspection, in that it seems to be one of the most flexible arrangements for acquiring data with high accuracy and resolution over a wide field of view. However for more detailed tasks requiring greater accuracy, it is not an ideal source since camera calibration problems seem to limit the accuracy of the system at present, and data from other visual sources could prove to be more accurate. Indeed, as we have discussed previously, other stages of the recognition and inspection process might benefit if additional visual information were present (for example see Chapter 4 on image segmentation and Section 5.3.2 on data fusion).

Thus for three-dimensional inspection to general geometric engineering tolerances, the use of the active vision system we have developed seems to be right choice.

13.3 Single Axis Inspection

Once an object has been placed on a turntable and its initial position has been accurately found then the inspection of any of the views which can be chosen by revolution of the turntable is possible. The inspection

13.3. SINGLE AXIS INSPECTION

of features in views subsequent to the first view is made simpler by the fact the object position can be safely and accurately hypothesised, and matching is not required for these subsequent views. It is assumed that the object is in a stable orientation on the turntable and does not slip when the turntable rotates.

In the last Chapter we discussed (Section 12.2) the best way of performing total inspection of an object. We noted that a method is required of choosing an optimum set of views which must satisfy the following conflicting requirements. On the one hand, a small number of views is desirable to save time. On the other hand, in any view, certain features will be at a glancing angle to the cameras, and so it will not be possible to accurately measure depth points for these features. A sufficiently large set of views must therefore be chosen to ensure that accurate depth readings can be taken for *every* feature.

The set of views of the object can be obtained by reasoning with the geometric object model offline. Clearly this set of views will be ordered so that rotations of the turntable through an increasing set of angles will provide each view in turn. The choice of each view is now described. Firstly we shall discuss how to group features into sets, one set per view, and secondly, how to obtain the best view of each set.

13.3.1 Choosing Sets of Features for Each View

In deciding how to choose views from which to inspect an object about a single axis of rotation of the turntable, there are many factors that need to be considered, and as we have already remarked many of these will have conflicting requirements. When inspecting features there are two basic components of the inspection task, namely to:

- inspect each feature individually for compliance to tolerance, and to

- inspect relationships between two or more features such as angles between planar surfaces. Inspection of this type can be achieved by considering *pairs* of features at a time. Clearly, if both features are visible in a single scene then their relationships can be verified immediately. However, in many cases, such as when considering whether two planar surfaces are parallel, they will generally not both be visible in a single scene. Here the accurate registration of the location and orientation between successive views of the object is very important so that feature inter-relationships can be determined from different views obtained in the inspection process.

Other problems may also occur when inspecting an object. Certain features may not be *completely* visible from any single view point and will thus require more than one view for complete inspection. This is obviously a problem for many curved surfaces, particularly cylindrical holes, for example, or any other surface where the range of surface normal directions occupies more than a hemisphere (or somewhat less in a stereo system). Similarly, a feature such as a large planar surface may be too large to be accommodated in a single view. However the latter type of problem can be avoided by imposing the constraint that the whole object must fit inside a given volume (a 0.5m cube for our system).

The former problem involving curved surfaces can be solved by dividing the surface up into pieces or *patches* that *can* be inspected in a single view. This can be easily achieved using our representation of surface patches, where for example, associated with a cylindrical surface is a pair of vectors that define the sector of the cylinder that the surface occupies (see Chapter 4). If the cylindrical surface spans more than 180° then the surface will clearly not be totally visible in any one view. Dividing the surface into smaller surface patches in this case is simply a matter of finding suitable intermediate vectors that break the surface into a series of adjacent sectors.

We now turn to the problem of choosing a suitable set of features for each view of the object and then the positions of each view for the single axis inspection stage.

In choosing the set of views of an object to use there is trade-off between inspecting every feature individually and inspecting suitably chosen sets of features in each view, as we have already noted. If each feature is inspected individually then it can guaranteed that the best possible view of the feature is used for inspection purposes. If we take sets of features then we shall have to sacrifice the best view for each individual feature for an adequate view for each feature in the set. If

13.3. SINGLE AXIS INSPECTION

we adopt the first approach then many views that are taken may be redundant in that a view chosen for inspecting one feature may be more than adequate for inspecting another but is not used for this purpose. Thus we adopt the approach of choosing sets of features and finding the best view for each set. We stress that the approach that we shall explain below is very basic and can be much improved upon. For instance, in choosing the views we the only knowledge taken into account about how the corresponding depth map is formed concerns the camera and turntable positions in the system. Much more serious consideration could be given to how the actual depth data is formed (*e.g.* how laser light is reflected by the object) to aid this process. Also much greater consideration could be made on how to cluster together faces in order to choose viewpoints.

The development of our method has been particularly influenced by the following factors:

- The feasibility of the general principle of inspection using our inspection system needs to be demonstrated before efforts to develop a more refined approach are worth considering.

- In an ideal case the turntable can simply be used to provide a new view in which the object's position and orientation can be deduced from the match provided by the initial view and the angle the turntable has turned through. Practical problems with camera calibration mean that this is no longer possible. This has implications when feature inter-relationships need to be measured and on the choice of the views that are to be used for single axis inspection.

In order to determine the set of features for each view and the consequent set of views the following are used:

- The stored geometric solid model of the object.

- The positions and orientations of the two cameras in the vision acquisition part of the inspection system. This information is obtained from the camera calibration routines.

- The position of the turntable's axis of rotation in the inspection system.

The first view that has to be inspected by the inspection system is the view that establishes the primary datum system. It is assumed for the methods we develop here that this view has indeed been found. We shall discuss how this is achieved in Section 13.4.

In an ideal system, each feature would be described by some method related to that of extended Gaussian images (see Section 5.5.2), to give the range of surface normals that the feature encompassed. This information would be used to cluster features into sets, and would take into account firstly the need to inspect each feature separately, and secondly, which pairs of features needed to be inspected together. These requirements would be separately weighted. We have not taken these ideas any further because of the practical problems mentioned earlier; techniques from the subject of *cluster analysis* [71] would be applied to perform this task. Some ideas on performing this task have recently been developed by Woo [242] and Requicha and Spyridi [202]. They have discused methods for computing the visibility of faces from a given solid model. Some of these methods are based on using the extended Gaussian image of the model. However, Requicha and Spyridi have introduced the idea of *global accessibilty cones* which describe all directions along which a face can be reached by a touch probe of a coordinate measuring machine. These ideas should be extensible to the stereo vision inspection techniques that we have described in the book.

In our practical system, because of the impossibility of accurately correlating positions and orientations between views, we have chosen a much more restricted and simplistic approach. Starting with the initial view, we use a dynamic method of choosing the next view. Basically, each new view adds the next feature in rotation order to the current set of features, while one or more features are removed from the set so that all features can be viewed by both cameras. In more detail, cylindrical and other curved surfaces are also dynamically split into patches for inspection at this stage. Although this sounds as though we have weighted the trade-off mentioned earlier complete towards a single view for each feature, the real reason we have done this is to maximise the number of possible feature inter-relationships that can be satisfactorily

13.3. SINGLE AXIS INSPECTION

measured, in light of the impossibility of accurately correlating positions and orientations between views. Note that we flag relationships as they are inspected to avoid inspecting the same relationship more than once. Furthermore, limited inspection of relationships between features which are not all visible in any single view is also performed by using secondary datums. For example, if planes A and B are visible in one scene, and planes B and C are visible in a second scene, the relationships of each of A and C to B may make it possible to infer under certain circumstances that the relationship between A and C does not meet tolerances. In summary, then, this method of grouping features into sets is very basic, and for example, the grouping does not consider *which* specific feature inter-relationships need to be measured.

13.3.2 Obtaining the Best View of a Set of Features

As mentioned, after the features have been grouped into sets, some method is needed to calculate the orientation of the turntable that gives the best view of each particular set of features. In this Section we describe such a method.

Let us commence by considering single features, however. In order to obtain the best view of a single feature for inspection purposes, the errors in depth readings taken for the feature should be kept to a minimum. For example, the best readings for a plane are obtained when the angles subtended by each of the cameras and the plane are equal. This can be readily computed from the plane normal. For curved surfaces, to a first approximation, an average surface normal can be used in a similar way. This will be discussed at more length later.

Let us now extend this idea to sets of features. The problem may be stated as: *Given two cameras of known position and orientation, and a set of surface normals, what is the angular orientation (θ) of the turntable which gives the best view of the surfaces to minimise errors in acquired depth information?*

In order to simplify this problem a virtual *middle camera* is assumed to exist to help in the estimation of θ. The results provided by a model using a single middle camera are approximately equivalent to those when two cameras are considered, but the calculations are much simpler to perform, as will be shown below. The middle camera is taken to be

CHAPTER 13. COMPLETE OBJECT GEOMETRIC INSPECTION

at the intersection of the base line of the two actual cameras and the angle bisector of the angle between the lines of sight of both cameras. In general the middle camera will be at the mid-point between the two cameras since the angle subtended by both cameras and their base line will be equal.

A note of caution is worth mentioning with regard to this simplification: this simplification may only be valid for positioning strategies for objects that are bounded by fairly large features that can reliably extracted from the depth data (*e.g.* planes and certain curved surfaces). If small details of an object or complex objects (with a lot of self-occlusion) are being inspected this approach may not be valid since the view obtained from each camera is different and only those features common to both views can be measured for depth by the stereo system. Nevertheless, this approach may be useful for hypothesising an initial viewpoint which can then be modified as the information from the two cameras is introduced to the reasoning process.

Let us now consider why the solution to choosing the optimum orientation of the turntable when there is one middle camera is equivalent to the problem with two cameras.

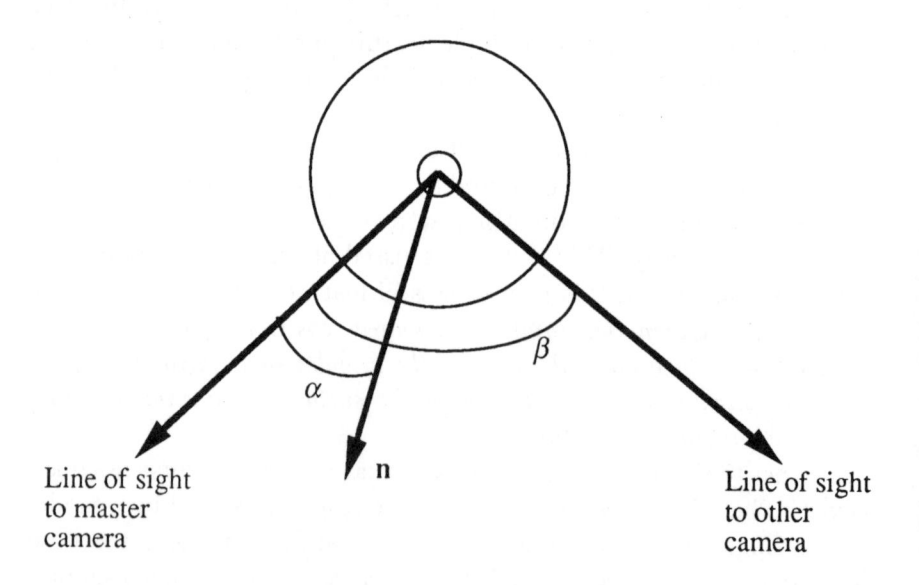

Figure 13.1: Equivalence of middle camera

13.3. SINGLE AXIS INSPECTION

In Fig. 13.1 the angle between the lines of sight of the cameras is β and the angle between some vector, $\mathbf{n}$ (*e.g.* a normal to some surface from which depth readings are to be taken) and the lines of sight to the master is α. In order to obtain the best possible depth readings of the plane, the component of $\mathbf{n}$ along the lines of sight to each camera should be equal.

Therefore the problem is to minimise the difference between the cosines of the angles between the lines of sight and $\mathbf{n}$ and that we want to minimise

$$D = \cos(\alpha) - \cos(\beta - \alpha) \tag{13.1}$$

with respect to α.

D is minimum when $dD/d\alpha = 0$, and therefore $\alpha = \beta/2$. This is exactly the orientation that will make the normal directly face a hypothetical middle camera.

Let us now consider the case the case where we wish to optimise the turntable orientation for a set of features. The estimation of the transformation that rotates a set of object features to provide the best possible overall view with respect to this middle camera may be found as follows (see Fig. 13.2).

Let $\mathbf{v}_c = (a_c, b_c, c_c)$ be the position of the middle camera, assuming the origin to be at the centre of the turntable. Let $\mathbf{v}_i = (a_i, b_i, c_i)$ be the normals associated with a set of features F_i. Let $\mathbf{R}$ represent a rotation through θ about the axis of the turntable. Furthermore, assume that the turntable's axis is parallel to the z-axis and that the positive z-axis points into the turntable. Assume also that all coordinate systems are right handed and rotation $\mathbf{R}$ is in a *clockwise* sense.

If the turntable is rotated through $\mathbf{R}$, each vector $\mathbf{v}_i$ is rotated to give $\mathbf{v}'_i$ where

$$\begin{aligned} \mathbf{v}'_i &= \mathbf{R}\mathbf{v}_i \\ &= (a_i \cos\theta + b_i \sin\theta, b_i \cos\theta - a_i \sin\theta, c_i) \end{aligned} \tag{13.2}$$

It has been shown in Chapters 4 and 8 that the larger the projected area of a viewed surface, the better the segmentation obtained for the surface will be. Thus, since the projected area of a surface depends on the cosine of the angle between $\mathbf{v}'_i$ and $\mathbf{v}_c$, we choose as the solution for θ the value which brings each $\mathbf{v}_i$ into view such that the sum of

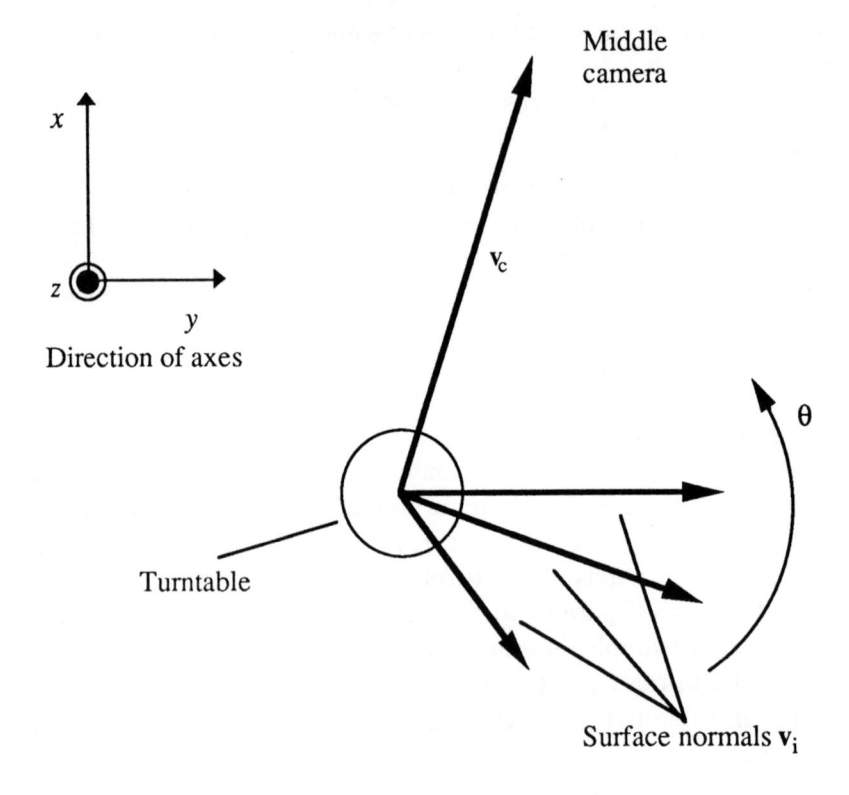

Figure 13.2: Estimation of best rotation

13.3. SINGLE AXIS INSPECTION

the cosines of angles between $\mathbf{v}'_i$ and $\mathbf{v}_c$ is maximised. The value of θ required to do this may be found analytically as below.

The cosine of the angle between vectors $\mathbf{v}'_i$ and $\mathbf{v}_c$ is given by

$$\mathbf{v}'_i.\mathbf{v}_c = a_c(a_i \cos\theta + b_i \sin\theta) + b_c(b_i \cos\theta - a_i \sin\theta) + c_c c_i \quad (13.3)$$

Note that that all vectors $\mathbf{v}'_i$ make an angle in the range $(-90°, 90°)$ to the line of sight of the middle camera, as they belong to surfaces which are visible, and so each of the terms in the sum is positive. Thus, we wish to maximise:

$$\begin{aligned} M &= \sum_i \mathbf{v}'_i.\mathbf{v}_c \quad (13.4) \\ &= \sum_i [a_c (a_i \cos\theta + b_i \sin\theta) + b_c (b_i \cos\theta - a_i \sin\theta) + c_c c_i] \end{aligned}$$

with respect to θ.

Now, M is maximised when $dM/d\theta = 0$, and

$$\frac{dM}{d\theta} = \sum_i [a_c (-a_i \sin\theta + b_i \cos\theta) + b_c (-b_i \sin\theta - a_i \cos\theta)]. \quad (13.5)$$

Thus, equating $dM/d\theta$ to zero and rearranging,

$$-\sin\theta (a_c \sum_i a_i + b_c \sum_i b_i) + \cos\theta (a_c \sum_i b_i - b_c \sum_i a_i) = 0,$$

(13.6)

giving for the required value of θ

$$\tan\theta = \frac{a_c \sum_i b_i - b_c \sum_i a_i}{a_c \sum_i a_i + b_c \sum_i b_i}. \quad (13.7)$$

Thus assuming that all these features are visible to both cameras at this orientation, the orientation of the turntable that provides the best possible view of these features is given by Eqn. 13.7.

Note that the minimisation problem as described above is to some extent arbitrary. For example, we could minimise the sum of the *squares* of $\mathbf{v}'_i.\mathbf{v}_c$ or minimise the maximum of value of $\mathbf{v}'_i.\mathbf{v}_c$ over all values of i. The formulation we chose results in simpler computations than the sum of squares method. The maximum value formulation does have some merit in that the end result of the minimisation would be to give more emphasis to reducing depth errors in the features which occur at the outer limits of our view. This issue probably deserves further consideration.

Earlier, we said that for curved surfaces, average surface normals could be used in this process. A more rigorous approach would be to be to integrate $\mathbf{v}'_i.\mathbf{v}_c$ over all surface normal directions encompassed by the surface patch in place of Eqn. 13.3, and to include these results in the sum in Eqn. 13.4.

13.4 The Single Axis Inspection Scheme

An initial approach to a single axis inspection scheme is summarised in Fig. 13.3. As can be seen the inspection scheme can be split into two phases,

- a registration phase, and
- an inspection phase.

These phases are described separately.

13.4.1 The Registration Phase

The purpose of this phase is to provide an accurate transformation that describes the spatial relationship between the coordinate systems of the observed object and the stored model. Once the transformation is known then the object can be rotated so that a set of views about the axis of the turntable may be inspected.

It is assumed that the object has been placed in a stable orientation on the turntable. The initial view of the object is taken by the vision system, segmented and matched to the corresponding object model as described in previous Chapters. The result obtained not only provides the transformation, but also a list of all model features matched to corresponding scene features. At this stage, it may be possible to rotate the turntable to the orientation which provides the best view of a set

13.4. THE SINGLE AXIS INSPECTION SCHEME

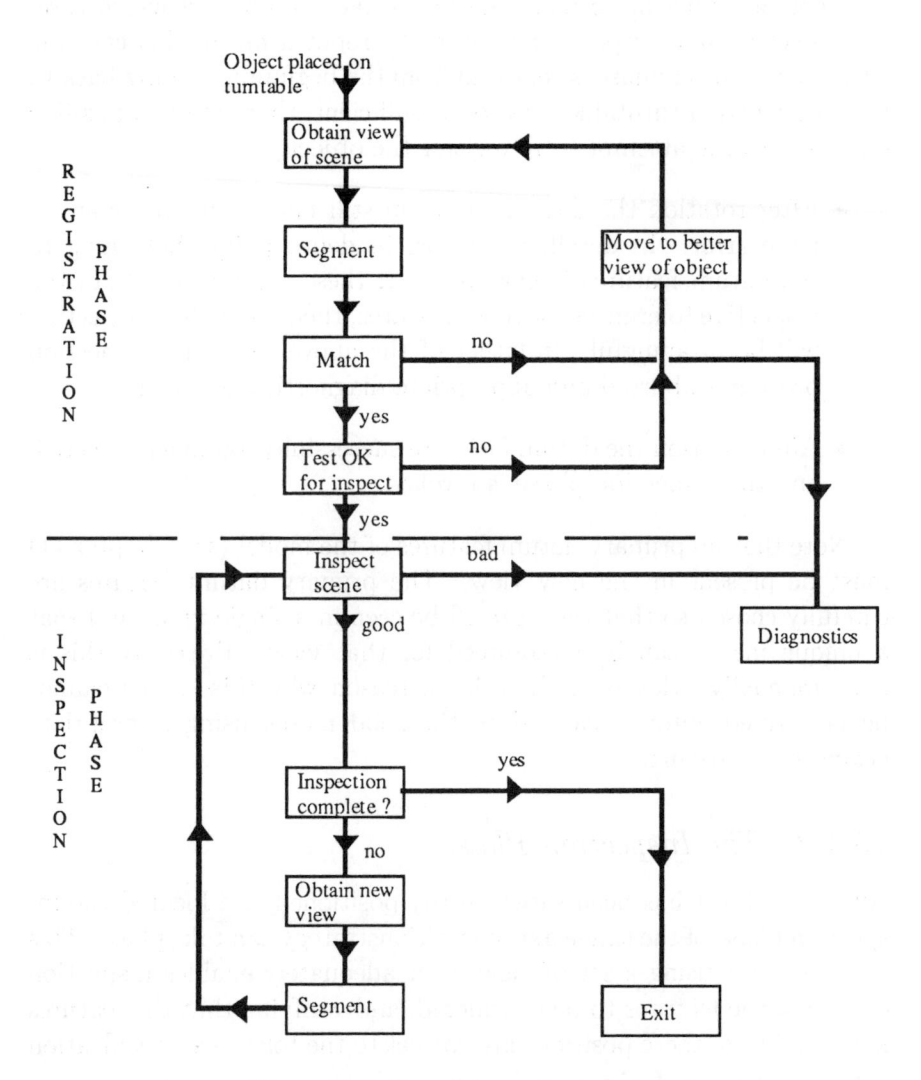

Figure 13.3: Overall inspection scheme

of features which includes all the primary datum features. The registration process is then applied to this view. However, a simple rotation may not be sufficient to make all datum faces visible, in which case, the object must be repositioned with the robot arm. In this case the whole registration phase is repeated from the beginning. Going back to the case where a turntable rotation is sufficient, there are two possible outcomes to the attempt to re-register the object.

- After rotation the datum faces can still not be found, or are in large error. In detail, the errors in depth points belonging to the datum features is examined. If these are not less than the respective tolerances for those features, then no further inspection will be meaningful. In either of the above cases the inspection process is aborted and appropriate diagnostics are given.

- After rotation the datum faces are successfully found as expected, and the inspection phase is invoked.

Note that *all* primary datum features of the model (see Chapter 11) must be present in the new view. The primary datum features are carefully chosen so that they *can* all be seen in a single view, and that a unique match can be guaranteed for that view. Presently this is done manually. However, there is no reason why this choice cannot be performed automatically from the solid model using appropriate geometric reasoning.

13.4.2 The Inspection Phase

After the object has been satisfactorily positioned and located the inspection phase of the single axis inspectionstrategy can take place. This stage involves using a set of views that adequately enables inspection of each of the features to be considered, and verifying that the features measured from these positions are correct to the tolerance specification for the given object.

The first process is the inspection of the initial scene. If the inspection of the features in this scene proves satisfactory then the following steps are repeatedly carried out for each of the desired viewpoints in turn.

13.4. THE SINGLE AXIS INSPECTION SCHEME

1. The turntable is rotated to position the object in the correct angular orientation to obtain the next view.

2. The new image of the object is acquired with the image acquisition system and is segmented (see Chapter 4).

3. The features in the observed view of the object are verified to be those predicted from a consideration of the viewpoint and the object model.

4. If verification is achieved then the features chosen for inspection in this view are inspected as described in Chapter 12 and Section 13.3.2. This includes individual features and their relationships.

5. If inspection shows that the object does not meets tolerances with respect to the features considered in this view then diagnostics are output, and the inspection process is aborted or allowed to continue depending on the severity of the error and whether it makes sense to continue the inspection. For example, if a hole has not been drilled then the object may still be fully functional after this has been remedied, so we will therefore wish to continue inspection. Conversely, if a hole has been found to be drilled in the wrong place or is found to be too large then the object can not be repaired, and the inspection process can be aborted and the object scrapped. It is an interesting question as to what extent such decisions can be made from the geometric model alone, and which require human assistance.

The basic inspection process was described in Chapter 12. We now describe the verification process of the inspection scheme.

Verification of the View

The verification process is basically a simplified matcher. The set of visible features in each new view is readily found when deciding on the set of views to be taken. Thus all that is required is to simply correlate this set of features predicted by the model with the segmented scene features. The combinatorics of this correlation is much simpler than that of the general matching problem, requiring only a single pass through the segmented set of data. Indeed, it should be noted that the general matching strategy cannot be safely applied since there is no guarantee that the set of visible features forms a unique view of the object.

Ideally, the verifier should just pair matching features since the rotation under registration should perfectly produce pairings of new scene features and hypothesised scene features. However, due to the fact that the segmentation of real data is not error free, we cannot guarantee the accuracy of the matching transformation, especially as these errors are compounded by rotation of the turntable. These errors can have a considerable effect on the tolerance limits that can be checked. This problem can be resolved by re-estimating the transformation of the matching using the set of match pairs formed in the verification process.

13.4.3 Diagnostics

Throughout the singe axis inspection process various diagnostics may be output that give information as to why inspection has failed. The diagnostics may contain varying degrees of information depending on the fault detected. In general, different inspection errors occur during two of the inspection processes:

Matching — It may not be possible to match all candidate segmented features to model features. This may be due to an incorrect object or model being used for a match (*i.e.* a user error), the object being incorrectly manufactured in gross detail, or the segmentation procedure not working correctly. The last case is unlikely as checks are made on the errors arising in the fitting of features and it is very rare to produce segmentations which are unreliable and yet have good feature approximations. It is difficult to decide which of the above errors may have caused the matching to fail and no attempt is made to do this in our practical system. Instead information about which features fail to match and the corresponding estimate of the translation and rotation and errors in this estimate are given.

Inspection — At this stage it is easier to provide more informative diagnostics. Information provided from this stage includes information on the three types of tolerance checks and the regions of features that fail. It is also possible to decide if a hole, say, has been omitted or incorrectly positioned by examining the out-of-tolerance regions in detail. If the points where a hole should be lie on the same plane as surrounding points then the hole is missing; if a cluster of misfitting points is analysed then it may be possible to deduce details about the nature of the cluster (for example, that they lie on a cylinder).

13.5 Complete Inspection

On completing single axis inspection certain features will still have not been inspected. In a complete inspection system the model would be used to work out motions for the robot arm to reposition the object. These tasks will involve complex planning and motion strategies for the robot arm. These subjects are productive research areas at present and discussion of such problems could form many books in their own right. We shall briefly highlight problems relevant to our task.

In order to manipulate the object in the workspace the robot arm will have to be able to grasp, pick up and move the object around. This involves much planning to decide the best faces of the object to grip to hold the object safely, and how to move the object within its environment without colliding with obstacles such as cameras and supports. These tasks clearly involve reasoning with geometric models not only of the object but of the environment as well. Such planning of robot motion comes under the heading of *path planning* which has been the subject of research in the artificial intelligence and robotics communities for many years (and will be for many years to come). Many texts have been published that address these areas. Good introductions are provided by the relevant Chapters in the books by Winston [239] and McKerrow [167]. For further reading a collection of articles in the book by Brady [32] is a good source (in particular the articles by Lozano-Perez [154] and Mason [164]).

Firstly, all of the objects in the workplace (including the robot) must be represented by a geometric model with attached tolerances. This allows the representation of *uncertainty* when planning robot tasks, which may be caused both by lack of precise original information about the sizes and positions of objects, and perhaps because of imperfect visual data. These must be taken into account when directing the robot to grasp and place objects. We have discussed how tolerances can be incorporated into geometric models in Chapter 11. We must also be able to adapt our models and actions in response to changes or errors found in our model of the workspace environment. We have discussed how to represent and use uncertainties in Section 5.3.2.

13.6 Results

The outlined single axis inspection procedure worked well on both artificial and real depth maps, enabling form, position and size tolerances to be tested to the tolerance accuracies discussed in Chapter 12 for single scenes. The objects used were the same test objects as previously, *i.e.* the cube, the widget and the widget with a cylindrical hole, as shown in Figs. 2.5 and 12.3, which contain planar and cylindrical features. When real data was used re-registration of the object was performed for each view as errors in the predicted and actual locations of features occurred each time. After initial re-orientation of the object for the first view used for inspection, the calculated rotations giving the best view of successive sets of features including successive faces of the widget were 90° as expected (see Section 13.3.2.

Sample views (as viewed from the master camera) of the the widget obtained by real and artificial means are shown in Figs. 13.4—13.7 and 13.8.

13.7 Conclusions

It has been shown that the single axis inspection of objects to typical inspection tolerancesis feasible using the very basic strategies outlined in this Chapter. We have demonstrated this on objects that consist mainly of planar, cylindrical and spherical features.

13.7. CONCLUSIONS

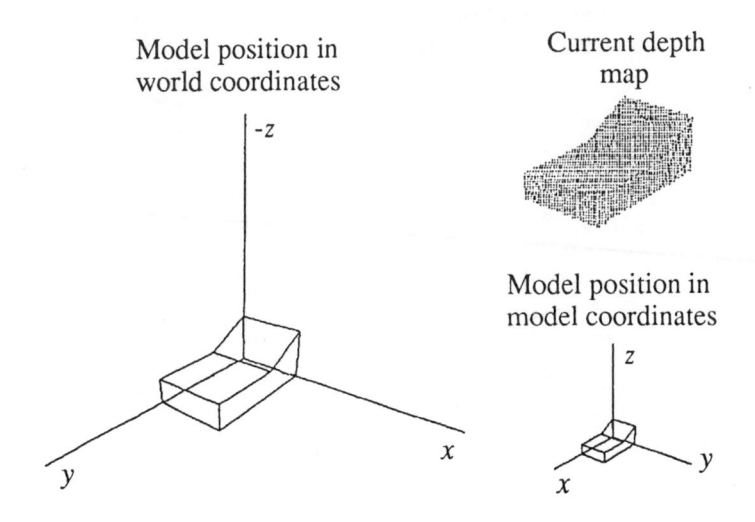

Figure 13.4: Single axis inspection using artificial data: initial view

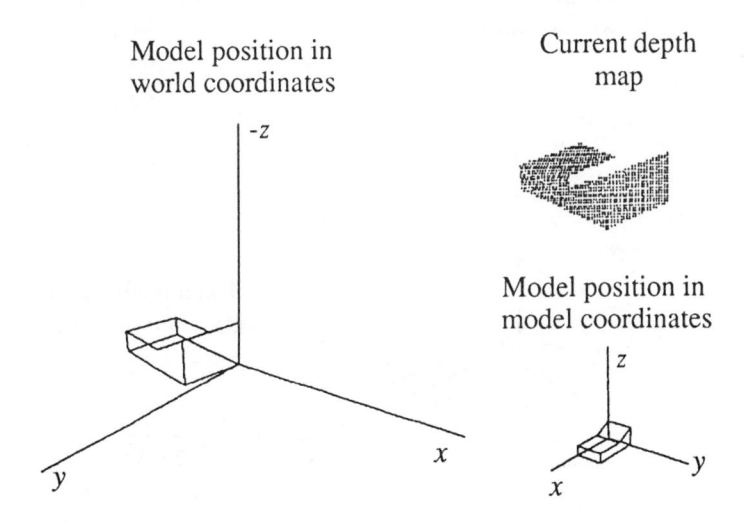

Figure 13.5: Single axis inspection using artificial data: second view

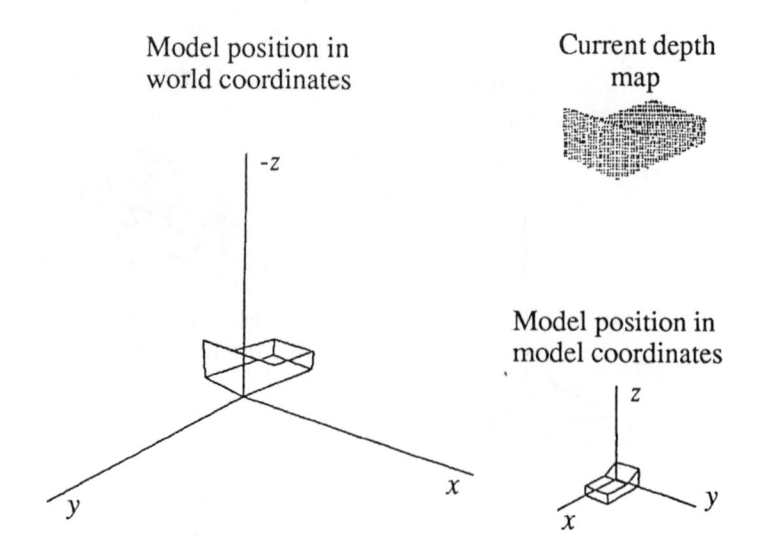

346 CHAPTER 13. COMPLETE OBJECT GEOMETRIC INSPECTION

Figure 13.6: Single axis inspection using artificial data: third view

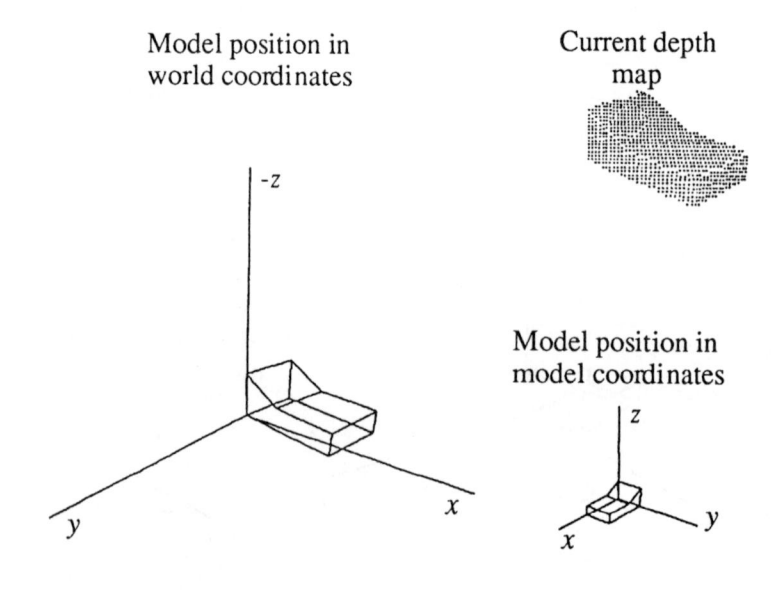

Figure 13.7: Single axis inspection using artificial data: fourth view

13.7. CONCLUSIONS

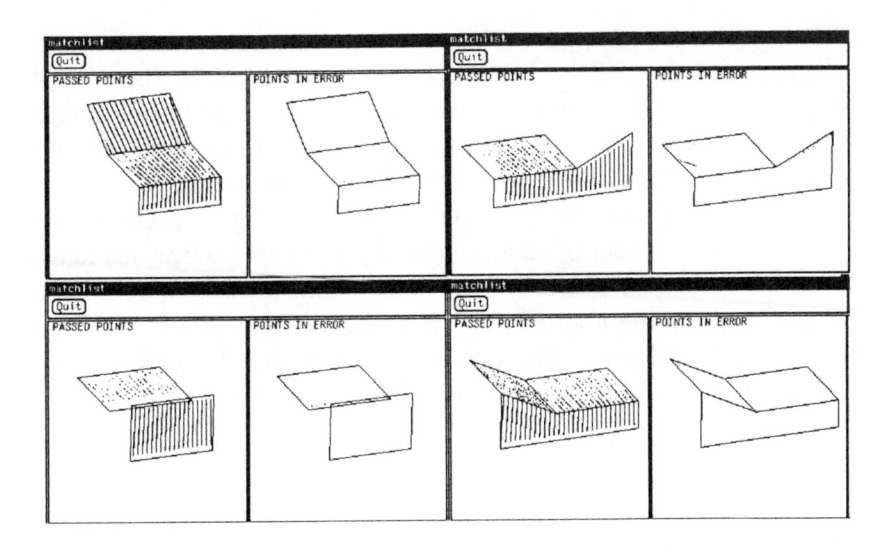

Figure 13.8: Single axis inspection using real data

When artificial depth map data were used it was possible to complete a single axis inspection to the required tolerances with only the need for a single initial registration phase. Again, it should be repeated that at present, the errors in real data arising from problems in calibrating the cameras make inspection to the desired tolerances infeasible. The best position tolerance inspectable so far is to a limit of 0.8mm (*position* tolerance is most affected by such errors). In order to obtain this inspection accuracy the pose of the object has to be re-estimated in each view otherwise the errors in the determined position of the object, caused by poor original position and orientation estimates, grow too large. However, once the camera calibration problems have been overcome we expect realistic inspection to be possible. We have managed to achieve inspection to within about 0.3mm for size and form tolerances which do not depend on the position estimate from the matching, but are only dependent on the segmentation.

The strategies outlined here are extremely rudimentary, although they have been shown to work for the small class of objects originally considered. Further work is needed to make strategies which are more efficient, robust and applicable to a larger class of objects. Some suggestions towards these ends have already been made in this Chapter. Also, we believe that further knowledge of depth image formation needs to incorporated to take into account scattering of laser light in the acquisition system which can give rise to poor depth readings.

In particular, more work is needed in reasoning with the model in order to choose the set of views of the object to use, and which features are to be inspected in each view. This could involve more advanced geometric reasoning about both the model and cameras . Many areas of artificial intelligence, robotics and computational geometry will be necessary to overcome all aspects of these problems.

Chapter 14

Other Types of Inspection

14.1 Introduction

So far we have concerned ourselves with one particular type of inspection problem, namely that of three-dimensional geometric tolerancing. There are many other important properties of objects that may need to be inspected, either separately, or as part of a complete inspection system. In this Chapter we shall briefly discuss how computer vision techniques can be applied to some of these other inspection tasks. The level of discussion in this Chapter is not as deep as in previous Chapters. This is because the tasks we discuss here involve lower level visual processes than those described for three-dimensional geometric inspection, making such tasks rather simpler to perform. In this Chapter we shall mention many of the ideas we introduced in the first Part of the book and have not touched on in subsequent Chapters. We have however decided to place this Chapter at this position in the book in order to keep the flow of the rest of the book intact. The purpose of this Chapter is mainly to bring to the readers attention some of the many different aspects that inspection entails and how computer vision techniques might be used for these purposes.

Nowadays no inspection process seems to have escaped at least an attempt to apply computer vision to it. One may find applications to many industrial manufacturing processes. However, it is also possible to find vision being applied to many other inspection processes involving for example foodstuffs and even live animals such as pigs.

Fig. 14.1 illustrates a very small sample of objects that have been inspected by some visual process. Fig. 14.1(a) shows a part of a typical mechanical component, Fig. 14.1(b) shows part of a printed circuit board, Fig. 14.1(c) shows part of the circuit on a semiconductor wafer (magnified many times) while Fig. 14.1(d) shows a cake that has not been iced properly. Manufacturing faults in these objects may or may not have serious consequences depending on how the items are to be used. However in the following four figures, undetected defects could directly endanger human health. Fig. 14.1(e) shows a glass bottle with a defect in it. A thread of glass has formed a *bird's swing* across the bottle as shown in close up in Fig. 14.1(f). Fortunately, this defect was detected in the bottling factory. However, Fig. 14.1(g) shows a *spike* glass filament found in a milk bottle that was actually delivered to a colleague's doorstep. Fig. 14.1(h) shows an X-ray image of a jar of jam that contains some glass fragments.

In this Chapter we shall examine various methods that can be used to inspect such objects and also consider related areas to which visual inspection systems could be applied. We shall firstly discuss a variety of inspection tasks and then go on to study two applications in more detail.

14.2 Inspection in Two Dimensions

In many cases it is easier to inspect objects using a two-dimensional profile or a series of profiles taken from selected views. The problems of acquiring three-dimensional data are eliminated and measurements are simply made with reference to the two-dimensional data. Examples of objects for which this type of inspection is suitable include those which are symmetric about an axis, those which may be regarded as being built up of two-dimensional *slices* or simply those for which all the important characteristics can be obtained from a two-dimensional view. Classic examples of such objects are printed circuit boards (Fig. 14.1), glass bottles or jars, pipes, box girders and certain mechanical components such as the connecting rod from a car engine shown in Fig. 14.4. Unless otherwise stated, we shall assume in this Chapter that any images discussed are greyscale images, acquired using a single camera.

14.2. INSPECTION IN TWO DIMENSIONS

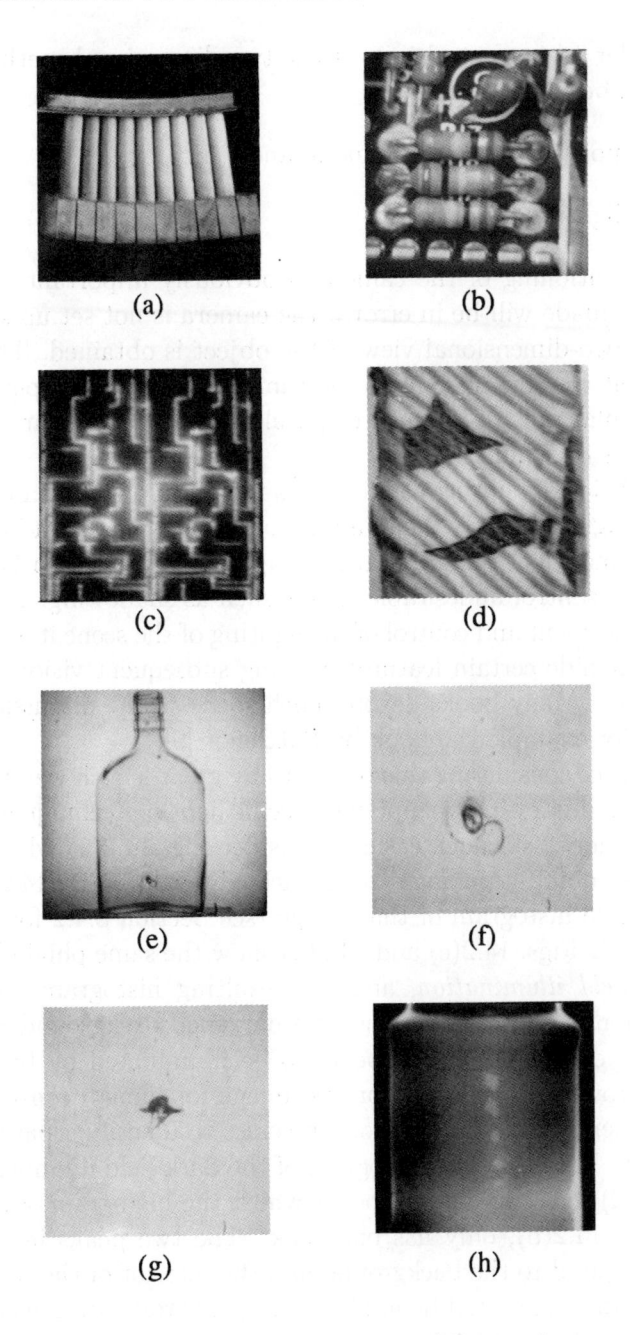

Figure 14.1: Examples of objects inspected by visual systems

In order to inspect objects using two-dimensional methods, great care must be taken over

- the positioning of the camera, and
- the lighting conditions.

The positioning of the camera is obviously important. Any measurements made will be in error if the camera is not set up so that the expected two-dimensional view of the object is obtained. This usually means that the line of sight of the camera should be perpendicular to the particular area of interest, or parallel or perpendicular as required to any axes of symmetry.

The choice of lighting conditions is more important than one might at first imagine. Clearly the illumination of the scene must be adequate to obtain sufficiently high quality images. We must also be watchful to avoid or control undesirable effects such as shadowing. However, by careful placement and control of the lighting of the scene it is possible to enhance or hide certain features, making subsequent vision processing tasks easier. Many books devote much space to the discussion of such matters, for example the book by Batchelor [9].

Instead of considering this topic at any great length we shall give an example to illustrate how applying specific lighting techniques can make the inspection task much easier. Consider Fig. 14.2(a) which shows a small glass phial illuminated by a backlight, while Fig. 14.2(b) shows the greyscale histogram of that image (see Section 3.4.2 for details on histograms). Figs. 14.2(c) and 14.2(d) show the same phial illuminated by *dark field illumination*, and the resulting histogram. Dark field illumination is a process producing a darkened background, but where the object of interest still appears to be illuminated by backlighting. Besides providing a vastly improved image for human consumption, it is also easier for computer vision processes to identify separate regions of interest. Note that the histogram of the dark field illuminated phial, Fig. 14.2(d), has two distinct peaks whilst the histogram of the backlit phial, Fig. 14.2(b), only has one peak. The two peaks in the former case correspond to the background and the interior of the phial. These can be readily separated by applying simple thresholding techniques as discussed in Section 3.4.2.

14.2. INSPECTION IN TWO DIMENSIONS

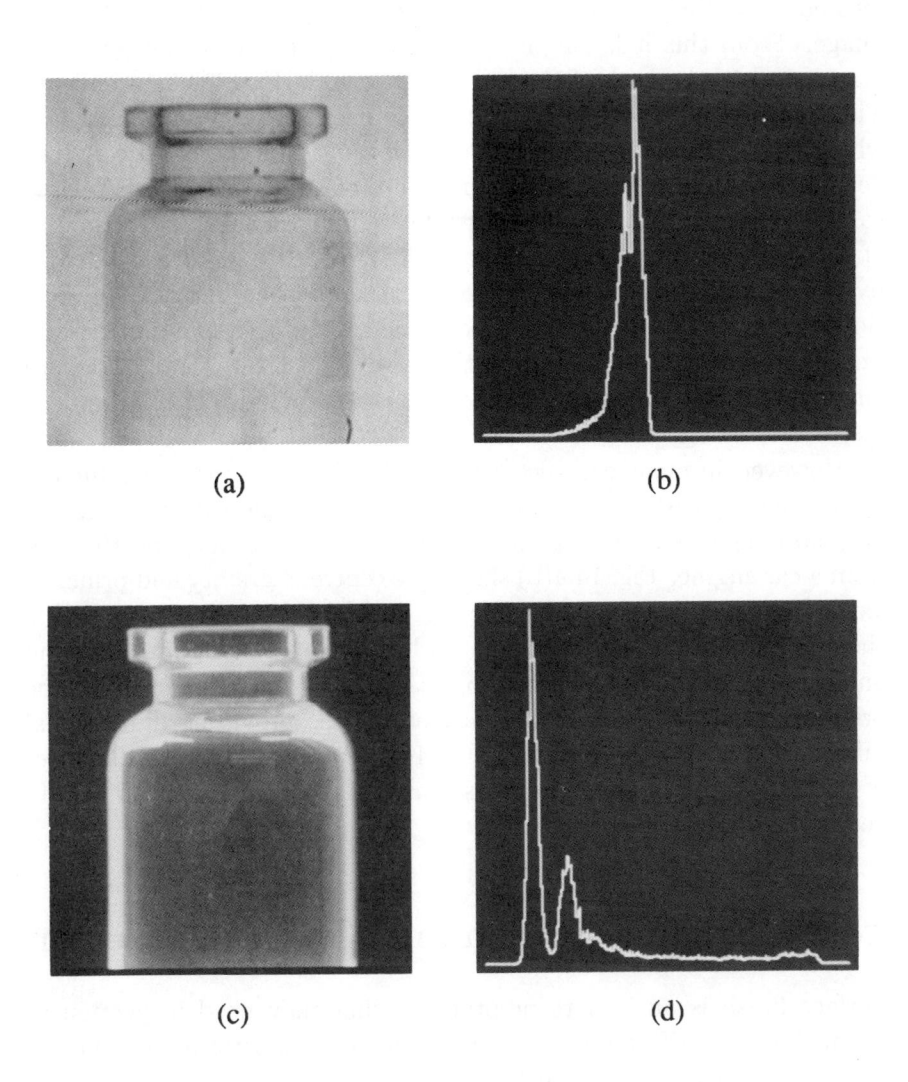

(a)

(b)

(c)

(d)

Figure 14.2: Using lighting to enhance an image

Other processing is also easier on the improved image, as illustrated by Fig. 14.3. Fig. 14.3(a) shows the edges of the phial detected in the image. From this it is simple to find the regions of the image which correspond to the neck of the phial, as shown in Figs. 14.3(b) and (c). These results allow the curvature of the neck to be determined, data which is very important for the inspection of this type of object. The neck of the phial has to satisfy a measure of strength which is related to this curvature. It is also possible to find the axes of symmetry of the phial as illustrated in Fig. 14.3(d), which can be used to verify the overall correctness of shape of the phial, as discussed in Sections 4.7 and 5.5. Thus, having obtained a suitable two-dimensional image of an object, inspection in this case is relatively straightforward employing image processing and inspection techniques that have been discussed throughout this book.

However, in many cases we can simplify the processing even further by dealing with binary images (by applying suitable thresholds to the original image). Fig. 14.4(a) shows a binary image of a connecting rod from a car engine. Fig. 14.4(b) shows the centre of gravity and principle axes of symmetry of the part while Fig. 14.4(c) shows the *convex hull* — the minimum bounding set of points which form a convex shape around an object — of the rod. These properties are useful in determining the orientation and location of the part as well as providing dimensional information about the part. Fig. 14.4(d) shows contours of distances of points within the shape from its centre of gravity, which can help to verify the dimensions of the object.

14.3 Surface Finish and Crack Detection

Surface finish is an important property that may need inspection as it can have an effect on the performance or cosmetic appearance of an object. Examples in the former category include the pistons and cylinder bores of a car engine, and the ball and socket of a replacement hip joint. Such surfaces need to be smooth to avoid friction during their working operation.

Inspection always requires comparison to an ideal model of some kind. So far we have mainly dealt with inspection of geometric data.

14.3. SURFACE FINISH AND CRACK DETECTION

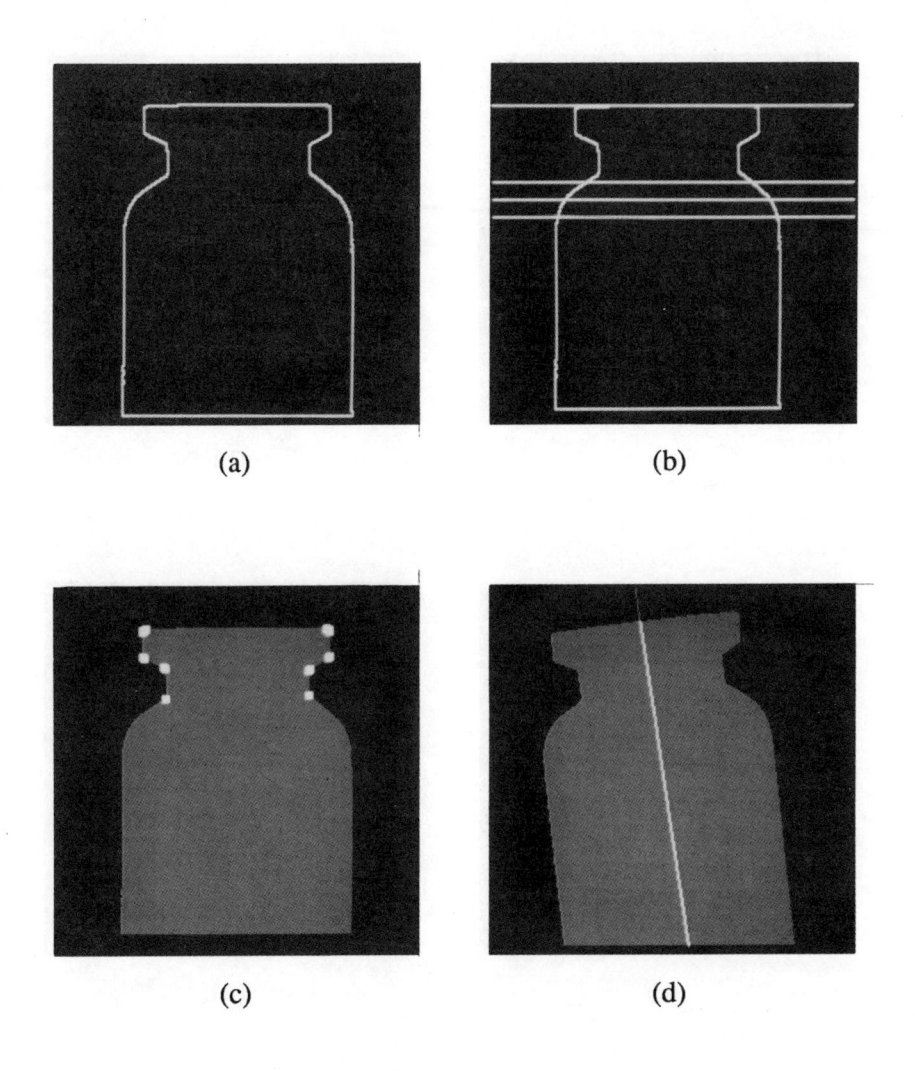

Figure 14.3: Inspecting a glass phial in two dimensions

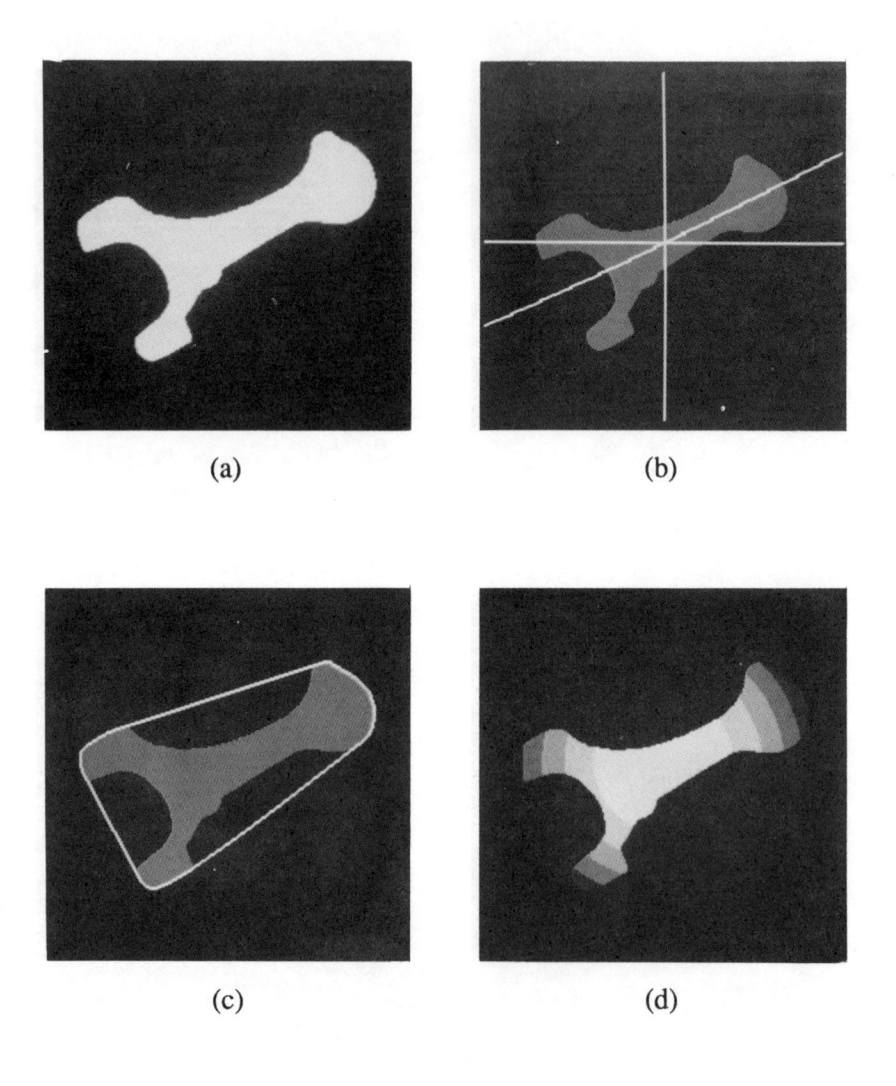

Figure 14.4: Analysing a connecting rod

14.3. SURFACE FINISH AND CRACK DETECTION

Although it may be possible to specify surface finish in geometric terms, such an approach is not really helpful. Instead for surface finish a different type of model is used. Surface finish depends on the surface material and the manufacturing process. The finish of a surface can be categorised by describing how rough, wavy, grainy, spotty or streaky it is. In order to inspect surface finish a known source of light is used to illuminate the surface and a camera records the reflected light from the surface. We have dealt with many techniques using projected light in Chapters 2 and 5 where we have studied image formation and extracting features from an image. In particular, we have discussed in Section 5.4 *shape from shading* techniques using models of lighting and reflection from a surface. Many techniques have been applied [171] to surface finish inspection using lighting models similar to those we have discussed.

Many surface finish properties are related to surface texture which we discussed in Section 5.7. We can clearly classify the roughness, waviness, spottiness *etc.* by means of a measure of texture. Many researchers have looked at ways of using texture for inspecting surface finish. In particular, Weszka and Rosenfeld [238] have identified a number of texture features related to surface finish and Don and his colleagues [66] have managed to classify metal surface samples into six different categories using texture. Others [211, 225] have used Fourier methods (see Section 3.2) for the analysis of texture. Some methods [211] have also used laser light as a source of illumination.

In general, however, it is difficult to measure surface finish visually because of the small size of the surface characteristics which affect its finish. Usually surface finish inspection requires the aid of high precision instruments such as *stylus profilers* which measure surface profiles by tracing a highly sensitive stylus along lines across the surface. Such instruments are very slow in operation and have the additional drawback of being contact sensing devices, so any excessive pressure in use will damage the object being tested. An alternative is to use a *light-section microscope*. Again, however, this requires skilled operation by a human operator. Both these instruments do not allow for an automated inspection strategy to be developed or even for modest size batches of objects to inspected manually. They clearly do not meet the ever increasing needs of industry for inspection. Vision systems to perform comparable tasks still have some way to go however, and much more work is needed on equipment, sensing and analysis techniques for vision systems to be able to perform such tasks.

Instead of trying to measure surface finish, it is much more common for a visual system to try to detect surface defects. This task is much easier and is based on detecting an abrupt change in the characteristics of the surface. Many systems have been developed to do this [9, 39, 139, 171]. These systems often involve deriving certain measures from a sequence of pixels along the surface and noting a change or deviation in the sequence. A simpler approach is to compare a series of images produced using very simple image processing operations, and look for any changes. To see how this idea works, consider Fig. 14.5.

Fig. 14.5(a) shows an iced cake which has a crack in the icing. Application of direct edge extraction techniques to the image is not appropriate due to the pattern of the icing. Instead a largest neighbour averaging technique is applied to the image to produce a new image. Each pixel is replaced by the largest pixel value of its neighbours, as shown in Fig. 14.5(b). If we now subtract the new image from the original image the crack becomes obvious, as shown in Fig. 14.5(c). Finally, Fig. 14.5(d) shows the crack superimposed on the original image.

One important application of crack detection is to printed circuit boards, where a crack can mean that a given connection is broken. There are many other properties that can be inspected on printed circuit boards, as will be discussed in the next Section.

Another approach that can be applied to the detection of surface defects is to use Moire fringes or other interference patterns. We have discussed Moire fringe image acquisition methods in Chapter 2. Moire fringe patterns give very accurate readings that can be used to check the shape of a surface, as any slight deviation in surface shape results in a different pattern. We may not necessarily be concerned with determining the exact shape of the surface but simply with detecting the presence of a defect. Alternatively, these methods may give accurate measurement of small changes in size. Many techniques have employed this type of approach to inspect the surface of objects. One recent example of such methods [46] has been used to detect and analyse surface wear on replacement knee and hip joints by comparing the interference pattern of an unworn joint with that of a worn joint. Subtracting the

14.3. SURFACE FINISH AND CRACK DETECTION

(a) (b)

(c) (d)

Figure 14.5: Example of crack detection

worn pattern from the unworn pattern results in a new pattern that shows the wear. This approach is also being developed to inspect a whole variety of objects including car headlamp reflectors, bearings, and fuel pump components.

14.4 Applications of Visual Inspection

In this Section we shall discuss the application of visual inspection to two particular areas, namely printed circuit board manufacture and the food and agricultural industries. From the study of these areas we hope to highlight many issues and problems relevant to inspection methodologies in general. Finally we shall discuss possible future applications for vision based inspection systems.

14.4.1 Printed Circuit Board Inspection

There are many different aspects of the process of printed circuit board manufacture that require inspection. For example, it is necessary to:

- Verify that components have been inserted in the correct places.
- Check circuit paths for any breaks.
- Check circuit paths for any short circuits due to solder debris.
- Check for defects (*e.g.* non-uniform bends) in the leads attaching components to the board.
- Verify that solder joints have been correctly formed.

Simple techniques for two-dimensional inspection and crack detection as already described in this Chapter can readily be applied to solve the first three problems. Testing for components being inserted in the correct place and for bent leads is not too difficult a two-dimensional inspection task. They both involve scanning the board for the occurrence of components and verifying their placement. Both these tasks can be derived from a model of the printed circuit board and its components. Further details of the tasks involved in these processes can be found in the papers by Chin and Harlow [56], Kaufman and his

14.4. APPLICATIONS OF VISUAL INSPECTION

colleagues [143] and Capson and Eng [49], and in the books by Batchelor [9] and Bretschi [36].

Checking for breaks in circuit boards is usually performed before components have been inserted, but checks for short circuits due to solder debris obviously have to be carried out after components have been inserted and soldered in place. However the basic processes used to detect such defects are similar. Again, the techniques are based on subtraction of an ideal image of the circuit from the image of the current board, much in the same way as crack detection was performed in the previous section. Clearly, before this can be done the ideal and actual images must be aligned. This can achieved by matching features from both images and estimating the relative position and orientation of the boards using techniques similar to those discussed in Chapters 8 and 9. Alternatively, the board can be positioned in a jig or similar device so that the two images should be perfectly aligned. Details of various methods that employ the above techniques can be found in Chin and Harlow's paper [56] and Bretschi's book [36].

The inspection of solder joints, however, presents a printed circuit board inspection system with a quite different set of problems unrelated to those posed by the other tasks mentioned above. Many researchers have looked into this problem [15, 49, 56, 166, 177, 213, 224, 233]. Fig. 14.6 show some typical solder defects. Fig. 14.6(a) shows a cross-section of a good joint. A good joint should have a uniform distribution of solder around the lead of the component and in the hole in the board, forming a concave fillet that lies part way up the lead. Fig. 14.6(b) shows a joint that has been missed during the soldering process. Fig. 14.6(c) shows a joint that has been formed with insufficient solder leading to a missing fillet on the lead. Too much solder has been applied to the joint in Fig. 14.6(d). Fig. 14.6(e) shows a joint that has been formed when the lead has not been *wetted* correctly — wetting is a process where the component leads and metal pad surrounding the hole in the board are coated with a small amount of solder before being soldered together to ensure a good bond between them.

It is important to note that many flaws of this type may easily escape the more traditional non-visual inspection methodologies such as *electrical continuity* testing since at the time of manufacturing these flaws do not necessarily exhibit any problems in the electrical properties

(a) Good Joint

(b) No Solder

(c) Insufficient Solder

(d) Too Much Solder

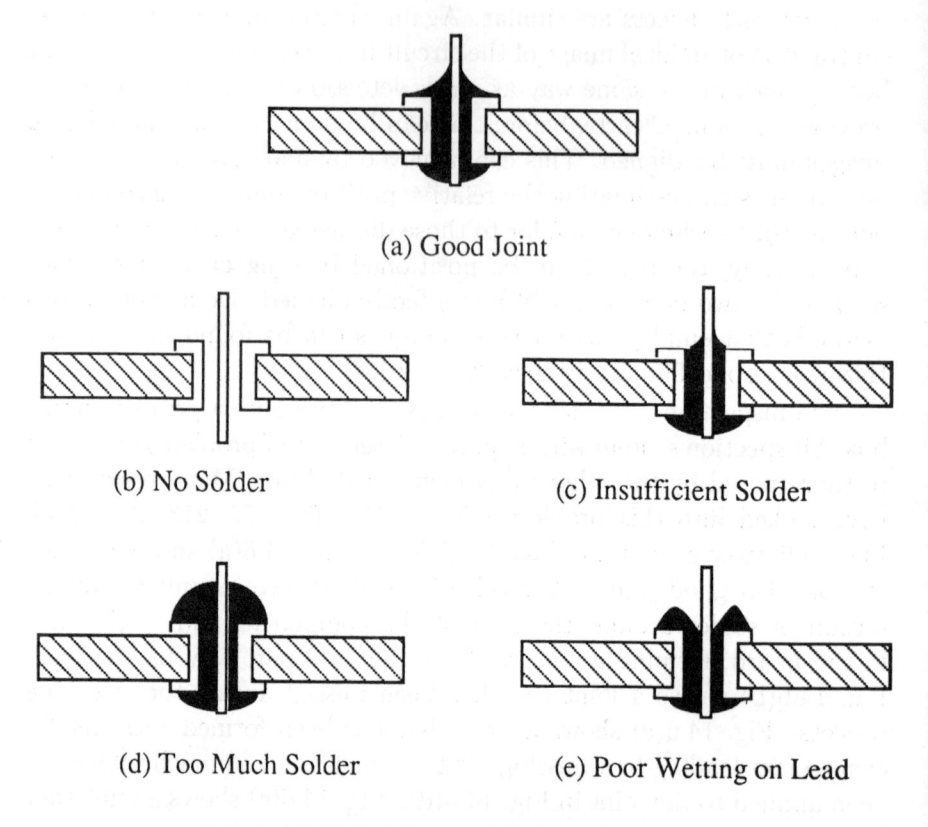

(e) Poor Wetting on Lead

Figure 14.6: Various solder joint defects

14.4. APPLICATIONS OF VISUAL INSPECTION

of the joint. However, when such flaws are subjected to everyday use they can easily break down and cause the board to fail.

Because of the invisibility or sometimes minute visual evidence of certain defects (such as *pin* or *blow holes* caused by contaminated solder) normal visual sensing methods may also be unsuitable for this type of task. Instead, both X-ray and thermal imaging methods have been applied to solder joint inspection. One popular method for thermal image inspection [213, 233, 232] directs a laser beam to heat each solder joint. The *thermal signature* — the recorded infrared image the heated joint makes — of the heated joint can then be tested. However, several types of defects can share similar thermal images, which can mean that classification of flaws is difficult. Furthermore, the heat emission patterns even for good joints can vary according to the soldering conditions, while the heating process can actually damage good joints and components.

In an alternative approach, Nakagawa [177] uses structured lighting to attempt to solve such problems. His system projects a stripe of light onto the solder joint and measures the shape profile of the solder joint. However, the high reflectance of certain types of solder joint can cause problems in obtaining sufficiently clear images of the stripe of light.

Other methods illustrate the point made previously in this Chapter about controlling the lighting of the scene to enhance specific properties of objects. McIntosh [166] and Kobayashi [146] suggested the use of coloured illumination to inspect solder joints. However, the use of colour information for each pixel can result in long image processing times. It can also be difficult to account for variations in perceived colours (red-green-blue intensities) between different colour images [224].

Consequently, *tiered coloured illumination* has been proposed [49] as a method for inspecting solder joints. The essence of such methods is to use two fluorescent light rings. The rings are mounted so that their centres lie on the same axis. One light ring which emits red light is mounted at a lower angle to the board whilst the other light ring is at a higher angle and emits blue light. The image is segmented into corresponding red, green and blue colour planes. The colour of the lighting is chosen so as to aid in the segmentation of the solder joints from the board itself which usually appears as a green background. A good solder joint (with a uniform concave fillet) appears in the image as a small red ring bounded by a blue ring when viewed from directly above.

Takagi and his fellow researchers [224] have also recently proposed a system of similar configuration. However, their lighting emits white light from both the low and high angle lights. The solder joints are inspected by considering the Lambertian and specular reflectance properties (see Section 5.4.3) of the solder joints in the image.

Finally, it is worth mentioning that Besl and his colleagues have also developed a system [15] using greyscale images and no special lighting arrangements to inspect solder joints for visible defects. Flaws are identified and classified based on the texture characteristics of the image of each joint.

Once more we have seen that no single method is able to solve all inspection needs and thus a variety of different approaches are useful for two-dimensional PCB inspection.

14.4.2 Inspection in the Food and Agriculture Industries

One sector that could reap huge benefits from automated inspection comprises the food and agricultural industries. However, until recently, these industries have not received much attention, probably because inspection problems in these areas are very hard to define due to the *non-deterministic* nature of their products. Indeed, inspection of such products will no doubt provide many interesting challenges for automated inspection for many years to come.

Visual sensing is potentially an ideal method for inspecting many foodstuffs and similar products since handling can easily damage or contaminate them. It is thus not surprising that today many institutions are considering computer vision inspection strategies for such products. For example, at a recent conference on computer vision and its applications [220], papers were presented on: foodstuff inspection using infrared illumination, cotton quality, inspection of cultivated plants for disease and rate of growth, inspection of the quality of apples, peanuts and carrots, inspection of the quality timber at sawmills and even the inspection of pigs and other live animals.

As can be seen, the possible range of things that might be visually

14.4. APPLICATIONS OF VISUAL INSPECTION

tested is immense and therefore so is the range of sensing and inspection techniques needed to solve these problems. However, the source of such problems can be broadly divided into two somewhat overlapping categories:

- production defects, and
- unwanted impurities.

Production defects can arise from many sources, including unreliable or faulty equipment, the wrong mixture of cooking ingredients, incorrect manufacturing processes such as incorrect cooking time or temperature, and defective packaging. Defects of this type can result in products suffering from, for example,

- a deformed shape or incorrect size,
- incorrect external texture,
- incorrect internal texture (such as large bubbles),
- unsatisfactory taste, and
- unsatisfactory colour.

Some of these defects are hard to detect with any inspection process, let alone a visual one. This problem is compounded by difficulties in quantifying what constitutes a particular defect.

For example, consider a loaf of bread. Two criteria it may have to satisfy in order to pass inspection might be

- Is it of a satisfactory overall shape?
- Has it been baked correctly?

Both of these criteria are highly subjective and will no doubt depend a lot on the baker's preferences. Measures of shape of the loaf can be determined by two-dimensional or three-dimensional techniques we have discussed previously in this Chapter and in Chapters 5 and 7. We can tell if the loaf has been baked correctly by examining the colour of the crust under controlled lighting conditions. Doney [67] has presented a method for *scoring* the quality of the baked loaf by considering such criteria, based on methods employed by human inspectors in bakeries. Essentially each property inspected is given a percentage score. The points for each test are totalled and the loaf passes inspection if this total is above a certain predetermined value.

As another example, let us reconsider the cake shown in Fig 14.5 which we used to illustrate crack detection. Besides cracks there are many other factors that need to be inspected. The consumer clearly expects the pattern on the cake to look appealing. However, substantial variations in the appearance and form of the pattern may occur which would still leave the cake clearly acceptable. For a vision system to deal with such large variations clearly requires a very different approach to the strict types of testing we have considered in other places in this book. For this particular application to baking and many others, classification of features by means of their texture, as discussed in Section 5.7, is useful [54]. In general, probability based algorithms, as discussed in Chapter 5, can be usefully applied to such problems of determining whether a product is defective or not [54, 67].

Production defects can also be one cause amongst several of the second category of *unwanted impurities* (foreign bodies). Detecting impurities is usually easier to perform by visual means than general production defect inspection, and has been the subject of research for some time [9]. It should be noted that impurities often have potentially much more serious consequences than production defects. If such foreign bodies are not detected there may well be a danger to human well-being as well to the commercial success of the producer. Occasional headlines in the newspapers make us aware that such problems still exist. For example, a small sample of headlines from British national newspapers taken over the last three years includes such unsavoury items as "Mouse in Milk", "Metal Pieces Found in Butter", "Glass in Crisps" and "More Glass in Baby Food". Some of the above occurrences may well be a result of industrial blackmail, where the product has been contaminated after it has left the factory, while in other cases employees have been known to attempt blackmail in this fashion. More usually, however, much more likely causes of these contaminants exist:

- Contaminants present in basic ingredients, such as nut shells that have not been separated from nuts, bones in meat, insects in

14.4. APPLICATIONS OF VISUAL INSPECTION

vegetation.

- Parts of faulty or broken machinery from the production process.

- Parts of packaging used, such as glass in bottled products.

- Human contaminants such as hair, jewellry or finger nails.

The need to detect such a wide range of foreign bodies presents an inspection system with many problems. Many of these impurities will not be visible to the naked eye as they will be contained within the product. Glass fragments in the jar of jam shown in Fig. 14.1 are one such example. Consequently alternative means of image production need to be employed, such as X-ray imaging. However, even with such techniques, certain foreign bodies may still not be visible. In particular human hairs are not easily detectable in most products using X-rays. Insects in cereals and perspex in most foodstuffs present similar problems [53, 54].

It is no surprise to find that if visual images are to be used strict control of lighting and image acquisition apparatus is required to achieve good results. In such circumstances, detecting foreign bodies from images becomes possible and techniques such as thresholding and subtraction of images, which we have discussed earlier, can be applied to produce the desired results [9].

Besides the application of a wide variety of imaging techniques, such as the use of greyscale and colour images, three-dimensional image data and X-ray images, various other techniques can also be applied to food inspection [54]. Many substances *fluoresce*, or emit visible light, when ultraviolet light is shone onto them. Certain foods exhibit a marked change in their fluorescence with age which can be used to ascertain their freshness. For example, fresh milk has a yellow fluorescence whilst sour milk appears white to grey violet under ultraviolet light. Ultrasound techniques, based on measuring the time or phase difference between transmitted and reflected ultrasound waves, can also be used to inspect food and other products. These have the advantage that they are very safe, and can be used to detect air bubbles or water in food [53]. However, ultrasound methods only give very low resolution images and cannot therefore be used for detailed inspection tasks.

In this Section, we have considered two very different case studies of printed circuit board inspection and food inspection. While perhaps not all of the techniques discussed are of immediate relevance to the inspection of engineering components, they do illustrate the point that many different inspection strategies need to be applied to completely satisfy the inspection requirements for any particular product.

14.4.3 Future Applications of Visual Inspection Systems

As we have seen in our brief consideration of visual inspection techniques in this Chapter, many objects have been inspected with such techniques. However, in many cases the solutions that have been developed are somewhat *ad hoc* in nature and thus are very specific in their application. Many developments are needed for inspection systems to progress further; a good discussion of this topic can be found in Freeman [90].

An *ad hoc* approach may in many ways be the best way to solve a particular inspection problem since such solutions are dealt with case-by-case and tend to be simpler and more compact. Industry has always tended to favour this approach [90]. Whilst this may be preferable in terms of development costs for industry, more general techniques are needed for the long term understanding of this subject. This will require re-examination of current algorithms as well as further quantification and identification of inspection problems. It is likely that this will result in a mixture of current and future techniques being used together to solve a particular problem, using methods based on the integration or fusion of data from multiple sources. We have already discussed this strategy in relation to general vision algorithms in Section 5.3.2.

Another factor that will no doubt have a vast effect on future vision systems will be advances in technology, especially in camera technology and basic computing power. Camera technology has advanced substantially in recent years and looks like continuing to do so for some time to come. It is expected that cameras with improved resolution, speed and sensitivity will be produced and with these better three-dimensional ranging systems will also be developed. Parallel and perhaps other advanced computer architectures coupled with generally faster processors and larger memory capabilities will also have an effect, allowing larger

14.4. APPLICATIONS OF VISUAL INSPECTION

or higher resolution images to be processed in a given time.

Most inspection systems to date, as we have just mentioned, solve only a particular task. For general purpose quality assurance in industry, such machines will not be cost effective in the future, as they will have high development costs. The potential exists for general purpose inspection machines to be constructed, capable of performing a variety of inspection tasks. Indeed, many such prototype machines are being built [209] or investigated at the present time. Many of these provide only a combination of visual techniques for performing inspection. However, not all inspection tasks will be able to be performed using solely visual means. For example, many manufactured parts have drilled or cast bores that can only really be inspected using conventional touch probe inspection methods. A realistic general purpose inspection cell would include a variety of different visual methods and a flexible lighting arrangement as well as robot arms for pick and place tasks and touch probes for more specific tasks. However, such a general system is probably some years away from being realised.

The development of multi-sensor controlled assembly systems will also occur in the future. These systems could and should be merged with inspection systems to produce overall systems that monitor the assembly and manufacturing processes as they are performed. Such ideas are still at embryonic stages of development at the moment with only a few having being implemented as laboratory models [209].

Chapter 15

Conclusions and Future Work

Firstly, we shall discuss the contributions that the work described in Parts 2 and 3 of the book has made towards achieving the goal of a totally automated visual inspection system. We shall then go on to consider various remaining problem areas. We shall look at these problems with reference to general computer vision tasks as well as to the inspection task.

15.1 Conclusions

In Part 1 of this book we presented an introduction to various methods used in computer vision for the acquisition of visual data and its subsequent processing. In particular, this processing is done in order to obtain an understanding of this data and ultimately to allow recognition of objects contained within a scene. In Part 2 we built upon the basis provided by Part 1 to show how model based matching strategies can recognise objects from three-dimensional data. Part 3 then extended these matching ideas with the aim of developing an automated inspection system capable of geometrically inspecting objects to typical engineering tolerances. We have also addressed some other important types of inspection task and how they can be performed.

The requirements and operation of vision-based inspection system have been outlined and the various issues arising at each stage of the vision and inspection processes have been addressed. The important points arising from each stage are briefly summarised below:

Acquiring three-dimensional depth data — The most flexible and accurate way of acquiring depth data is to use the principal of active stereoscopy. (Moire fringe methods may be more accurate, but are less flexible.) Active stereoscopy involves illuminating the object with a point or strip light source, noting the positions of corresponding points in a pair of images, and using these to determine the three-dimensional position of each point on the object. The preferred method of achieving this is to eliminate all external light from the vision system and to project stripes of laser light onto the object. This was discussed in Section 2.5.

In order that the depth of each illuminated point can be calculated the positions and orientations of the pair of stereo cameras has to be calibrated relative to some world coordinate system. Various internal camera parameters, representing for example lens distortion, must also be estimated. Such a method for calibrating the cameras has been outlined in Section 2.5. In practice errors from this stage still constitute a problem in our system as we have seen in studies of segmentation in Section 9.6, matching in Section 9.7 and inspection in Sections 12.5 and 13.6.

Extracting information from depth data — In order that efficient and reliable inspection can be achieved, higher level descriptions than simple arrays of depth readings of points on the surface of the viewed object are required. Currently the only higher level descriptions we can produce that are capable of providing accurate and reliable enough data, for object recognition, positioning and inspection purposes, are planar descriptions of the faces of the object. A method for extracting planar regions from the depth maps has been given. Other non-planar surface information can be inspected with or without extraction from the data at this stage.

Geometric modelling of the objects — Since the result of the previous step is to describe the viewed object in terms of a set of

15.1. CONCLUSIONS

surface regions, it is desirable that similar descriptions of the object should be readily available from a stored geometric model. The boundary representation modelling scheme is thus the most promising method of representing solid objects when comparing viewed and modelled object features. The issue of including geometric tolerance specifications in this representation has also been addressed.

Matching scene and model descriptions — Before the inspection of an object can be performed, the precise location and orientation of the object under test must be determined. This is achieved by matching the scene description of the viewed object to the corresponding stored model of the object. Various current model based matching strategies have been assessed and one has been adopted as best suited to the inspection task. For application of the method to the inspection problem, a range of improvements have been made, both to the accuracy of the estimation of the location and orientation provided, and to the efficiency. The basic matching strategy, employing geometric constraints to control the search space, has proved to be very efficient as also noted by Grimson [109]. Such strategies are extremely efficient at finding the correct model to scene match when there is only one object present, which is the expected case in our inspection problem.

Inspecting the object — A method for automatically inspecting a series of views of an object taken about a single axis has been outlined. The basic method for inspecting the object is to inspect information extracted from each individual view of the object separately and then to rotate the object on a turntable to the next best view which brings in features that have not yet been inspected. The set of views is compiled offline using the object model and geometric information about the positions of the turntable and cameras.

It is also suggested how these methods may be extended to perform complete inspection, incorporating manipulation with a robot arm to reposition the object on the turntable.

Throughout each of the above stages tests have been performed using both synthetically created depth map data and real depth map data provided from the working vision system.

Results from depth map data synthetically created to simulate the typical performance of the vision system, based on a theoretical assessment of errors inherent in such a system, have shown that the outlined vision and inspection strategy is capable of geometrically inspecting objects to typical engineering tolerances of 0.254mm.

Results on the actual depth data produced by the real vision system have shown that currently inspection to such tolerances is not quite possible. However, errors in the calibration of the cameras have been shown to be the most likely cause of the inaccuracies causing this problem. Once these errors have been remedied it is expected that inspection to the required tolerances will be possible. Work is currently being carried out to further investigate the exact source of the errors. This could necessitate improving the calibration model, improving the calibration procedures, or both. The following observations may be relevant to improving the camera calibration procedure and calibration results.

- Certain camera parameters are assumed to be constant for similar cameras after one initial calibration (see Chapter 2), for example the effective focal length, lens distortion parameters and the uncertainty scale factor discussed in Section 2.3.2. These should be calibrated for each camera.

- The camera calibration chart as used at present, and as shown in Fig 2.4, only occupies a two-dimensional slice of the viewing volume visible to the cameras. Thus, the lens distortion parameters are not entirely representative of points which are to be acquired during inspection. The use of calibration points distributed throughout the three-dimensional viewing volume could overcome such problems. The use of several calibration charts placed in different positions and orientations throughout the viewing volume is currently being investigated [132, 133].

- More realistic camera models may be required, although these will require careful consideration. For example, we currently ignore tangential lens distortion, on the grounds that the calibration procedure will become more numerically unstable if it is included, which may outweigh any theoretical improvements (see Section 2.3.2).

- The use of more accurate sensing equipment may be desirable. The accuracy and general performance of sensing equipment is constantly being improved. Therefore, we can expect a slightly better performance from the sensing equipment. However, there are some fundamental problems that can not be overcome. One major problem is in the hardware timing between devices (see Section 2.3.2) resulting from device clocks being out of synchronisation. One possible way to avoid this problem could be to use *digital cameras* [147]. Such cameras output precise x, y pixel locations as well as the greyscale value for that pixel output by conventional CCD cameras. The use of such devices is also currently being investigated [132, 133].

15.2 Future Work

This book has demonstrated that geometric inspection to standard engineering tolerances is feasible using a three-dimensional vision system employing a range of computer vision techniques. However, many areas requiring further work have been noted throughout the book. Some good sources on the state of the art of computer vision which deal specifically with the future development of the science are the paper by Brady and his colleagues [31], the series of dialogues by leading computer vision researchers in a recent issue of *Computer Vision, Graphics and Image Processing* [63] and the recent books containing collections of papers and panel discussions edited by Freeman [90, 91].

These discussions of further work are mainly concerned with obtaining better results for the particular computer vision tasks they are concerned with. However, as we have seen throughout this book, many vision tasks are interrelated, particularly when incorporated into complete systems. Inspection is a particular case in which high degrees of accuracy are required for each stage of the overall process. The main areas which we expect to benefit from further developments will now be considered, both from the point of view of computer vision problems in general, and geometric inspection in particular.

Camera calibration — The main source of errors in our vision process, developed in this book, has been shown to lie in camera calibration problems. Clearly such errors need to be reduced. We have discussed the likely sources of error and some immediate remedies for these calibration problems in the previous Section. However, in the long run, further theoretical work is probably needed to develop better camera models and calibration procedures.

Segmentation — Extracting high level primitives from depth data is a difficult task as we have seen in Chapter 4. Throughout the book, and in particular in Chapter 5 we have discussed techniques that could make this task easier. In particular, the use of information from sets of images obtained from more than one visual source seems to be the most promising approach. Methods are required to integrate or *fuse* the data from these multiple sources together. Presently this is an area of active research with a whole host of applications. One example is the development of mobile robots or vehicles which are to navigate in fairly simple environments while avoiding collisions. Another example of data fusion is in the area of segmentation when edge information can be used to improve region segmentation techniques as those in Chapter 4. Taking this a stage further, potentially more accurate methods, such as Moire fringe techniques, could also be integrated to achieve greater accuracy of surface information for inspection and other tasks. Various ways of achieving data fusion have been discussed in Chapter 5.

Data fusion techniques need not only be concerned with different sources of information but also some use can be made of different representations of the same data. For example, many tasks can be performed on the frequency representation of an image provided by its Fourier transform as discussed in Chapter 3. In his book, Wechsler [236] provides a good account of the simultaneous use of spatial and frequency representations (*cojoint representations*)

15.2. FUTURE WORK

for a variety of computer vision tasks. Indeed, evidence exists that the human visual system uses such methods [156, 236].

Coming closer to the theme of this book, there is no reason why stored object model information cannot be used in the segmentation process. Most of the model based recognition strategies we have discussed have used a *free* fitting strategy for surface reconstruction followed by a model based matching strategy. Thus, at the segmentation stage, no assumptions are made as to the surfaces which are likely to be found in the scene. In certain circumstances, such as those typified by inspection, this is clearly not very sensible. Great benefits could arise from using the stored geometric model to drive the segmentation from the actual surfaces expected to be visible. In detail, the number of parameters that describe the primitive surface can be restricted, so we can search for restricted classes of surfaces. Thus the numerical approximations used to fit data to the surface can be made simpler, and, as values for less free parameters are to be fitted, the results are likely to be more accurate. We have used similar strategies for cylindrical and spherical surfaces in the inspection stages described in Section 11.4. An hypothesis and test strategy is used to predict where each primitive is, and its type. The inspection process then tests whether the point data fits those surfaces as expected.

Object Recognition — We have described many recognition algorithms and demonstrated the application of a modified version of one of them in Section 9.7. However, nearly all of these recognition methods suffer from the same drawbacks. They can only recognise objects that have a strict or rigid geometric representation. An object that cannot be represented easily by current types of solid models as discussed in Chapter 7 cannot be readily recognised. However, some progress is being made with objects that can be *parametrised* in terms of scale and position of certain key features [34, 109]. Here whole classes of objects can be represented and hence recognised by a single object model as opposed to requiring a separate model for every instance of the object within the class. This will improve the efficiency of the model based matching strategy in general, as fewer models will need to be stored and matched to, although comparisons with such a model will be more complex.

Secondly, the matching strategies we have described are quite efficient when applied to relatively uncluttered scenes. When there are many objects in a scene the efficiency of such methods suffers. Some methods try to counter this by choosing subsets of features to initially recognise an object. They then hypothesise (using the object model) sets of features that should be in the scene and apply a strategy to test the hypothesis, backtracking and trying another hypothesis if necessary. Further work is needed on methods that can efficiently and reliably choose relevant subsets of data that can be used at this initial stage.

Another approach that could be used to avoid a combinatoric explosion during the matching process is to use the object model to guide the search. This will involve developing more robust geometric reasoning techniques, such as those reported in [85, 86, 87, 109, 139, 142]. When a large database of objects exists that are candidates for matching, efficient searches of the database are required. This problem does not really concerned us in inspection as only a single model needs to be retrieved from the database. Nevertheless, for recognition tasks it may be a real and serious problem. Many standard database retrieval methods have been developed to address this problem such as linear searches, feature indexing (the feature focus methods of the *3DPO* and *TINA* vision systems described in Chapter 8 fall into this category) and hashing [109]. However, more research into the choice and use of features for matching is needed for complex scene understanding.

One further useful advance for an object recognition system would be for systems to be able to *learn* new objects that do not fit previously stored models [109].

Automated Inspection Strategies — The inspection strategies detailed in this book were primarily developed to demonstrate the feasibility of automatic inspection using a vision system. Further work is needed to develop these and more advanced strategies so

15.2. FUTURE WORK

that more general geometric inspection of objects can be achieved. Strategies will need to be developed to enable complete inspection of the object rather than just a single set of views obtained by rotating a turntable. This will typically involve object manipulation with a robot arm. Also, strategies need to be developed that can reason with the geometric model of the object so that the inspection tasks can be planned and performed more efficiently. At present our method of selecting the views is crude. Methods of selecting a better set of views are needed which provide control over both the accuracy of inspection, and the number of different views used (and hence the time taken), a pair of conflicting requirements.

From the above analysis it is clear that many advances are required before computer vision systems can be developed for general purpose vision tasks, as opposed to the often *ad-hoc* applications generally found at the moment. We have identified two main areas in which we feel significant progress can be made in the short term, namely data fusion and geometric reasoning. Furthermore, there is also a need to reason about the physics of the image formation processes and the effects that various image processing and segmentation processes have on the resultant data. We also need to be able to deal with uncertainties resulting from inaccurate data and false hypotheses [139]. We therefore need to be able to deal with uncertainties both in logical and geometric data. The application of probabilistic approaches to this area seems most appropriate [68].

All of the above will no doubt provide the computer vision community with much work for many years to come.

Appendices

Appendices

Appendix A

Matrix Definitions

In this appendix we shall give some basic definitions and properties of matrices that are required in this book. We assume that the reader has a basic grounding in elementary matrix and vector concepts such as addition and multiplication, the identity matrix and determinants. A good introduction to all of these concepts, with details of many other matrix computations, is the book by Golub and van Loan [104].

A.1   Matrix Signature and Rank

Let $\mathbf{A}$ be an $n \times m$ matrix with elements denoted by a_{ij}, $1 \leq i \leq n, 1 \leq j \leq m$.

The *signature* or *trace*, s, of a matrix is simply the sum of all the *leading diagonal* elements of the matrix, which must be square, so $m = n$. Thus

$$s = \sum_{i=1}^{n} a_{ii}. \tag{A.1}$$

The *rank* of a matrix $\mathbf{A}$ is defined as follows. Let us form determinants of all possible square submatrices of $\mathbf{A}$ which are formed by striking out zero or more arbitrary rows or columns of $\mathbf{A}$. Let $r \times r$ be the size of the largest non-zero determinant (or determinants). Then r is the rank of $\mathbf{A}$. In particular, if the matrix is the null matrix its rank is zero. We have given a method in Section 9.4.6 to determine the rank of a matrix using techniques based on singular value decomposition (which is defined in the next Section).

A.2 Eigenvectors and Eigenvalues

The eigenvalues and associated eigenvectors of a matrix have many useful mathematical properties. We have used eigenvalues and eigenvectors when solving systems of linear equations, determining orthogonal coordinate systems from a set of vectors and determining the rank of a matrix. These are easily realised from the definition of eigenvalues and eigenvectors. Let us now define the eigenvalues and eigenvectors of a matrix $\mathbf{A}$.

Consider the matrix equation

$$\mathbf{A}\mathbf{x} = \lambda \mathbf{x} \qquad (A.2)$$

where $\mathbf{A}$ is an $n \times n$ square matrix, $\mathbf{x}$ is column vector of n components and λ is a scalar value.

We can rewrite Eqn. A.2 as

$$(\mathbf{A} - \lambda \mathbf{I}_{n \times n})\mathbf{x} = 0 \qquad (A.3)$$

where $\mathbf{I}_{n \times n}$ is the $n \times n$ identity matrix.

Clearly a *trivial* solution to this exists when $\mathbf{x}$ is the column vector $\{0, 0, \ldots, 0\}$. Besides this solution a number of *non-trivial* solutions exist that satisfy

$$|\mathbf{A} - \lambda \mathbf{I}_{n \times n}| = 0. \qquad (A.4)$$

To see this, if $|\mathbf{A} - \lambda \mathbf{I}_{n \times n}|$ were not zero, it would be possible to take the inverse of the matrix $\mathbf{A} - \lambda \mathbf{I}_{n \times n}$, leading to the conclusion that $\mathbf{x} = (\mathbf{A} - \lambda \mathbf{I}_{n \times n})^{-1} 0$, which is clearly a contradiction for $\mathbf{x}$ non-zero.

Eqn. A.4 is called the *characteristic* equation of the matrix $\mathbf{A}$. The n roots of the equation, denoted by $\lambda_1, \lambda_2, \ldots, \lambda_n$ are called the *eigenvalues* of $\mathbf{A}$.

Corresponding to each eigenvalue there is a solution vector $\mathbf{x}_i$ to Eqn. A.2. These solutions are called the *eigenvectors* of $\mathbf{A}$.

A.3 Singular Value Decomposition

The *singular value decomposition* (SVD) of a matrix provides a method of solving linear systems of equations and also of determining the rank

A.3. SINGULAR VALUE DECOMPOSITION

of a matrix [104]. The SVD method is a particularly numerically stable method of computation. We have discussed methods for computing the SVD of a matrix in Section 9.3.1 and its use to determine the rank of a matrix in Section 9.4.6. Here we shall simply provide a brief summary of the relevant definitions.

The SVD of a matrix $\mathbf{A}$ (not necessarily square) is formed if it is possible to factorise $\mathbf{A}$ into

$$\mathbf{A} = \mathbf{U}\mathbf{\Sigma}\mathbf{V}^T \quad (A.5)$$

where $\mathbf{U}^T\mathbf{U} = \mathbf{V}^T\mathbf{V} = \mathbf{V}\mathbf{V}^T = \mathbf{I}$, the identity matrix of appropriate size, and $\mathbf{\Sigma}$ is a diagonal matrix whose elements are $(\sigma_1, \sigma_2, \sigma_3, \ldots)$. The latter are the singular values. (^{T} denoted matrix transpose.)

In fact, the matrix $\mathbf{U}$ consists of the orthonormalised eigenvectors of $\mathbf{A}\mathbf{A}^T$, while the matrix $\mathbf{V}$ consists of the orthonormalised eigenvectors of $\mathbf{A}^T\mathbf{A}$. The singular values are the square roots of the eigenvalues of $\mathbf{A}^T\mathbf{A}$..

Appendix B

Least Squares Approximation

In this appendix we give details of least squares approximations used in the segmentation methods described in Chapter 4. In the context of this book, least squares approximations are used for surface fitting. Let us suppose some particular type of surface is described by a number of parameters. In general, a certain number of points in space will serve to uniquely fix those parameters. For example, three non-collinear points in space will define a plane. Many applications, however, produce a large set of points which are all supposed to lie on the same surface, where there are now many more points than the minimum number required to define the surface. Errors in the positions of these points mean that in practice they do not all lie exactly on the expected surface. Least squares approximation methods find the surface of the given type which best represents the data points. In the case of planes this is in the sense that the sum of the squares of the distances of the data points to the fitted plane (*i.e.* the errors) is minimised.

B.1 Approximation of a Planar Surface

This Section shows how the method of least squares fitting may be used to find the plane which best fits a set of three-dimensional data points. Suitable precautions for avoiding numerical errors will be discussed.

Let us suppose we are given a set of points in the form $z = z(x, y)$, i.e. we are given z_i values for the points lying at a set of known (x_i, y_i) positions. The equation of a plane can be written (as discussed in Section 4.6.1) as

$$z = ax + by + c. \tag{B.1}$$

For a given point $P_i = (x_i, y_i, z_i)$ the error in the fit is measured by

$$\sigma_i = z_i - (ax_i + by_i + c). \tag{B.2}$$

(Although this measures the error in the z direction, and not the *perpendicular* distance to the plane directly, the two are in fact proportional to each other by a factor of $\sin \theta$ where θ is the angle between the plane and the z axis.)

The best fitting plane to a set of n points $\{P_i\}$ occurs when $\chi^2 = \sum \sigma_i^2$ is a minimum, where this and all subsequent sums are to be taken over all points, $P_i, i = 1, \ldots, n$. This occurs when

$$\frac{\partial \chi^2}{\partial a} = 0 \tag{B.3}$$

and hence

$$\sum x_i z_i - a \sum x_i^2 - b \sum x_i y_i - c \sum x_i = 0, \tag{B.4}$$

when

$$\frac{\partial \chi^2}{\partial b} = 0 \tag{B.5}$$

and hence

$$\sum y_i z_i - a \sum x_i y_i - b \sum y_i^2 - c \sum y_i = 0, \tag{B.6}$$

and when

$$\frac{\partial \chi^2}{\partial c} = 0 \tag{B.7}$$

and hence

$$\sum z_i - a \sum x_i - b \sum y_i - nc = 0. \tag{B.8}$$

Now, some of these sums can become quite large and any attempt to solve the equations directly can lead to poor results due to rounding errors. One method of reducing the rounding errors is to re-express the coordinates of each point relative to the centre of gravity of the set of points, (x_g, y_g, z_g). All subsequent calculations are performed under this translation with the final results requiring the inverse translation to return to the original coordinate system.

The translation also simplifies our minimisation problem since now

$$\sum_{\forall i} x_i = \sum_{\forall i} y_i = \sum_{\forall i} z_i = 0. \tag{B.9}$$

Therefore, using the above fact in Eqns. B.4, B.6 and B.8 and solving them gives:

$$a = \frac{\sum x_i z_i \sum y_i^2 - \sum y_i z_i \sum x_i y_i}{\sum x_i^2 \sum y_i^2 - (\sum x_i y_i)^2}, \tag{B.10}$$

$$b = \frac{\sum y_i z_i \sum x_i^2 - \sum x_i z_i \sum x_i y_i}{\sum x_i^2 \sum y_i^2 - (\sum x_i y_i)^2}, \tag{B.11}$$

$$c = 0. \tag{B.12}$$

B.2 Approximation of a Quadric Surface

A general quadric surface can be expressed in the form (note that this is slightly different to that used in Chapter 4)

$$a_1 x^2 + a_2 y^2 + a_3 z^2 + \sqrt{2} a_4 xy + \sqrt{2} a_5 xz + \sqrt{2} a_6 yz + a_7 x + a_8 y + a_9 z + a_{10} = 0, \tag{B.13}$$

or alternatively,

$$\mathbf{x}^T \mathbf{A} \mathbf{x} + \mathbf{x} \mathbf{v} + d = 0 \tag{B.14}$$

where

$$\mathbf{A} = \begin{bmatrix} a_1 & a_4/\sqrt{2} & a_5/\sqrt{2} \\ a_4/\sqrt{2} & a_2 & a_5/\sqrt{2} \\ a_5/\sqrt{2} & a_5/\sqrt{2} & a_3 \end{bmatrix},$$

$$\mathbf{v} = (a_7 \; a_8 \; a_9)^T,$$

$$d = a_{10},$$

$$\mathbf{x} = (x, y, z)^T$$

The matrix $\mathbf{A}$ gives information about the geometric form of the quadric. The eigenvectors of $\mathbf{A}$ give the directions of the principal axes of the quadric, while the determinant of **A** gives the class of the quadric [149] (*e.g.* if $|A| = 0$ then the quadric is a cone or a cylinder), as discussed in Section 4.6.1.

The natural geometric interpretation of error in least squares fitting is the distance of each point to the surface, as stated earlier. However, in the case of quadric surfaces, using such a measure would cause the mathematics of the approximation to become very awkward. Instead we use quantities similar to those used in the case of planes (but with no direct geometric meaning) to keep the approximation simple. Thus the error in the fit of the set of points $\{x_i\}, i = 1, \ldots, n$ to a quadric surface is defined [80], using Eqn. B.14, as

$$\chi^2 = \sum (x_i^T A x_i + x_i v + d)^2, \qquad (B.15)$$

where again the sum is taken over all data points x_i.

The minimisation of Eqn. B.15 must be constrained in order to avoid the trivial solution $\{a_j = 0; j = 1 \ldots 10\}$. However no natural geometric constraints now exist, unlike in the case for planes, where we can assume in our formulation that the coefficient of z is 1, because the plane must point at least partly towards the camera. Note that care must be taken in choosing constraints that are invariant with respect to geometric transformations. Faugeras and Hebert [80] suggest using the constraint that

$$\mathrm{Tr}(AA^T) = \sum_{j=1}^{6} a_j^2 = 1 \qquad (B.16)$$

where Tr is the trace of a matrix (see Appendix A). This constraint is invariant to both translation and rotation, because **A** is invariant to translation and rotation [80, 219]. Other possible choices of constraints such as $\sum_{j=1}^{10} a_j^2 = 1$ are not invariant to rotation or translation and cause problems in use.

We will now describe Faugeras and Hebert's method further.

Let us define the three vectors $p = (a_1, \ldots, a_{10})^T$, $p_1 = (a_1, \ldots, a_6)^T$ and $p_2 = (a_7, \ldots, a_{10})^T$. Then the constraint of Eqn. B.16 may be rewritten as

$$\|p_1\| = 1. \qquad (B.17)$$

Eqn. B.14 is a quadratic function of the components of **p** and so it

B.2. APPROXIMATION OF A QUADRIC SURFACE

can be written
$$\chi^2 = \sum \mathbf{p}^T \mathbf{M}_i \mathbf{p}, \qquad (B.18)$$
where the summation is over all data points $\mathbf{x}_i = (x_i, y_i, z_i)$. $\mathbf{M}_i$ is the symmetric matrix
$$\mathbf{M}_i = \begin{bmatrix} \mathbf{B}_i & \mathbf{C}_i \\ \mathbf{C}_i^T & \mathbf{D}_i \end{bmatrix};$$
by comparing coefficients of Eqns. B.15 and B.18 it can be seen that matrices $\mathbf{B}_i, \mathbf{C}_i$ and $\mathbf{D}_i$ are of the form

$$\mathbf{B}_i = \begin{bmatrix} x_i^4 & x_i^2 y_i^2 & x_i^2 z_i^2 & \sqrt{2} x_i^3 y_i & \sqrt{2} x_i^3 z_i & \sqrt{2} x_i^2 y_i z_i \\ x_i^2 y_i^2 & y_i^4 & y_i^2 z_i^2 & \sqrt{2} x_i y_i^3 & \sqrt{2} x_i y_i^2 z_i & \sqrt{2} y_i^3 z_i \\ x_i^2 z_i^2 & y_i^2 z_i^2 & z_i^4 & \sqrt{2} x_i y_i z_i^2 & \sqrt{2} x_i z_i^3 & \sqrt{2} y_i z_i^3 \\ \sqrt{2} x_i^3 y_i & \sqrt{2} x_i y_i^3 & \sqrt{2} x_i y_i z_i^2 & 2 x_i^2 y_i^2 & 2 x_i^2 y_i z_i & 2 x_i y_i^2 z_i \\ \sqrt{2} x_i^3 z_i & \sqrt{2} x_i y_i^2 z_i & \sqrt{2} x_i z_i^3 & 2 x_i^2 y_i z_i & 2 x_i^2 z_i^2 & 2 x_i y_i z_i^2 \\ \sqrt{2} x_i^2 y_i z_i & \sqrt{2} y_i^3 z_i & \sqrt{2} y_i z_i^3 & 2 x_i y_i^2 z_i & 2 x_i y_i z_i^2 & 2 y_i^2 z_i^2 \end{bmatrix},$$

$$\mathbf{C}_i = \begin{bmatrix} x_i^3 & x_i^2 y_i & x_i^2 z_i & x_i^2 \\ x_i y_i^2 & y_i^3 & y_i^2 z_i & y_i^2 \\ x_i z_i^2 & y_i z_i^2 & z_i^3 & z_i^2 \\ \sqrt{2} x_i^2 y_i & \sqrt{2} x_i y_i^2 & \sqrt{2} x_i y_i z_i & \sqrt{2} x_i y_i \\ \sqrt{2} x_i^2 z_i & \sqrt{2} x_i y_i z_i & \sqrt{2} x_i z_i^2 & \sqrt{2} x_i z_i \\ \sqrt{2} x_i y_i z_i & \sqrt{2} y_i^2 z_i & \sqrt{2} y_i z_i^2 & \sqrt{2} y_i z_i \end{bmatrix},$$

$$\mathbf{D}_i = \begin{bmatrix} x_i^2 & x_i y_i & x_i z_i & x_i \\ x_i y_i & y_i^2 & y_i z_i & y_i \\ x_i z_i & y_i z_i^2 & z_i & z_i \\ x_i & y_i & z_i & 1 \end{bmatrix}.$$

Eqn. B.18 can be written as
$$\chi^2 = \mathbf{p}^T \mathbf{M} \mathbf{p}, \qquad (B.19)$$
where $\mathbf{M} = \sum \mathbf{M}_i$ taken over all $\mathbf{x}_i$.

The minimisation of χ^2 can be solved by using the method of Lagrangian multipliers [100, 123]. The problem reduces to finding the minimum λ and corresponding vector $\mathbf{p}$ such that
$$\mathbf{M}\mathbf{p} - \lambda \mathbf{p}_1 = 0. \qquad (B.20)$$

Since $\mathbf{p} = (\mathbf{p}_1, \mathbf{p}_2)$, the system of Eqns. B.20 can be split into two parts:

$$\mathbf{B}\mathbf{p}_1 + \mathbf{C}\mathbf{p}_2 - \lambda\mathbf{p}_1 = 0, \qquad (B.21)$$

$$\mathbf{C}^T\mathbf{p}_1 + \mathbf{D}\mathbf{p}_2 = 0, \qquad (B.22)$$

where $\mathbf{B} = \sum \mathbf{B}_i, \mathbf{C} = \sum \mathbf{C}_i$, and $\mathbf{D} = \sum \mathbf{D}_i$, again summing over all data points.

The minimum value of λ, λ_{min}, from Eqn. B.21 is the lowest eigenvalue of the matrix $\mathbf{B} - \mathbf{C}\mathbf{D}^{-1}\mathbf{C}^T$. The unit eigenvector corresponding to λ_{min} is the solution for $\mathbf{p}_1$. Knowing $\mathbf{p}_1$, the solution for $\mathbf{p}_2$ can be obtained from Eqn. B.22 by solving the linear system of equations

$$\mathbf{D}\mathbf{p}_2 = \mathbf{C}^T\mathbf{p}_1. \qquad (B.23)$$

Once again, rounding errors can be reduced by performing all of the calculations about the centre of gravity of the points, as in case of the approximation of planes discussed in Section B.1.

Appendix C

Quaternions

The theory of quaternions is used in the estimation of the rotation component of the transformation between model and object coordinate systems in Chapter 9. The elements of quaternion algebra are presented in this appendix together with some relevant results.

A quaternion may be regarded as pair $\mathbf{q} = (\mathbf{v}, s)$ where $\mathbf{v}$ is vector of $\Re^3$ and s is a scalar. Many other definitions exist but this is the one most suited to our problem of representing the rotation.

The product ($*$) of two quaternions is defined by:

$$(\mathbf{v}, s) * (\mathbf{v}', s') = (\mathbf{v} \times \mathbf{v}' + s'\mathbf{v} + s\mathbf{v}', ss' + \mathbf{v}.\mathbf{v}') \qquad \text{(C.1)}$$

where $\times$ is the vector cross product and . is the vector dot product.

The conjugate $\bar{\mathbf{q}}$ of a quaternion $\mathbf{q} = (\mathbf{v}, s)$ is defined by

$$\bar{\mathbf{q}} = (-\mathbf{v}, s). \qquad \text{(C.2)}$$

The norm of a quaternion is defined by

$$|\mathbf{q}|^2 = \mathbf{q} * \bar{\mathbf{q}} = \|\mathbf{v}\|^2 + s^2. \qquad \text{(C.3)}$$

The norm is also multiplicative

$$|\mathbf{q} * \mathbf{q}'|^2 = |\mathbf{q}|^2 * |\mathbf{q}'|^2. \qquad \text{(C.4)}$$

In our discussions we always deal with unit quaternions. However in general the inverse of a quaternion, denoted by $\mathbf{q}^{-1}$ is often more useful than its conjugate. For a quaternion, **q**, the inverse is defined by:

$$\mathbf{q}^{-1} = \frac{\bar{\mathbf{q}}}{|\mathbf{q}|}. \tag{C.5}$$

C.1 Rotation by Quaternions

Rotations may be performed in two dimensions by using complex numbers. Quaternions may be regarded as a generalisation of complex numbers to three-dimensional space with the vector part **v** corresponding to the imaginary part. In detail, complex numbers of modulus 1 are used to represent two-dimensional rotations. The equivalent holds for quaternions in three-space.

Let **R** be a rotation represented by the quaternion **q**, and suppose we wish to rotate a vector **u** by **R**. If

$$\mathbf{u}' = \mathbf{R}\mathbf{u} \tag{C.6}$$

then

$$\mathbf{u}'_q = \mathbf{q} * \mathbf{u}_q * \bar{\mathbf{q}} \tag{C.7}$$

where the quaternion representing any vector **v** is $\mathbf{v}_q = (\mathbf{v}, 0)$. Also, if **a** is the axis of the rotation and θ is the angle of rotation, then the quaternion **q** performing the rotation is $(\mathbf{s}, c)$, where

$$\mathbf{s} = \sin(\theta/2) \quad \mathbf{a}, \tag{C.8}$$
$$c = \cos(\theta/2). \tag{C.9}$$

Appendix D

Outliers

The theory of *outliers* is appropriate when trying to decide whether a given data point lies within the naturally expected range of values for a set of data points, or whether it represents a value outside the expected variation. In this appendix we give the appropriate test and statistical table for determining whether a data point is classed as such an outlier. This test is used when inspecting an object for compliance to tolerance limits as discussed in Chapter 11.

D.1 Significance Test For a Single Outlier

Let us be given a sample of n data values x_i with sample mean μ and sample variance s^2. The mean and variance are calculated as follows:

$$\mu = \frac{1}{n}\sum_i x_i,$$
$$s^2 = \frac{1}{n-1}\sum_i (x_i - \mu)^2. \qquad (D.1)$$

Let us suppose we wish to test if a given data value x_j is an outlier which may be unexpectedly larger or smaller than the sample mean. In particular we consider the case of a *single* outlier, which implies that we are only considering one point at a time as being an outlier. The following statistic is calculated:

$$\left|\frac{x_j - \mu}{s}\right|. \tag{D.2}$$

This statistic is now compared to a standard value depending on the sample size, and a standard significance level. If the calculated statistic is less than the appropriate standard value then the point is considered to belong to the sample otherwise it is considered to be an outlier. The significance level tells us how certain we are of this classification. Thus, if we use a 5% significance level then we would expect only 5% of the points identified as outliers to be incorrectly classified, while a significance level of 1% reduces the chances of mistakenly calling a point an outlier even further.

Details of the calculation of the appropriate values to test our statistic against are given in [8]; this statistic is expected to follow the *internally Studentized* distribution from the mean. Since the sample sizes used in this book are of a small and usually fixed size (for example a region of 5×5 pixels), the values may be precomputed and stored in a look-up table for our purposes.

Table D.1 is based on Table VIII from Barnett and Lewis's book [8], which shows the critical values for the outlier test for 5% and 1% significance levels for different values of n. Wherever we have used outliers in this book, we have used a significance test based on the 5% significance level.

n	5%	1%
3	1.15	1.15
4	1.46	1.49
5	1.67	1.75
6	1.82	1.94
7	1.94	2.10
8	2.03	2.22
9	2.11	2.18
10	2.18	2.41
12	2.29	2.55
14	2.37	2.66
15	2.41	2.71
16	2.44	2.75
18	2.50	2.82
20	2.56	2.88
25	2.66	3.00
36	2.82	3.19
49	2.95	3.33
64	3.05	3.43
81	3.13	3.52
100	3.21	3.60
121	3.27	3.66

Table D.1: Table of significance levels for outlier test

Bibliography

[1] E. H. Al-Hujazi, Integration of Edge and Region Based Techniques for Range Image Segmentation, *Proceedings of SPIE OE/BOSTON 90 Conference No.* **1381** *on Intelligent Robots and Computer Vision IX: Algorithms and Techniques*, Ed. D. P. Casasent, SPIE, Washington, USA, pp 589-600, (1990).

[2] J. A. D. W. Anderson, G. D. Sullivan and K. D. Baker, Constrained Constructive Solid Geometry: A Unique Representation of Scenes, *Proceedings of the 4th Alvey Vision Conference*, pp 91-96, (1988).

[3] Anorod Precision Tables, *Sales Literature*, (1991).

[4] N. Ayache and O. D. Faugeras, HYPER:A New approach for the Recognition and Positioning of 2-D Objects, *IEEE Transactions on Pattern Analysis and Machine Intelligence*, **8** (1), pp 2-14, (1986).

[5] N. Ayache, O. D. Faugeras, B. Faverjon and G.Toscani, Matching Depth Maps Obtained by Passive Stereo, *Proceedings 3rd Workshop on Computer Vision: Representation and Control*, pp 197-204, (1985).

[6] D. H. Ballard and C. M. Brown, *Computer Vision*, Prentice Hall, Englewood Cliffs, New Jersey (1982).

[7] D. H. Ballard and D.H. Sabbah, On Shapes, *Proceedings of 7th International Joint Conference on Artificial Intelligence*, pp 607-612, (1981).

[8] V. Barnett and T. Lewis, *Outliers in Statistical Data*, J. Wiley and Sons, Chichester, (1984).

[9] B G. Batchelor, D. A. Hill and D. C. Hodgson, *Automated Visual Inspection*, North Holland, Amsterdam, (1985).

[10] F. L. Bauer, Elimination with Weighted Row Combinations for Solving Linear Equations and Least Squares Problems, *Numerische Mathematik*, **7**, pp 338-352, (1965).

[11] B. G. Baumgart, Geometric Modelling for Computer Vision, *Stanford Artificial Intelligence Lab. Report STAN-CS-74-463*, (1974).

[12] R. C. Beach, *An Introduction to the Curves and Surfaces of Computer Aided Design*, Van Nostrand Reinhold, New York, (1991).

[13] P. J. Besl, Active Optical Range Imaging Sensors, in *Advances in Machine Vision*, Ed. J. L. C. Sanz, Springer-Verlag, New York (1988).

[14] P. J. Besl, *Surfaces in Range Image Understanding*, Spriner-Verlag, New York, USA, (1988).

[15] P. J. Besl, E. J. Delp and R. K. Jain, Automatic Visual Solder Joint Inspection, *IEEE Journal of Robotics and Automation*, **1** (1), pp 42-56, (1985).

[16] P. J. Besl and R. K. Jain, Three-Dimensional Object Recognition, *ACM Computing Surveys*, **17** (1), pp 75-145, (1985).

[17] P. J. Besl and R. K. Jain, Invariant Surface Characteristics for 3D Object Recognition of Range Images, *Computer Vision, Graphics and Image Processing*, **33**, pp 33-80, (1986).

[18] P. J. Besl and R. K. Jain, Segmentation Through Variable-Order Surface Fitting, *IEEE Transactions on Pattern Analysis and Machine Intelligence*, **10** (2), pp 167-192, (1988).

[19] B. Bhanu, Representation and Shape Matching of 3-D Objects, *IEEE Transactions on Pattern Analysis and Machine Intelligence*, **6** (3), pp 340-351, (1984).

BIBLIOGRAPHY

[20] T. O. Binford, Visual Perception by Computer, *Proceedings of IEEE Conference on Sytems and Controls, December 1971*, Miami, (1971).

[21] T. O. Binford, Inferring Shapes from Images, *Artificial Intelligence*, **17**, pp 205-244, (1981).

[22] S. P. Black, Utilisation of 3D Coordinate Machines, *International Metrology Conference NELEY 80*, paper 2.3, 7-9 October, (1980).

[23] R. E. Blahut, *Fast Algorithms for Digital Signal Processing*, Addison Wesley, Reading, Massachusetts (1985).

[24] A. Blake, Surface Reconstruction from Stereo and Shading, *Alvey Computer Vision and Image Interpretation Meeting Proceedings*, University of Sussex, September 15-18, (1986).

[25] R. M. Bolle and B. C. Vemuri, On Three-Dimensional Surface Reconstruction Methods, *IEEE Transactions on Pattern Analysis and Machine Intelligence*, **13** (1), pp 1-13, (1991).

[26] R. C. Bolles, P. Horaud and M. J. Hannah, 3DPO: A Three-dimensional Part Orientation System, *Proceedings of 8th International Joint Conference on Artificial Intelligence*, pp 1116-1120, (1983).

[27] R. C. Bolles and P. Horaud, 3DPO: A Three-dimensional Part Orientation System,*The International Journal of Robotics Research*, **5** (3), pp 3-26, (1986).

[28] P. Boulanger, K. B. Evans, M. Rioux and P. Ruhlmann, Object Input for CAD / CAM using a 3D Laser Scanner, *National Research Council of Canada Laboratory for Intelligent Systems*, NRCC Memo No. 25446, (1985).

[29] A. Bowyer and J.R. Woodwark, *A Programmers Geometry*, Butterworths, London, UK (1983).

[30] B. A. Boyter and J. K. Aggarwal, Recognition of Polyhedra from Range Data, *IEEE Expert (Spring)*, pp 47-59, (1986).

[31] J. P. Brady, N. Nandhakumar and J. K. Aggrawal, Recent Progress in Object Recognition from Range Data, *Image and Vision Computing*, pp 295-307, (1989).

[32] M. Brady (Editor), *Robotics Science*, MIT Press, Cambridge, Massachusetts, USA, (1989).

[33] M. Brady, J. Ponce, A. Yuille and H. Asada, Describing Surfaces, *Proceedings of Second International Symposium on Robotics Research*, MIT Press, Cambridge, Massachusetts, USA, (1985).

[34] M. Brady, I. Reid, R. Rixon and B. Knight, Design of a Range Finder and the Recognition of Parameterised Objects, *Proceedings of the IEE Colloquium on Active and Passive Techniques for 3D Vision*, (Savoy Place, London, 19 th February), Ed. M. Brady, (1991).

[35] J. Bremner, Private Communications, *Quality Assurance Dept., British Aerospace*, 1988-90.

[36] J. Bretschi, *Automated Inspection Systems for Industry*, IFS Publications Ltd., UK, (1981).

[37] C. B. Brice and C. L. Fennema, Scene Analysis using Regions, *Artificial Intelligence*, **1** (3), pp 205-226, (1970).

[38] British Standards **BS 308**, Parts 2 and 3:1972. *Dimensioning and Tolerancing of Size*, and *Geometric Tolerancing*, (1972).

[39] R. A. Brook, An Experimental Automatic Surface Inspection System, *Metron*, **3** (8), pp 219-223, (1971).

[40] R. A. Brooks, Symbolic Reasoning Among 3D Models and 2D Images, *Artificial Intelligence*, **17**, pp 285-384, (1981).

[41] R. A. Brooks, Model-based Three-Dimensional Interpretations of Two-dimensional Images, *IEEE Transactions on Pattern Analysis and Machine Intelligence*, **5** (2), pp 140-149, (1983).

[42] P. Brou, Using the Gaussian Image to Find the Orientation of Objects, *International Joutnal of Robotics Research*, **3** (4), pp 89-125, (1984).

[43] R. A. Brooks, R. Griener, and T. O. Binford, The ACRONYM Model-based Vision System, *Proceedings of 6th International Joint Conference on Artificial Intelligence*, pp 105-113, (1979).

[44] P. Brunet and D. Ayala, Extended Octtree Representation of Free Form Surfaces, *Computer Aided Geometric Design*, **4**, pp 141-154, (1987).

[45] P. Brunet and I. Navazo, Geometric Modelling Using Exact Octree Representation of Polyhedral Objects, in *Eurographics '85*, Ed. C. E. Vandoni, North-Holland, pp 159-169, (1985).

[46] D. R. Burton, M. J. Lalor and J. T. Atkinson, The Growth of Modern Interferometry for Industrial Inspection, *Proceedings of IEE Colloquim on Active and Passive Techniques for 3D Vision*, (Savoy Place, London, 19th February), Ed. M. Brady, (1991).

[47] M. C. Cakir, *The Reconstruction of Measured Engineering Components and Their Comparison with Solid Models*, Ph.D. Thesis, University of Bath, (1989).

[48] J. F. Canny, A Computational Approach to Edge Detection, *IEEE Transactions on Pattern Analysis and Machine Intelligence*, **8** (6), pp 679-698, (1986).

[49] D. W. Capson and S. Eng, A Tiered-Color Illumination Approach for Machine Inspection of Solder-Joints, *IEEE Transactions on Pattern Analysis and Machine Intelligence*, **10** (3), pp 387-393, (1988).

[50] I. Carlbom, I. Chakravarty and D. Vanderschel, A Heirarchical Data Structure for Representing the Spatial Decomposition of 3-D Objects, *IEEE Computer Graphics and Applications*, **5** (4), pp 24-31, (1985).

[51] P. Carnavali, L. Coletti and S. Patarnello, Image Processing by Simulated Annealing, *IBM Journal of Research and Development*, **29** (6), pp 569-579, (1985).

[52] D. W. Castelow, D. W. Murray, G. L. Scott and B. F. Buxton, Matching Canny Edgels to Compute the Principal Components of Optical Flow, *Proceedings of Third Alvey Vision Conference* (Cambridge, U.K. 14-17 Sept), Ed. J. F. Frisby, Sheffield University Press, pp 193-200, (1987).

[53] J. P. Chan, *Automated X-ray Inspection of Foodstuffs*, M.Sc. Thesis, University of Wales College of Cardiff, (1988).

[54] J. P. Chan, B. G. Batchelor, I. P. Harris and S. J. Perry, Intelligent Visual Inspection of Food Products, *Proceedings of SPIE OE/BOSTON '90: Machine Vision Systems Integration in Industry, Conference No.* **1386**, Eds. B. G. Batchelor and F. M. Waltz, SPIE, Washington, USA, pp 171-179, (1990).

[55] R. T. Chin and C. R. Dyer, Model Based Recognition in Robot Vision, *ACM Computing Surveys*, **18** (1), pp 67-108, (1986).

[56] R. T. Chin and C. A. Harlow, Automated Visual Inspection: A Survey, *IEEE Transactions on Pattern Analysis and Machine Intelligence*, 4 (6), pp 557-573, (1982).

[57] R. J. Clarke, *Transform Coding of Images*, Academic Press, London, (1985).

[58] G. R. Cross and A. K. Jain, Markov Random Field Texture Models, *IEEE Transactions on Pattern Analysis and Machine Intelligence*, **5** (1), pp 25-39, (1983).

[59] P. Davies and D. Hutber, A Survey of Possible Methods for Automatic Inspection of Three Dimensional Objects, *British Aerospace, Sowerby Research Centre Internal Report No. JS 10631*, (1986).

[60] L. Davis, *Genetic Algorithms and Simulated Annealing*, Pitman, London, (1987).

[61] K. DeJong, Learning with Genetic Algorithms, *Machine Learning*, **3**, pp 127-163, (1988).

[62] H Derin and H Elliot, Modeling and Segmentation of Noisy and Textured Images Using Gibbs Random Fields, *IEEE Transactions on Pattern Analysis and Machine Intelligence*, **9** (1), pp 39-55, (1987).

[63] Dialogues on Ignorance, Myopia and Naivete in Computer Vision Systems, *Computer Vision, Graphics and Image Processing*, **53** (1), (1991).

[64] W. Dickson, Feature Grouping in a Hierarchical Probabilistic Network, *Image and Vision Computing*, **9** (1), Butterworth-Heinemann, UK, (1991).

[65] G. Doemens, R. Buerger, W. W. Gaebel, G. Haas and R. Schneider, A Fast 3D Sensor with High Dynamic Range for Industrial Applications *Proceedings of ROVISEC Conf*, (1986).

[66] H. S. Don, K. S. Fu, R. Liu and W. W. Lin, Metal Surface Inspection Using Image Processing Techniques, *IEEE Transactions on Systems, Man and Cybernetics*, **14** (1), pp 139-146, (1984).

[67] T. A. Doney, Production Quality Control Products, *Proceedings of SPIE OE/BOSTON '90: Intelligent Robotics and Computer Vison IX: Algorithms and Techniques, Conference No.* **1381**, Ed. D. P. Casesant, SPIE, Washington, USA, pp 9-21, (1990).

[68] H. F. Durrant-Whyte, *Integration, Coordination and Control of Multi-Sensor Robot Systems*, Kluwer Academic Publishers, Boston, USA, (1988).

[69] M. J. Dürst and T. L. Kunii, Integrated Polytrees: A Generalized Model for the Integration of Spatial Decomposition and Boundary Representation, in *Theory and Practice of Geometric Modeling*, Ed. W. Straßer and H.-P. Seidel, Springer, Berlin, pp 329-348, (1989).

[70] C. R. Dyer, Computing Euler Number of an Image from its Quadtree, *Computer Graphics and Image Processing*, **13** (3), pp 270-276, (1980).

[71] B. Everitt, *Cluster Analysis*, Halstead Press, New York, (1980).

[72] T. G. Fan, *Describing and Recognising 3-D Objects using Surface Properties*, Springer Verlag, New York, (1990).

[73] T. G. Fan, G. Medioni and R. Navatia, Description of Surfaces from Range Data Using Curvature Properties, *Proceedings of Computer Vision and Pattern Recognition Conference* (Miami, June 22-26), pp 86-91, (1986).

[74] G. Farin, *Curves and Surfaces for Computer Aided Geometric Design*, Academic Press, San Diego, California, (1988).

[75] O.D. Faugeras and M. Berthod, Improving Consistency and Reducing the Ambiguity in Stochastic Labeling: an Opimization Approach, *IEEE Transactions on Pattern Analysis and Machine Intelligence*, **3** (4), pp 412-424, (1980).

[76] O. D. Faugeras, F. Germain, G. Kryze, J. D. Boissonnat, M. Hebert, J. Ponce, E. Pauchon and N. Ayache, Towards a Flexible Vision System, in *Robot Vision*, Ed. A. Pugh, Springer-Verlag, New York, pp 129-147, (1983).

[77] O.D. Faugeras and M. Hebert, A 3-D Recognition and Positioning Algorithm Using Geometric Matching Between Primitive Surfaces, *Proceedings of the Eighth International Joint Conference on Artificial Intelligence*, Karlsruhe, Germany, pp 996-1002, (1983).

[78] O.D. Faugeras and M. Hebert, Object Representation, Identification and Positioning From Range Data, *Proceedings of First International Symposium on Robotics Research*, pp 425-446, (1984).

[79] O.D. Faugeras and M. Hebert, The representation, recognition and positioning of 3-D shapes from range data, in *Machine Intellegince and Patttern Recognition 3: Techniques for 3-D machine perception*, Ed. A. Rosenfield, North-Holland, pp 13-53, (1986).

[80] O.D. Faugeras and M. Hebert, The Representation, Recognition and Locating of 3-D Objects, *International Journal on Robotics Research*, **5** (3), pp 27-52, (1986).

[81] O. D. Faugeras, M. Hebert, and E. Pauchon, Segmentation of Range Data into Planar and Quadric Patches, *Proceedings of Third Computer Vision and Pattern Recognition Conference*, Arlingtion, VA, pp 8-13, (1983).

[82] R. B. Fisher, Using Surfaces and Object Models to Recognise Partially Obscured Object, *Proceedings of the Eighth International Joint Conference on Artificial Intelligence*, Karlsruhe, Germany, pp 989-995, (1983).

[83] R. B. Fisher, Representing Three-dimensional Structures for Visual Recognition, *Artificial Intelligence Review*, **1**, pp 183-200, (1987).

[84] R. B. Fisher, SMS:A Suggestive Modelling System For Object Recognition, *Image and Vision Computing*, **5**, pp 98-104, (1987).

[85] R. B. Fisher, *From Surfaces to Objects*, J. Wiley and Sons, Chichester, (1989).

[86] R. B. Fisher, Geometric Constraints from Planar Patch Matching, *Image and Vision Computing*, **8** (2), pp 148-154, (1990).

[87] R. B. Fisher, The Design of the Imagine II Scene Analysis Program, in *3D Model Recognition From Stereoscopic Cues*, Eds. J. E. W. Frisby and J. P. Frisby, MIT Press, pp 239-243, (1991).

[88] A. Fleming, Geometric Relationships Between Toleranced Features, *Artificial Intelligence*, **37**, pp 403-412, (1989).

[89] A. Fleming, A Representation for Geometrically Toleranced Parts, *Geometric Reasoning* Ed. J. Woodwark, Oxford University Press, Oxford, (1989).

[90] H. Freeman, Is Industry Ready for Machine Vision, in *Machine Vision for Inspection and Measurement*, Academic Press, San Diego, CA, USA, (1988).

[91] H. Freeman, *Machine Vision for Three-dimensional Scenes*, Academic Press, San Diego, CA, USA, (1990).

[92] J. Funda and R. P. Paul, A Comparison of Transforms and Quaternions in Robotics, *Proceedings of IEEE Conference on Robotics and Automation,* **2**, pp 886-891, (1988).

[93] P. C. Gaston and T. Lozano-Perez, Tactile Recognition and Localization using Object Models, *M.I.T. A.I. Memo no. 705,* (1983).

[94] M. Gay, Private Communications, *Advanced Image Processing Dept., Sowerby Research Centre, British Aerospace, Filton, Bristol,* 1988/9.

[95] M. Gay, Segmentation Using Region Merging With Edges, *Proceedings of the 5th Alvey Vision Conference,* (Reading Univ. 25-28 September), pp 115-120, (1989).

[96] D. Geman and S. Geman, Stochastic Relaxation, Gibbs Distributions and the Baysian Restoration of Images, *IEEE Transactions on Pattern Analysis and Machine Intelligence* **6** (6), pp 721-741, (1984).

[97] M. Gervautz and W. Purgathofer, A Simple Method for Color Quantization: Octree Quantization, in *Graphics Gems,* Ed. A. S. Glassner, Academic Press, San Diego, California, USA, (1990).

[98] E. Gilheany and E. T. Treywin, Development in Three-Coordinate Measuring Machines and Associated Software, *International Metrology conference NELEY 80,* paper 2.2, 7-9 October, (1980).

[99] K. Gill, Case Study — The Automotive Industry, *Proceedings of The Machine Vision Conference,* June 7-8, Queen Elizabeth Conference Centre, London, UK, (1990).

[100] P. E. Gill, W. Murray and M. Wright, *Practical Optimisation,* Academic Press, New York, USA, (1981).

[101] G. Golub and W. Kahan, Calculating the Singular Values and Pseudo-inverse of a Matrix, *Society of Industrial and Applied*

Mathematics Journal of Numerical Analysis, **2** (2), pp 205-224, (1965).

[102] G. Golub, Numerical Methods for Solving Linear Least-Squares Problems, *Numerische Mathematik*, **7**, pp 206-216, (1965).

[103] G. Golub and C. Reinsch, Singular Value Decomposition and Least Squares Solutions, *Numerische Mathematik*, **14**, pp 403-420, (1970).

[104] G. Golub and C. Van Loan, *Matrix Computations*, North Oxford Academic Publishing Co., Oxford, (1983).

[105] R. C. Gonzalez and P. Wintz, *Digital Image Processing (2nd Edn.)*, Addison Wesley, Reading, Massachusetts (1987).

[106] N. Goto and T. Kondo, An Automatic Inspection System for Printed Wiring Board Masks, *Pattern Recognition*, **12**, pp 443-455, (1980).

[107] W. E. L. Grimson, The Combinatorics of Local Constraints in Model-Based Recognition and Localization from Sparse Data, *Journal of the Association of Computing Machinery*, (1986).

[108] W. E. L. Grimson, Sensing Strategies for Disambiguating Among Multiple Objects of Known Poses, *IEEE Journal of Robotics and Automation*, **2** (4), pp 196-213, (1986).

[109] W. E. L. Grimson, *Object Recognition by Computer: The Role of Geometric Constraints*, MIT Press, Cambridge, Massachusetts, USA, (1990).

[110] W. E. L. Grimson and T. Lozano-Perez, Model Based Recognition and Localization From Sparse Range or Tactile Data, *International Journal of Robotics Research*, **3** (3), pp 3-35, (1984).

[111] Y. Hara, N. Akiyama and K. Karasaki, Automatic Inspection System for Printed Circuit Boards, *IEEE Transactions on Pattern Analysis and Machine Intelligence*, **5** (6), pp 623-630, (1983).

[112] R. M. Haralick, H. Joo, C. Lee, X. Zhuang, V .G. Vaidya and M. B. Kim, Pose Estimation From Corresponding Point Data, in *Machine Vision for Inspection and Measurement*, Ed. H. Freeman, Academic Press, London, pp 1-84, (1989).

[113] M. Hebert and J. Ponce, A New Method For Segmenting 3-D Scenes Into Primitives, in *Proc. 6th International Conference on Pattern Recognition*, (Munich, W. Germany, Oct 19-22), IEEE New York, pp 836-838, (1982).

[114] M. Hebert and T. Kanade, The 3-D Profile Method for Object Recognition, *IEEE Computer Vision and Pattern Recognition Conference*, San Fransisco, USA, pp 458-463, (1985).

[115] P. S. Heckbert, Color Image Quantization for a Frame Buffer Display, *ACM Computer Graphics*, **16** (3), pp 297-307, (1982).

[116] T. C. Henderson, Efficient 3-D Object Representations For Industrial Vision Systems, *IEEE Transactions on Pattern Analysis and Machine Intelligence*, **5** (6), pp 609-617, (1983).

[117] R. C. Hillyard, *Dimensions and Tolerances in Shape Design*, Ph.D. Thesis, Cambridge University, (1978).

[118] C. M. Hoffmann, *Geometric and Solid Modeling: An Introduction*, Morgan Kaufmann, San Mateo, California, USA, (1989).

[119] R. Hoffman and R. K. Jain, Segmentation and Classification of Range Images, *IEEE Transactions on Pattern Analysis and Machine Intelligence*, **9** (5), pp 608-620, (1987).

[120] J. H. Holland, *Adaptation in Natural and Artificial Systems*, University of Michigan Press, (1975).

[121] P. Horaud and R. C. Bolles, 3DPO's Strategy for Matching Three-dimensional Objects in Range Data, *Proceedings of the International Conference on Robotics*, (Atlanta, Georgia, March 13-15), pp 78-85, (1984).

[122] B.K.P. Horn, Extended Gaussian Images, *Proceedings of IEEE*, **72** (12), pp 1656-1678, (1984).

[123] B. K. P. Horn, *Robot Vision*, MIT Press, Cambridge, Massachusetts, USA, (1986).

[124] B.K.P. Horn and K. Ikeuchi, The Mechanical Manipulation of Randomly Oriented Parts, *Scientific American*, **251** (2), pp 100-111, (1984).

[125] B. K. P. Horn and B. G. Schunck, Determining Optical Flow, *Artificial Intelligence*, **17** (1-3), pp 185-203, (1981).

[126] S. L. Horowitz and T. Pavlidis, Picture Segmentation by a Directed Split-and-Merge Method, *Proceedings of Second International Joint Conference on Pattern Recognition*, pp 424-433, (1974).

[127] M. K. Hu, Visual Pattern Recognition by Moment Invariants, *IEEE Transactions on Information Theory*, **8**, pp 179-187, (1962).

[128] R. A. Hummel and S. W. Zucker, On the Foundations of Relaxation Labelling Processes, *IEEE Transactions on Pattern Analysis and Machine Intelligence.* **5** (3), pp 267-287, (1983).

[129] S. Hurley, *Solving Certain Least Squares Problems with Rank Deficiency*, M.Sc. Dissertation, University of Reading, (1985).

[130] D. Hutber, Automatic Inspection of 3-D Objects using Stereo, in Proceedings of SPIE Conference No. **850** on Optics, Illumination and Image Sensing For Machine Vision II, SPIE, Washington, USA, pp 146-151, (1987).

[131] D. Hutber, Improvement of CCD Camera Resolution using a Jittering Technique, in *Proceedings of SPIE Conference No.* **849** on Automated Inspection and High Speed Vision Architectures, SPIE, Washington, USA, pp 11-17, (1987).

[132] D. Hutber, Active Stereo Vision and Its Application to Industrial Inspection, *Proceedings of IEE Colloquium on Active and Passive Techniques for 3D Vision*, (Savoy Place, London, 19 th February), Ed. M. Brady, pp 1-4, (1991).

[133] D. Hutber, A. D. Marshall, M. Gay and A. Page, Multiple Scene Inspection of Objects with Accurate Depth Data, in Preparation, (1991).

[134] D. Hutber and S. M. Wright, A Digital Camera and Real-Time Geometric Correction for Images, *Proceedings of Third Alvey Vision Club Conference*, Cambridge, U.K., Sept 15-17th, Ed. J. Frisby, Sheffield University Press, Sheffield, pp 151-157, (1987).

[135] M. Idesawa, Y. Yatagai and T. Soma, A Method for the Automatic Measurement of 3D Shapes by a New Type of Moire Topography, *Proceedings of Third International Conference on Pattern Recognition*, Cornado, California, USA, pp 708-712, (1976).

[136] K. Ikeuchi, Recognition of Objects Using the Extended Gaussian Image, *Proceedings of 7th International Joint Conference on Artificial Intelligence* (Vancouver, B.C., Canada, Aug 24-28), pp 595-600, (1981).

[137] K. Ikeuchi, Generating an Interpretation Tree from a CAD Model to Represent Object Configurations for Bin-Picking Tasks, *Dept. of Computer Science, Carnegie-Mellon University, Report, CMU-CS-86-144*, (1986).

[138] K. Ikeuchi, B. K. P. Horn, S. Nagata, T. Callahan and O. Feimgold, Picking Up an Object From a Pile of Objects, *MIT AI Memo No. 726*, (1983).

[139] R. C Jain and A. K. Jain, *Analysyis and Interpretation of Range Images*, Springer-Verlag, New York, USA, (1990).

[140] J. F. Jarvis, A Method for Automating the Visual Inspection of Printed Circuit Boards, *IEEE Transactions on Pattern Analysis and Machine Intelligence*, **2** (1), pp 77-82, (1980).

[141] K. Kanatani, *Group-Theoretical Methods in Image Understanding*, Springer Verlag, Berlin, (1990).

[142] D. Kapur and J. L. Mundy, Geometric Reasoning and Artificial Intelligence: A Special Volume of Artificial Intelligence, *Artificial Intelligence*, **37**, (1988).

[143] P. Kaufmann, G. Medioni and R. Nevatia, Visual Inspection using Linear Features, *Pattern Recognition*, **17** (5), pp 485-491, (1984).

[144] S. Kirkpatrick, C. D. Gebalt and M. P. Vecchi, Optimisation by Simulated Annealing, *Science*, **220**, pp 671-680, (1983).

[145] J. J. Koenderink, *Solid Shape*, MIT Press, Cambridge, Massachusetts, USA, (1990).

[146] S. Kobayashi, Automatic Solder Joint Inspection Utilizing Color Illumination, *Journal of the Society for Production Engineering*, **56** (8), pp 1375-1380, (1990).

[147] Kodak-Eastman, *Digital Camera Sales Literature*, (1989).

[148] D. T. Kuan and R. J. Drazovich, Model Based Interpretation of 3-D Range Data, in *Machine Intelligence: Techniques for 3-D Machine Perception*, Ed. A. Rosenfield, North-Holland, (1986).

[149] J. Levin, A Parametric Algorithm for Drawing Pictures of Solid Objects Composed of Quadric Surfaces, *Communications of the ACM*, **19** (10), pp 555-563, (1976).

[150] D. G. Lowe, *Perceptual Organization and Visual Recognition*, Kluwer, Norwell, Massachusetts, (1985).

[151] D. G. Lowe, Three-Dimensional Object Recognition from Single Two-dimensional Images, *Artificial Intelligence*, **31**, pp 355-395, (1987).

[152] D. G. Lowe and T. O. Binford, Segmentation and Aggregation: An approach to Ground Figure Phenomena, in *Readings in Computer Vision:Issues, Problems, Principles and Paradigms*, Eds. M. A. Fischler and O. Firschein, Morgan Kaufmann, Los Altos, California, USA, (1987).

[153] T. Lozano-Perez and W. E. L. Grimson, Recognition and Localization of Overlapping Parts from Sparse Data, *IEEE Transactions on Pattern Analysis and Machine Intelligence*, **9** (4), pp 469-482, (1987).

[154] T. Lozano-Perez and R. H. Taylor, Geometric Issues in Planning Robot Tasks, in *Robotics Science*, Ed. M. Brady, MIT Press, Cambridge, Massachusetts, USA, (1989).

[155] M. Mäntylä, *An Introduction to Solid Modeling*, Computer Science Press, Rockville, Maryland, USA, (1988).

[156] D. Marr, *Vision*, Freeman, San Francisco, (1982).

[157] D. Marr and E. Hildreth, Theory of Edge Detection, *Proceedings of the Royal Society*, **B 207**, pp 187-217, (1980)

[158] D. Marr and H. K. Nishihara, Representation and Recognition of the Spacial Organisation of Three-Dimensional Shapes, *Preceedings of the Royal Society*, **B 200**, pp 269-294, (1978).

[159] A. D. Marshall, *The Automatic Inspection of Machined Parts Using Three-Dimensional Range Data and Model Based Matching Techniques*, Ph.D. Thesis, University of Wales College of Cardiff, (1990).

[160] A. D. Marshall, Automatically Inspecting Gross Features of Machined Objects using Three-dimensional Depth Data, *Proceedings of SPIE OE/BOSTON '90 Conference No.* **1386** on *Machine Vision Systems Integration in Industry*, Eds. B. G. Batchelor and F. M. Waltz, SPIE, Washington, USA, pp 243-254, (1990).

[161] A. D. Marshall, Inspecting Geometric Features of Machined Objects Using Three-Dimensional Depth Data, submitted to *Image and Vision Computing*, (1991).

[162] A. D. Marshall and R. R. Martin, 3-D Model Based Matching for Automatic Inspection From Depth Data, *Proceedings of IMA conference on Robotics: Applied Mathematics and Computational Aspects*, (Loughborough Univ., 12-14 th July, 1989), to appear, Oxford University Press.

[163] A. D. Marshall, R. R. Martin and D. Hutber, Automatic Inspection of Mechanical Parts Using Geometric Models and Laser Range Finder Data, *Image and Vision Computing*, **9** (6), (1991).

[164] M. T. Mason, Robotic Manipulation: Mechanics and Planning, in *Robotics Science*, Ed. M. Brady, MIT Press, Cambridge, Massachusetts, USA, (1989).

[165] J. E. W. Mayhew and J. P. Frisby (eds.), *3D Model Recognition from Stereoscopic Cues*, MIT Press, Cambridge, Massachusetts, USA, (1991).

[166] W. E. McIntosh, Automating the Inspection of Printed Circuit Boards, *Robotics Today*, June Edition, pp 75-78, (1984).

[167] P. J. Mckerrow, *Introduction to Robotics*, Addison Wesley, Sydney, Australia, (1991).

[168] G. Medioni and R. Navatia, Description of 3D Surfaces Using Curvature Properties, *Proceedings of Image Understanding Workshop DARPA*, (New Orleans, October 3-4), pp 291-299, (1984).

[169] M. C. Morrone, J. Ross, D. C. Burr and R. Owens, Mach Bands are Phase Dependent, *Nature*, **324** (6094), pp 250-253, (1986).

[170] M. C. Morrone, and R. A. Owens, Feature Detection from Local Energy, *Pattern Recognition Letters*, **6**, pp 303-313, (1987).

[171] J. L. Mundy, Visual Inspection of Metal Surfaces, *Proceedings of the Eighth International Conference on Pattern Recognition*, pp 227-231, (1982).

[172] D. W. Murray, Model-based Recognition Using 3-D Structure from Motion, *Image and Vision computing*, **5** (2), pp 85-90, (1987).

[173] D. W. Murray, Model-based Recognition Using 3-D Shape Alone, *Computer Vision, Graphics and Image Processing*, **40**, pp 250-266, (1987)

[174] D. W. Murray and B. F. Buxton, *Experiments in the Machine Interpretation of Visual Motion*, MIT Press, Cambridge, Massachusetts, USA, (1990).

[175] D. W. Murray, D. A. Castelow and B. F. Buxton, From an mage Sequence to a Recognized Polyhedral Object, *Proceedings of the Third Alvey Vision Conference* (Cambridge, U.K. 15-17 Sept), Ed. J. F. Frisby, Sheffield University Press, Sheffield, pp 201-210, (1987).

[176] D. W. Murray and D. B. Cook, Using the Orientation of Fragmentory 3-D Edge Segments for Polyhedral Object Recognition, *International Journal of Computer Vision*, **3** (3), pp 107-120, (1988).

[177] Y. Nakagawa, Automatic Visual Inspection of Printed Circuit Boards, *Robot Vision*, **336** pp 676-682, (1982).

[178] R. Nevatia and T. O. Binford, Description and Recognition of Curved Objects, *Artificial Intelligence*, **8** (1), pp 77-98, (1977).

[179] A. W. Nutbourne and R. R. Martin, *Differential Geometry Applied to Curve and Surface Design, (Volume 1: Fundamentals)*, Ellis Horwood, Chichester, UK, (1988).

[180] E. E. Osborne, On Least Squares Solutions of Linear Equations, *Journal of the ACM*, **8**, pp 628-636, (1961).

[181] M. Oshima and Y. Shirai, A Scene Description Method Using Three-Dimensional Information, *Pattern Recognition*, **11**, pp 9-17, (1979).

[182] M. Oshima and Y. Shirai, Object Recognition Using 3-D Information, *IEEE Transactions on Pattern Analysis and Machine Intelligence*, **5** (4), pp 353-361, (1983).

[183] R. Owens, S. Venkatesh and J. Ross, Edge Detection is a Projection, *Pattern Recognition Letters*, **9**, pp 233-244, (1989).

[184] A. Page, Segmentation Algorithms, *British Aerospace, Sowerby Research Centre Internal Report No. AOI/TR/BASR/880201*, (1988).

[185] T. Pavlidis and S. L. Horowitz, Segmentation of Plane Curves, *IEEE Transactions on Computing*, **23** (8), pp 860-870, (1974).

[186] J. Pearl, *Probabilistic Reasoning in Intelligent Systems: Networks of Plausible Inference*, Morgan Kaufmann, Palo Alto, California, USA, (1988).

[187] M. Pietkianen, A. Rosenfeld and I. Walter, Split-and-Link Algorithms for Image Segmenatation, *Pattern Recognition*, **15**, (1982).

[188] S. B. Pollard, J. E. W. Mayhew and J. P.Frisby, PMF: A Stereo Correspondance Algorithm Using A Disparity Gradient Limit, *Perception*, **14**, pp 449-470, (1985).

[189] S. B. Pollard, J. Porrill, J. E. W. Mayhew and J. P.Frisby, Matching Geometric Descriptions in Three Space,*AVC '86: Proceedings of the 2nd Alvey Vision Conference*, (University of Bristol, UK, 22-25 September), Ed. M. Brown, (1986).

[190] S. B. Pollard, T. P. Pridmore, J. Porrill, J. E. W. Mayhew and J. P. Frisby, Geometric Modelling From Multiple Stereo Views, *Sheffield Univ., AIVRU No. 02*, (1987).

[191] J. Porrill, S. B. Pollard, T. P. Pridmore, J. E. W. Mayhew and J. P. Frisby, TINA: The Sheffield AIVRU Vision System, *Sheffield Univ., AIVRU No. 031*, (1987).

[192] M. Potmesil, Generating Models of Solid Objects by Matching 3D Surface Segments, *Proceedings of International Joint Conference on Artificial Intelligence*, (Karlsruhe, West Germany, Aug 8-12), pp 1089-1093, (1983).

[193] G. Reid, R. Rixon and H. Messer, Absolute and Comparative Measurements of Three Dimensional Shape by Phase Measuring Moire Topography, *Optics and Laser Technology*, **16**, pp 315-319, (1984).

[194] Reinshaw Transducer Systems Ltd., Precision Metrology and Inspection Equipment, *Sales Literature*, (1991).

[195] A. A. G. Requicha, Mathematical Models of Rigid Solid Objects, *University of Rochester, College of Engineering Science, Production Automation Project, Technical Memorandum 28*, (1977).

[196] A. A. G. Requicha and H. B. Voelcker, Solid Modeling: Current Status and Research Directions, *IEEE Computer Graphics and its Applications*, **3** (7), pp 25-37, (1983).

[197] A. A. G. Requicha, Toward a Theory of Geometric Tolerancing, *International Journal of Robotics Research*, **2** (4), pp 45-60, (1983).

[198] A. A. G. Requicha, Representation of Tolerances in Solid Modeling: Issues and Alternative Approaches, in *Solid Modeling by computers: From Theory to Applications*, Eds. J. W. Boyse and M. S. Pickett, Plenum, New York, pp 3-22, (1984).

[199] A. A. G. Requicha and S. C. Chan, Representation of Geometric Features, Tolerances and Attributes in Solid Modelers Based on Constructive Geometry, *IEEE Robotics and Automation*, **2** (3), pp 156-166, (1986).

[200] A. A. G. Requicha, Representations for Rigid Solids: Theory, Methods and Systems, *ACM Computer Surveys*, **12** (4), pp 437-464, (1980).

[201] A. A. G. Requicha and H. B. Voelcker, Solid Modeling: A Historical Summary and Contemporary Assessment, *IEEE Computer Graphics and its Applications*, **2** (2), pp 9-24, (1982).

[202] A. A. G. Requicha and A. J. Spyridi, Computing Global Accessibility Directions for the Faces of a Solid, *Proceedings of the Society for Industrial and Applied Mathematics Second Conference on Geometric Design*, November 4-8, Tempe Arizona, USA, abstract only, (1991).

[203] L G Roberts, Machine Perception of Three Dimensional Solids, in *Optical and Electro-optical Information Processing*, Ed. J P Tipper, MIT Press, Cambridge, Massachusetts, USA, (1965).

[204] D. F. Rogers and J. A. Adams, *Mathematical Elements for Computer Graphics (2nd edition)*, Mcgraw-Hill, New York, USA, (1990).

[205] J. Rooney, A Survey Of Representations Of Spatial Rotation About A Fixed Point, *Environment and Planning*, **B 4**, pp 185-210, (1977).

[206] J. Rooney and P. Steadman (Eds.), *Principles of Computer-Aided Design*, Pitman, London, (1987).

[207] A. Rosenfeld, Machine Vision for Industry: Concepts and Techniques, *Robotics Today*, (1985).

[208] A. Rosenfeld, R. Hummel and S. Zucker, Scene Labelling by Relaxation Operations, *IEEE Transactions on Systems, Man and Cybernetics*, **6**, pp 420-433, (1979).

[209] P. Rummel, Applied Robot Vision: Combining Workpiece Rocognition and Inspection, in *Machine Vision for Inspection and Measurement*, Ed. H. Freeman Academic Press, San Diego, California, USA, (1988).

[210] F. A. Sadjadi and E. L. Hall, Three-Dimensional Moment Invariants, *IEEE Transactions on Pattern Analysis and Machine Intelligence*, **2** (2), pp 127-136, (1980).

[211] R. Safabakhsh, *Automated Visual Inspection for Detection, Characterization and Measurement of Machined Metal Surfaces*, PhD Thesis, University of Tennesse, Knoxville, USA, (1986).

[212] B. G. Schunck and B. K. P. Horn, Constraints on Optical Flow Computation, *Proceedings of Pattern Recognition and Image Processing Conference*, Dallas, Texas, USA, pp 205-210, (1981).

[213] M. P. Seah and C. Lea, Certainty Measurement using an Automated Infrared Laser Inspection System for PCB Solder Joint Integrity, *Journal of Physical, Electronic and Scientific Instrumentation*, **18**, pp 676-682, (1985).

[214] G. Shafer, *A Mathematical Theory of Evidence*, Princeton University Press, (1976).

[215] G. Shafer and J. Pearl (Eds.), *Readings in Uncertain Reasoning*, Morgan Kaufmann, Palo Alto, California, USA, (1990).

[216] Y. Shirai, *Three-Dimensional Computer Vision*, Springer-Verlag, Berlin, (1987).

[217] E. Shortliffe and B. Buchanan, *Rule Based Expert Systems*, Addison-Wesley, Reading, Massachussetts, (1984).

[218] T.M. Silberberg, D.A. Harwood and L.S.Davis, Object Recognition Using Oriented Model Points, *Computer Vision Graphics and Image Processing*, **35**, pp 47-71, (1985).

[219] D. M. Sommerville, *Analytical Geometry of Three Dimensions*, Cambridge University Press, Cambridge, UK, (1947).

[220] Proceedings of *SPIE International Symposia OE/BOSTON '90: Applications in Optical Science and Engineering*, Nov. 4-9, Boston, MA, USA, (1990).

[221] V. Srinivasan, H. C. Liu and M. Halioua, Automated Phase Measuring Profilometry: A Phase-Mapping Approach, *Applied Optics*, **24** (2), pp 185-188, (1985).

[222] G. Stiny and J. Gips, *Algorithmic Aesthetics: Computer Models for Criticism and Design in the Arts*, University of California Press, Berkeley, California, USA, (1972).

[223] H. Szu and R. Hartley, Fast Simulated Annealing, *Physics Letters*, **A 122**, pp 157-162, (1987).

[224] Y Takagi, S. Hata and S. Hibi, Visual Inspection Machine for Solder joints using Tiered Illumination, *Proceedings of SPIE OE/BOSTON 90 Conference No.* **1386** *on Machine Vision Systems Integration in Industry*, Eds. B. G. Batchelor and F. M. Waltz, SPIE, Washington, USA, pp 21-29, (1990).

[225] M. Takeda and I. Mutoh, Fourier Transform Profilometry for the Automatic Measurement of 3D Object Shapes, *Applications for Optics*, **22**, (1983).

[226] J. M. Tannenbaum and H. G. Barrow, Experiments in Interpretation-Guided Segmentation, *Artificial Intelligence*, **8** (3), pp 241-274, (1977).

[227] K. C. E. Truslove, The Implications of Tolerancing for Computer-Aided Design, *Computer-Aided Engineering Journal*, **7**, pp 79-85, (1988).

[228] R. Y. Tsai, A Versatile Camera Calibration Technique for High Accuracy Machine metrology using Off the Shelf TV Cameras and Lenses, *IBM Research Report RC 51342, October 1*, (1985).

[229] R. Y. Tsai, An Efficient and Accurate Camera Calibration Technique for 3-D Machine Vision, *Proceedings of IEEE Conference on Computer Vision and Pattern Recognition*, pp 364-374, (1986).

[230] F. Tuijnman, The Use of 3D CAD Models in Robot Vision, in *Geometric Modeling for CAD Applications*, Eds. M. Wozny, H. W. Mclaughlin and J. L. Encarnacao, North Holland, pp 307-314, (1988).

[231] J. U. Turner and M. J. Wozny, A Mathematical Theory of Tolerances, *Geometric Modeling for CAD Applications*, Eds. M. Wozny, H. W. Mclaughlin and J. L. Encarnacao, North Holland, pp 163-187, (1988).

[232] R. Vanzetti and A. C. Traub, Thermal Imaging as a Diagnostic Tool, *Sensor Review*, **2** (3), pp 17-29, (1985).

[233] R. Vanzetti, A. C. Traub and J. S. Ele, Hidden Solder Joint Defects by Laser Infrared System, *Proceedings of IPC 24th Annual Meeting*, pp 1-15, (1981).

[234] B. C. Vemuri, A. Mitchie and J. K. Aggarwal, Curvature Based Representation of Objects from Range Data, *Image and Vision Computing*, **4** (2), pp 107-114, (1986).

[235] D. L. Waltz, Generating Semantic Descriptions from Drawings of Scenes with Shawdows, in *The Psychology of Computer Vision*, Ed. P. H. Winston, McGraw-Hill, New York (1972).

[236] H. Wechsler, *Computational Vision*, Academic Press, San Diego (1990).

[237] K. J. Weiler, *Topological Strutures for Geometric Modeling*, Ph.D. Thesis, Rensselaer Polytechnic Institute, (1986).

[238] J. S. Weszka and A. Rosenfeld, An Application of Texture Analysis to Materials Inspection, *Pattern Recognition*, **8**, pp 195-199, (1976).

[239] P. H. Winston, *Artificial Intelligence (2nd Edition)*, Addison Wesley, Reading, Massachusetts, (1984).

[240] P. R. Wolf, *Elements of Photogrammetry (2nd Edition)*, McGraw Hill, New York, (1983)

[241] R. Y. Wong and K. Hayrepetian, Image Processing with Intensity and Range Data, *Proceedings of Pattern Recognition and Image Processing Conference* (Las Vegas, Nevada, June 14-17), pp 518-520, (1982).

[242] T. C. Woo, Linear and Circular Visibility for Workpiece Orientation and Operations, *Proceedings of the Society for Industrial and Applied Mathematics Second Conference on Geometric Design*, November 4-8, Tempe Arizona, USA, abstract only, (1991).

[243] J. R. Woodwark, *Computing Shape*, Butterworths, London, (1986).

[244] W. A. Wright, A Markov Random Field Approach to Data Fusion and Colour Segmentation, *Image and Vision Computing*, **7** (2), pp 144-150, (1989).

[245] I. Zeid, *CAD/CAM Theory and Practice*, McGraw-Hill, New York, (1991).

[246] S. W. Zucker, R. A. Hummel and A. Rosenfeld, An Application of Relaxation Labelling to Line and Curve Enhancement, *IEEE Transactions on Computing*, **26** (4), pp 394-403, (1977).

Index

3DPO Vision System, 216, 225, 237, 378

Accuracy of Inspection, 286
ACRONYM Vision System, 217, 236, 293
Active Stereo, 12, 36, 291, 313, 328, 372
Adjacency of Features, 209
Aerial Photography, 5
Agricultural Inspection, 360, 364
Algebraic Surfaces, 114, 136
Apples, 364
Arc of Graph, 209
Artificial
    Depth Maps, 263, 264, 272, 316, 347, 374
    Intelligence, 5, 7, 8, 343, 348
Assembly Systems, 7, 369
Autonomous Guided Vehicles, 7, 154, 186, 376
Axis of Symmetry, 350
Azimuth, 159

B-Spline Patches, 112
Baby Food, 366
Backlighting, 352
Backtracking, 148, 179
Baking, 365
Ball and Socket Joint, 354

Basis Functions, 112
Bayesian
    Networks, 155
    Statistics, 155
Belief Models, 156
Bernstein Basis, 112
Bézier Patches, 112
Bicubic Surface, 131
Bidirectional Reflectance Distribution Function, 158
Binary Images, 354
Biquadratic Surface, 131
Bird's Swing Defect, 350
Blackmail, 366
Blow Holes, 363
Blurring, 64, 72, 74
Boltzmann Distribution, 152
Bones, 366
Bottles, 350, 367
Boundary Representation, 198, 202, 206, 217, 304, 373
Box Girders, 350
Bread, 365
British Aerospace, 11
Bubbles in Products, 365
Butter, 366
Butterworth Filter, 67

CAD, 4, 7, 9, 187, 189, 203, 207, 298

Cake, 350, 358
Calculus of Variations, 173
Calibration, 18, 19, 22, 38, 73, 77, 263, 269, 272, 291, 317, 320, 322, 326, 328, 347, 372, 374, 376
    Accuracy, 43
    Analysis, 25, 43
    Chart, 23, 24, 374
Camera, 265, 269, 348, 368
    Calibration, see Calibration
    CCD, 6, 18, 375
    Digital, 375
    Errors, 6, 72
    Geometry, 18, 19
    Master, 38, 269
    Middle, 333
    Model, 18, 19, 291, 376
    Pin-Hole, 19
    Position, 352
    Set Up, 315
    Television, 16, 72
Canny Edge Detector, 101
Car
    Body Panel, 3
    Connecting Rod, 350, 354
    Cylinder Bores, 354
    Engine, 350, 354
Carrots, 364
CCD Cameras, 6, 18, 375
Centre of Gravity, 138, 262, 389
Cereals, 367
Certainty Factors, 156
Characteristic Equation, 384
Chromaticity, 127
Classification by Texture, 366
Cluster Analysis, 332

Clustering of Surfaces, 231
CMMs, 285–287, 332
Coalescing, 129, 263
Cojoint Representations, 376
Collision Avoidance, 7, 186, 376
Colour
    Images, 18, 81, 264, 363, 367
    Look-up Table, 81
    of Products, 365
    Space Compression, 81
Compactness, 138, 140, 209
Component Insertion, 360
Compression
    Colour Space, 81
    Lossless, 78
    Lossy, 78
Computational
    Geometry, 348
    Solid Geometry, 195, 201
Computer
    Aided Design, 4, 7, 9, 187, 189, 203, 207, 298
    Graphics, 7, 8, 164, 187, 203
Concave Edges, 144, 151, 208, 216
Condensation, 129
Cones, 117, 196, 212
    Generalised, 120
    Global Accessibility, 332
Connecting Rod, 350, 354
Connectivity, 139
Constraints, 256, 276
    Local, 220, 222
    Propagation of, 149
    Rigidity, 227
    Similarity, 121, 126
Continuity Testing, Electrical, 361

INDEX

Contrast, 7
    Enhancement, 68
Control Points, 112
Controlled Lighting, 352, 365
Convex
    Edges, 144, 151, 216
    Hulls, 354
    Objects, 169, 294
Convolution, 55, 75, 101
    Kernels, 95
    Masks, 95
    Theorem, 57, 75
Cooking, 365
Coordinate Measuring Machines, 285–287, 332
Correspondence Problem, 32, 290
Costs, 284
Cotton, 364
Crack Detection, 354, 358, 361
Crease Edge, 233
Crisps, 366
CSG, 195, 201
Cube, 198, 275, 344
Curvature, 354
    Gaussian, 132, 140
    Mean, 132, 140
    Surface Normal, 132
Cut-off Frequency, 66
Cylinder, 110, 118, 200, 212, 255, 287, 325, 330, 344, 377
    Bores, 354
    Generalised, 120, 185, 209

Dangling
    Edge, 197
    Face, 196
Dark Field Illumination, 352

Data Acquisition, 289
Data Fusion, 137, 154, 322, 328, 344, 376, 379
Datum, 285, 289, 299, 304
    Systems, 340
Deblurring, 74
Deconvolution, 58
Defects
    Bird's Swing, 350
    in Production, 365, 366
    in Surfaces, 358
Delta Function, 58, 75
Dempster-Shafer Models, 156
Dents, 308
Depth Map, 38, 49, 127, 228, 240, 262, 272, 291, 297, 315, 331, 372, 374, 376
    Artificial, 263, 264, 272, 316, 347, 374
    Construction, 38
    Dense, 27
    Errors, 40
    Simulation, 45
Devil's Fork, 189
Difference (of Sets), 196
Differential
    Coding, 79
    Geometry, 131
Digital Cameras, 375
Digitising Images, 16
Dirac Delta Function, 58, 75
Discrete Fourier Transform, 54, 60
Distortion, Lens, 21, 22, 25, 73, 374
DOG Operator, 99
Dynamic Programming, 104

Eccentricity, 138, 139
Edge, 13, 91, 143, 185, 216, 264, 295
  Bar, 94, 101
  Concave, 144, 151, 208, 216
  Convex, 144, 151, 216
  Crease, 233
  Dangling, 197
  Data, 27
  Detection, 230
  Detector
    Canny, 101
    DOG, 99
    LOG, 98
    Morrone and Owens, 101
    Roberts, 96
    Sobel, 96, 230
  Discontinuity
    Gradient, 94
  Extraction, 91
  Jump, 94, 233
  Leaking, 125
  Linking, 102, 151, 230
  Matching, 219
  Occluding, 144, 151, 228
  of Graph, 209
  Point Classification, 93
  Points, 93
  Primitives, 91
  Ramp, 94
  Roof, 94
  Shadow, 148
  Silhouette, 109
  Step, 94, 216
Edgels, 93
Eigenvalues, 245, 384, 392
Eigenvectors, 245, 384, 389

Electrical Continuity Testing, 361
Ellipse, 218
Ellipsoid, 117, 212
Elliptic
  Cylinder, 117
  Paraboloid, 117
Elongation, 138
Engine, 354
Engineering Drawings, 296
Errors
  in Depth Map, 40
  in Depth Readings, 313
  in Stereo, 40, 334, 348
  in Vision System, 315
  Rounding, 388
Euler
  Angles, 243
  Number, 139
  Operators, 202
  Rotation Formula, 245
Euler-Poincaré Formula, 202
Extended Gaussian Images, 167, 224, 235, 236, 294, 332
External Parameters, 19, 21
Extrinsic Parameters, 19, 21

Faces, *see* Surfaces
Fan's Matching Method, 233, 295
Fast Fourier Transform, 60
Faugeras and Hebert Matching Method, 226, 240, 246, 251, 273, 277, 295
Feasible Interpretations, 220
Feature, 86, 201, 209
  Adjacency, 209
  Classification Network, 217
  Clusters, 216

Description, 88
Detection, 216
Determination of, 264
Extraction, 88
Focus Matching Methods, 378
Indexing, 378
Planar, 317
Representation, 85
Sets
   Choice of, 329
   Choice of Best View, 333
Feet, 3
Finger Nails, 367
Fitting
   Least Squares, *see* Least Squares Fitting
   of Surfaces, 127, 137, 377, 387
   Planar, 387
      Errors, 388
   Quadric, 389
Fluorescence, 367
Focal Length, 21, 25
Food Inspection, 3, 360, 364
Forcing, 129, 264
Foreign Bodies, 366
Fork, Devil's, 189
Form Tolerance, 301, 307, 317, 320, 347
Fourier
   Descriptors, 140
   Smoothing, 65
   Space, 50
   Theorems, 59
   Transform, 50, 65, 75, 80, 101, 140, 376
      Discrete, 54, 60

      Fast, 60
      Inverse, 50
Fractures, 10
Frame, 16
   Grabber, 18
   Store, 18
Frequency, 50
   Component, 50
Freshness, 367

Gaussian
   Curvature, 132, 140
   Distribution, 45, 93, 98, 307; 319
   Images, 332
   Noise, 265
   Smoothing, 97
   Sphere, 168
General Position, 145, 213
Generalised
   Cones, 120
   Cylinders, 120, 185, 209
   Function, 58
Genetic Algorithms, 154
Genus, 202
Geodesic Torsion, 132
Geometric
   Correction, 73
   Inspection, 4, 10, 11, 283–286, 328, 344, 379
   Modelling, 4, 7, 13, 186, 187, 189, 207, 217, 232, 273, 284, 296, 298, 312, 327, 329, 338, 344, 372, 377, 379
   Primitives, 118
   Reasoning, 285, 329, 333, 340,

348, 378, 379
  Tolerance, 10, 11, 283, 295–297, 300, 311, 349
  Validity, 189
Geometry, 200, 203
Gibbs Distribution, 152
Glass, 367
  Fragments, 350, 366
Global Accessibility Cones, 332
Gradient, 85, 163
  Direction, 95
  Edge Discontinuity, 94
  Magnitude, 95
Graininess, 357
Graph, 209
  Relational, 209, 212, 217, 232
  Search, 104
Greyscale Images, 16, 264, 350, 367
Grimson and Lozano-Perez Matching Method, 219, 236, 243, 258, 295

Hadamard Transform, 80
Hair, 367
Half-Spaces, 197
High-Level Data, 85
Hilbert Transform, 101
Hip Joints, 354, 358
Histogram, 352
  Equalisation, 68
  Filtering, 126
  Modification, 71
Hole
  Cylindrical, 330
  Loop, 202
Hough Transform, 104, 235, 294

Householder Transformation, 248
Hue, 127
Human Visual System, 3, 229
Hyperbolic
  Cylinder, 117
  Paraboloid, 117
Hyperboloids, 117, 212
Hypothesis and Test
  Matching, 231, 257
  Strategy, 329, 342

Icing, 350
Illumination, 290
Image
  Acquisition, 5, 12, 15, 16, 18, 26, 32, 289, 327, 368, 372
  Binary, 354
  Coherence, 77
  Colour, 264, 367
  Compression, 8, 77
    Colour Space, 81
  Encoding, 78
  Greyscale, 16, 264, 350, 367
  Input Devices, 16
  Preprocessing, 49
  Processing, 7, 49
  Two-Dimensional, 6
IMAGINE Vision System, 231, 236, 295
Imaging
  Geometry, 18
  Stereo, 19, 23, 333, 368
  Thermal, 5, 49, 363
  Three-Dimensional, 5, 6
Implicit
  Segmentation, 319

Surfaces, 112, 197, 200
Impulse
    Function, 58
    Response Function, 75
Impurities, 365, 366
Indoor Scenes, 5
Industrial
    Inspection, 9, 284, 285
    Vision Systems, 286
Infrared
    Inspection, 364
    Light, 5
Input Devices, 16
Insects, 366, 367
Insertion of Components, 360
Inspection, 3, 4, 7, 13, 49, 137, 186, 187, 239, 264, 272, 283, 292, 296, 314, 325, 341–343, 349, 369, 373–375
    Accuracy of, 286
    Agricultural, 360, 364
    Automatic, 378
    Cell, 369
    Diagnostics, 342
    Food, 3, 360, 364
    Geometric, 4, 10, 11, 283–286, 295, 297, 328, 344, 379
    Industrial, 9, 284, 285
    Infrared, 364
    Manual, 285
    of Complete Object, 343
    of Multiple Scenes, 325
    of Single Scenes, 311
    Preferred Vision System for, 328
    Registration Phase, 338
    Reliability of, 287
    Single Axis, 328, 338, 340
    Speed of, 287
    Strategies, 312
    System, 11, 325, 326, 373
    Tasks, 283
    Two-Dimensional, 4, 286, 350
    Verification Phase, 341
    Visual, 286
        Advantages of, 287
        Disadvantages of, 287
Intelligent Vision, 4
Intensity, 127
Internal Parameters, 19, 20
Interpretation Tree, 220, 226, 234, 240, 259, 294
Intersection (of Sets), 196
Intrinsic Parameters, 19, 20
Invariant
    Measures, 140
    Moments, 138
Inverse Fourier Transform, 50
Irradiance, 158

Jars, 350
Jewellry, 367
Jig, 361
Joints, 211
Jump Edge, 94, 233
Junction, 146

Kernels
    Convolution, 95
    Matching, 213
Knee Joints, 358

Lagrangian Multipliers, 172, 391

Lambert's Law, 162
Lambertian Reflectance, 160, 364
Laplacian Operator, 96
Laser
    Light, 291, 348, 372
    Rangefinder, 32, 290
    Striping, 291, 372
Leading Diagonal, 383
Learning, 378
Least Squares Fitting, 25, 110, 118, 136, 243, 253, 287, 291, 306, 342, 387, 389
Lens Distortion, 21, 22, 25, 73, 374
Lighting
    Conditions, 352
    Controlled, 352, 365
    Models, 357
Line
    Extraction, 230
    Junction, 146
    Labelling, 143, 151
    Representation, 91
Linear Operator, 59, 74
Link of Graph, 209
Local
    Constraints, 220, 222
    Energy Function, 101
    Feature Focus Methods, 216, 225
LOG Operator, 98, 230
Lossless Compression, 78
Lossy Compression, 78
Low-Level Data, 85
Lowpass Filter, 66

Machinery, 367

Manifold Objects, 199
Manipulation, 186
Manual Inspection, 285
Markov Random Field, 152
Masks
    Convolution, 95, 101
Master Camera, 38, 269
Matching, 13, 167, 185, 187, 204, 251, 264, 272, 273, 276, 278, 283, 314, 315, 317, 326, 327, 329, 338, 341, 342, 347, 361
    Constraints, 220, 222, 227, 373
    for Inspection, 292
    Hypothesis and Test, 231, 257
    Kernel, 213
    Model Based, 4, 13, 186, 187, 207, 230, 239, 292, 373, 377
    Pairs
        Order of, 257
    Transformation, 256
Material Condition
    Maximum, 303
    Minimum, 303
Matrix
    Conditioning, 247
    Definitions, 383
Matt Surface, 160
Maximum Material Condition, 303
Mean Curvature, 132, 140
Measuring Instrument, 55
Meat, 366
Median
    Cut Algorithm, 82
    Filtering, 64

Merging, 121
Middle Camera, 333
Milk, 350, 366, 367
Minimum Material Condition, 303
Model
   Based
      Matching, 4, 13, 186, 187, 207, 230, 239, 292, 373, 377
      Recognition, 7
      Vision System, 5
   Compilation from Multiple Views, 312
Moire Fringes, 33, 34, 290, 322, 358, 372, 376
Moments
   Invariant, 138
   of Inertia, 138, 140
   Central, 138
Morrone and Owens Edge Detector, 101
Motion, 170
Mouse, 366
Murray Matching Method, 251, 295

Navigation, 7, 186
Neighbourhood, 197
   Averaging, 63
Nevatia and Binford Matching Method, 209, 236, 251, 293
Newspapers, 366
Node of Graph, 209
Noise, 7, 63, 65
   Filtering, 153
Nominal Surface Features, 303

Non-Manifold Objects, 144
Normal
   Curvature, 132
   Surface, 325, 333
Null Branch, 259
Nuts, 366

Object
   Description From Multiple Views, 312
   Model, 167, 272, 312, 327, 338, 348, *see also* Geometric Modelling
   Parametric, 377
   Properties For Vision, 204
   Orientation, 187
   Position, 187
   Recognition, 4, 185, 187, 377
Observation, 55
Occluding Edge, 144, 151, 228
Occlusion, 94, 208, 222, 225, 236, 305, 313, 334
   Partial, 87
   Self, 208
OCR, 8
Octree, 123, 192, 201
   Quantization, 82
Optical
   Character Recognition, 8
   Flow, 169, 170
Optics, 7, 20
Orientation
   from Texture, 175
   of Objects, 187
Oriented Surface Points, 13, 89, 128, 219
Orthonormal Matrices, 243

Oshima and Shirai Matching Method, 212, 293
Outdoor Scenes, 5
Outlier Test, 307, 308, 319, 322, 395
Oversegmentation, 268

Parabolic Cylinder, 117
Paraboloids, 212
Parallel Computers, 125, 154, 264, 368
Parametric
    Object Model, 377
    Surfaces, 200, 206, 231
Partial Occlusion, 87
Passive Stereo, 35, 225, 290
Patches
    B-Spline, 112
    Bézier, 112
    Surface, 112, 231, 233, 330
Path Planning, 186, 328, 343, 344
Pattern Recognition, 8
    Statistical, 174
Peanuts, 364
Perceptual Grouping, 229, 230
Phase, 51
    Components, 101
Phial, 352
Photometric Stereo, 34, 165
Pick and Place Tasks, 343, 344
Pigs, 349, 364
Pin Holes, 363
Pin-Hole Camera, 19
Pipes, 350
Pistons, 354
Pixel, 16, 49, 192

Planar
    Features, 317
    Surface, 110, 131, 136, 185, 200, 240, 255, 257, 268, 272, 275, 287, 292, 333, 344, 372
    Fitting, 387
    Segmentation, 128, 135
Plants, 364, 367
Point, 185, 295
    Extraction, 89
    Features, 89
    Primitives, 89
    Spread, 75
Polar Angle, 159
Polyhedra, 144, 185, 208, 220
Polytree, 194
Popularity Algorithm, 82
Pose, 4
Position
    General, 145, 213
    of Objects, 187
    Recovery from Insufficient Data, 259
    Tolerance, 302, 306, 317, 320, 347
Prediction Graph, 218
Primary Datum Systems, 340
Primitive, *see* Feature
    Surfaces, 118
Principal
    Axes, 139, 390
    Curvatures, 132
    Directions, 132
Printed Circuit Boards, 286, 350, 358, 360
Probabilistic Segmentation, 126

Probe, 332, 369
Production Defects, 365, 366
Productivity, 288
Profiles, 350
Projections, 6, 218
Pseudo-Inverse Method, 246, 247
Psychology, 7
Psychophysics, 7

QR Algorithm, 248
Quadric
    Fitting, 389
    Surfaces, 114, 131, 136, 140, 185, 226, 255, 325, 389
        Classification of, 115, 116, 390
Quadtree, 123, 192
Quality Assurance, 4, 284, 369
Quaternions, 243, 393

Radar, 5
Radiance, 158
Ramp Edge, 94
Random Textures, 174
Rank of a Matrix, 261, 383
Real Space, 49
Reasoning, 143
Recognition, 167, 187, 204, 239, 283, 328
Redundant Views, 313
Reflectance, 158, 290
    Lambertian (Matt), 160
    Map, 164
    Specular, 160, 364
Region
    Description
        Statistical, 137
    Growing, 121, 123

    Simultaneous, 125
    Map, 129
    Segmentation, 110, 120
    Splitting, 120, 121
Registration
    of Multiple Views, 313
    of Object Position, 287
Regularised Set Operators, 197, 201
Relational Graph Structure, 209, 212, 217, 232, 233
Relaxation Labelling, 104, 149, 236
    Matching Methods, 228, 293
    Methods, 151
Reliability
    of Manufactured Products, 284
    of Segmentation, 87
    of Visual Inspection, 287
Resolution
    Depth, 289
    Spatial, 289
Restriction Graph, 217
Ribbons, 211, 218
Rigidity Matching Constraint, 227
Roberts Edge Operator, 96
Robot, 3
    Arm, 326–328, 343, 344, 369, 373, 379
    Mobile, 376
    Motion, 343
Robotics, 343, 348
Rogue Faces, 258
Roof Edge, 94
Roots of Unity, 61
Roughness, 357

Rounding Errors, 388
Run Length Coding, 78

Saturation, 127
Scan Line, 16
SCERPO Vision System, 156, 229, 251, 293
Search
    Constraints, see Constraints
    Control, 256
    Tree, 220
Seed Pixel, 123
Segmentation, 12, 85, 86, 137, 154, 167, 186, 216, 240, 257, 263, 264, 272, 273, 277, 278, 293, 315, 317, 326, 328, 335, 338, 341, 342, 347, 363, 372, 376
    Goals, 87
    Implicit, 319
    of Surfaces, 110, 120, 129, 135
    Poor, Consequences of, 256
    Probabilistic, 126
    Reliability of, 87
    Stability of, 293
Self-Occlusion, 208
Semiconductor Wafer, 350
Set Operators, 195, 196
    Regularised, 197, 201
Set-Theoretic Modelling, 195, 201, 304
Shadow, 94, 352
    Edges, 148
Shape
    from Shading, 34, 157, 290, 357
    from Texture, 175
    Grammars, 175
Sheet Metal Components, 286
Shoes, 3
Signal Processing, 7
Signature of a Matrix, 383
Silhouette, 146
    Edges, 109
Similarity
    Constraints, 121, 126
    Function, 215
Simulated
    Annealing, 153
    Depth Maps, see Depth Map, Artificial
Simultaneous Region Growing, 125
Sinc Function, 53
Single Axis Inspection, 328, 338, 340, 344
Singular Value Decomposition, 248, 261, 384
Size Tolerance, 301, 306, 317, 320, 347
Slices, 350
Sobel Edge Operator, 230
Solder
    Debris, 360
    Joints, 360, 361
Solid Modelling, see Geometric Modelling
Sonar, 5
Sound, 50
Spatial
    Enumeration, 192, 201
    Resolution, 289
Specular Reflectance, 160, 364

INDEX

Speed, of Inspection, 287
Spheres, 114, 119, 197, 255, 344, 377
Spike, 350
Split and Merge, 121
Spottiness, 357
SRC Vision System, 26, 36, 37, 43, 128
Standards, 296
Statistical
    Methods, 149
    Pattern Recognition, 174
    Region Description, 137
Statistics, 8
Step Edge, 94, 216
Stereo, 27, 30
    Active, 12, 36, 291, 313, 328, 372
    Correspondence Problem, 32
    Imaging, 19, 23, 333, 368
        Errors in, 40
    Passive, 35, 225, 290
    Photometric, 34, 165
Stochastic Labelling, 150, 228
Streakiness, 357
Structured
    Lighting, 32, 290, 363
    Textures, 174
Stylus Profilers, 357
Surface
    Algebraic, 114, 136
    B-Spline, 112
    Bézier, 112
    Bicubic, 131
    Biquadratic, 131
    Clustering, 231
    Completion, 231
    Curvature, 132
    Curved, 191, 197, 212, 272, 333
    Defects, 358
    Features
        Actual, 303
        Nominal, 303
        of Object, 13
    Finish, 4, 10, 296, 297, 354, 357
    Fitting, 127, 137, 377, 387
    Higher Order, 185
    Implicit, 112, 197, 200
    Matt, 160
    Normal, 89, 132, 168, 325
    Normals, 333
    of Car Body, 4
    Parametric, 200, 206, 231
    Patches, 112, 231, 233, 330
    Planar, 110, 131, 136, 185, 200, 240, 255, 257, 268, 272, 275, 287, 292, 333, 344, 372
        Fitting, 387
    Quadric, 114, 136, 140, 185, 226, 255, 325, 389
        Classification of, 115, 116, 390
        Fitting, 389
    Representation, 110
    Rogue, 258
    Segmentation, 110, 120, 129, 135
SVD, 384
Symbolic Reasoning, 8
Symmetric Function, 53
Symmetry, 254, 350

Tactile Probe, 285, 332, 369
Taste, 365
Television Camera, 16, 72
Texels, 179
Text, 51
Texture, 297, 305, 357, 365, 366
    Classification by, 366
    Orientation from, 175
    Structured, 174
Thermal
    Imaging, 5, 49, 363
    Signature, 363
Three-Dimensional
    Image Data, 288
        Acquisition, 18, 26, 32, 289, 368, 372
    Imaging, 5, 6
Thresholding, 64, 352, 367
Tiered Coloured Illumination, 363
Timber, 364
TINA Vision System, 27, 36, 224, 243, 251, 295, 378
Tolerance, 284, 289, 317, 320, 322, 340, 341, 343, 347, 374
    Conventional, 298
    Form, 301, 307, 317, 320, 347
    Geometric, 10, 11, 283, 295–297, 300, 311, 349
    Position, 302, 306, 317, 320, 347
    Representation, 298
    Size, 301, 306, 317, 320, 347
    Specification, 303, 373
    Zones, 299, 301
Top Hat Function, 52, 55, 58, 66
Topology, 198, 202

Touch Probe, 285, 332, 369
Trace of a Matrix, 383
Transform Encoding, 79
Transformation Estimation, 242
Translation Estimation, 246
Transmission, 78
Tree Search, 104, 148, 179, 224, 226, 234, 294
Triangulation, 19, 25, 27, 30, 40
Turntable, 13, 326–328, 335, 340, 341, 373, 379
Two-Dimensional
    Image, 6
        Acquisition, 16
    Inspection, 4, 286
    Vision, 5, 286

Ultrasound, 367
Ultraviolet Light, 5, 367
Umbilic, 132
Uncertainty Scale Factor, 21
Union (of Sets), 196

Validity, 189, 201
Variance, 395
Variational Graph Structure, 304
Views
    Choice of, 329
    Multiple, 312
Vision
    System, 11, 341, 348
        Industrial, 286
    Two-Dimensional, 5, 286
Voids, 367
Voxel, 192, 201

Walsh Transform, 80
Wavyness, 357

Widget, 253, 256, 265, 268, 275–278, 317, 320, 344
Wireframe Models, 189
World Coordinates, 18, 21

X-Rays, 5, 350, 363, 367

Zero-Crossings, 99, 216, 230